Evolutionary Biology – New Perspectives on Its Development

Evolutionary biology has been a remarkably dynamic area since its foundation. Its true complexity, however, has been concealed in the last 50 years under an assumed opposition between the "Extended Evolutionary Synthesis" and an "Alternative to the Evolutionary Synthesis". This multidisciplinary book series aims to move beyond the notion that the development of evolutionary biology is structured around a lasting tension between a Darwinian tradition and a non-Darwinian tradition, once dominated by categories like Darwinian Revolution, Eclipse of Darwinism, Evolutionary Synthesis, and Post-Synthetic Developments.

The monographs and edited volumes of the series propose an alternative to this traditional outlook with the explicit aim of fostering new thinking habits about evolutionary biology, a multifaceted area composed of changing and interacting research entities and explanatory levels. Contributions by biologists and historians/ philosophers are welcomed. Topics covered in the series span from (among many other possibilities):

- An Overview of Neutralist Theories in Evolutionary Biology
- Developmental Biology: From Reductionism to Holism and Back
- Selection Theories Beyond Hard and Soft Inheritance
- Divergent, Parallel, and Reticulate Evolution: Competing or Complementary Research Programs?
- The Rise of Molecular Biology: Between Darwinian and Non-Darwinian
- Biologizing Paleontology: A Tradition with Deep Historical Roots
- The Darwinian Revolution and the Eclipse of Darwinism: Blurring the Historiographical Lines
- Darwinism, Lamarckism, Orthogenesis: Can We Really Define Them by Their Hard Explanatory Cores?
- The Evolutionary Synthesis: A Fabricated Concept?
- The Opposition to the Evolutionary Synthesis: Criticizing a Phantom?
- A Reversed Perspective: Approaching Charles Darwin from the Pre-1859 Period
- The Long Development of the Multilevel Paradigm in Evolutionary Biology
- Self-Organization: A Research Tradition from Morphology to Cosmology
- Human Evolution: Sociobiological or Sociocultural?

More information about this series at http://www.springer.com/series/16175

Richard G. Delisle

Charles Darwin's Incomplete Revolution

The *Origin of Species* and the Static Worldview

Springer

Richard G. Delisle
Department of Philosophy and
School of Liberal Education
University of Lethbridge
Lethbridge, AB, Canada

ISSN 2524-7751 ISSN 2524-776X (electronic)
Evolutionary Biology – New Perspectives on Its Development
ISBN 978-3-030-17202-2 ISBN 978-3-030-17203-9 (eBook)
https://doi.org/10.1007/978-3-030-17203-9

Cover photo by GLady is licensed under CC0
https://pixabay.com/photos/mosaic-fish-tile-art-ceramic-200864

This Springer imprint is published by the registered company Springer Nature Switzerland AG.
The registered company address is: Gewerbestrasse 11, 6330 Cham, Switzerland

Acknowledgements

This monograph has been long in the making. It has been difficult to unlearn what I thought I knew about the *Origin of Species*, especially for someone teaching a yearly course on the topic for over 10 years. During that unlearning process, I read with great profit a significant part of the huge corpus of the literature produced by what is called the "Darwin industry." That is why I should like to dedicate this book to scholars of this industry who have collectively made a remarkable contribution to intellectual history. I understand the expression "Darwin industry" in its widest possible sense to comprise Darwin's contemporaries. Indeed, this monograph is also an invitation to (re)discover what some of them had to say when the *Origin* was freshly published. I, for one, have rediscovered them only at the very end of this project. Only when liberated from the traditional view was I able to grasp the profound relevance of their reflections. Reference to them, therefore, is more often than not an afterthought. It is my impression that the authors of the "first Darwin industry" have much more to offer, a research topic worth revisiting in the near future.

This monograph also benefited from the expertise of James Tierney, Language Program Director in the English Language Programs from the Center for Language Study at Yale University. For more than 10 years, Jim has been my voice to the English-speaking world by giving my English texts the necessary polish. I wish to express my warmest appreciation for this fruitful collaboration.

This book project was kindly reviewed by two anonymous referees. My gratitude is also extended to them.

I am glad to acknowledge my debt to Dr. Sabine Schwarz, Executive Editor: Biomedicine, Genetics & Microbiology, who welcomed this monograph in the newly created book series "Evolutionary Biology—New Perspectives on its Development,"

Finally, I should like to apologize to my wife, Anik, for having spent too much time with Charles Darwin. We all know that three's a crowd.

Contents

Chapter 1
Introduction: Darwin in the Larger Intellectual Context

Abstract Our image of Charles Darwin's work has been modernized to excess by often presenting it as a contribution to modern evolutionism. Uncovering the real Darwin requires a recalibration of his image: seeing him afresh as a scholar in transition between two revolutions (the Scientific Revolution and the Transformist Revolution), two worldviews (the static and the evolutionary), and two epistemologies. In order to recast the *Origin of Species* against this complex canvas, an "intellectual grid" presenting *fundamental ideas* and *dominant systems of thought* is sketched. Our view of Darwin is situated in reference to the views of major Darwin scholars like John C. Greene, Michael Ruse, and Robert J. Richards. It is argued that the historiography needs to be completed by paying closer attention to Darwin's profound debts to the static worldview developed in the seventeenth and eighteenth centuries.

Keywords Traditional historiography · Static worldview · Evolutionary worldview · Scientific Revolution · Transformist revolution · Atomism · Monism · Teleology · Chain of being · Organicism · Mechanism · Natural theology · Romantic biology · John C. Greene · Michael Ruse · Robert J. Richards

1.1 Recalibrating Darwin's Image

The name of Charles Darwin (1809–1882) is familiar to many people today. Leaving aside the deforming lens of a picture of a Darwin constructed by pop culture that depicts him as the mythical founder of the idea of evolution itself, the familiarity of his name among scholars is largely related to the fate of his ideas in twentieth-century science. While Darwin's books of the 1860s and 1870s became widely known, especially following the initial publication and reception of the *Origin of Species* in 1859 (Desmond and Moore 1991: 477–481), his ideas were often judged insufficient to explain biological evolution, hence an assumed "eclipse" of

R. G. Delisle, *Charles Darwin's Incomplete Revolution*, Evolutionary Biology – New Perspectives on Its Development 1, https://doi.org/10.1007/978-3-030-17203-9_1

Darwinism in the late nineteenth and early twentieth centuries (Bowler 1983, 1988).[1] Yet, Darwin's name really came of age in twentieth-century evolutionary biology, when a series of conceptual and theoretical developments between the 1920s and the 1960s in such areas as population genetics, zoology, and paleontology indicated that evolution might be consistent with key Darwinian concepts such as small variation, random distribution, selective pressure, gradualism, extrapolationism (microevolution = macroevolution), and the like. These developments eventually came to be known as the Evolutionary Synthesis.[2] Evolution was explained as a two-step process: the rise of small and random genetic variations subsequently oriented (selected) at the organismic level in accordance to the relative advantage they provided in differential environmental settings. Darwin's place in this story was made ever more official when the ongoing professionalization of the fields of history and philosophy of biology—which really made itself felt at the Centennial in 1959 and after (Greene 1975; Churchill 1982; Oldroyd 1984; Kohn 1985)—generated a literature analyzing the rise of this Evolutionary Synthesis with presumably Darwinian roots.[3]

Historiographically speaking, a mid-nineteenth-century scholar like Darwin was thereby pulled into the twentieth century. While this interpretation is not entirely deprived of some support, it comes with the important disadvantage of overlooking Darwin's debt to previous centuries. It is the main goal of this book to contribute to the historiographical "recalibration" of Darwin's image, by showing him not to be as modern as the traditional historiography would have it. By so doing, we will walk in the footsteps of other historians and philosophers who have already significantly contributed to connecting Darwin with his own past. The issue is of some importance, because it impacts how evolutionary biology is envisioned today. If Darwin is as modern as is sometimes claimed, it is then legitimate for today's biologists to see themselves as Darwinians in the full sense of the expression. If he is not, the presumed connections between Darwin and modern biologists become more problematic. In this regard, the situation in the field of evolutionary biology seems rather unique. As John Horgan (1996: 114) remarked: "No other field of science is as burdened by its past as is evolutionary biology... [It] can be defined to a large degree as the ongoing attempt of Darwin's intellectual descendants to come to terms with his overwhelming influence." No surprise, then, that Darwin is sometimes used as the final arbiter of contemporary debates. In the words of Robert J. Richards (1988: 146): "Among contemporary evolutionary theorists Darwin functions as an icon, an image against which theories may receive approbation or reprobation. To select from the historical Darwin those features that best comport with one's own predilections

[1]This is not the place to contest the historiographical value of the notion of an "eclipse of Darwinism." For a critic, see Largent (2009, 2013) and Delisle (2017).

[2]Among key publications often associated with this theoretical corpus, see Fisher (1930), Wright (1931), Haldane (1932), Dobzhansky (1937, 1951), Huxley (1942), Mayr (1942, 1963), Simpson (1944, 1953), Jepsen et al. (1949), Stebbins (1950), and Rensch (1960).

[3]Chapter 8 of this book will hint at reasons for questioning this traditional narrative.

in the contemporary scientific debate is to have those predilections sanctioned by the master." The implication underlying this attitude has been succinctly summarized by Timothy Shanahan (2004: 293), from whom I have borrowed the preceding analysis: "One way of coming to terms with Darwin's influence is to appropriate him in support of one's own views." Among contemporary evolutionists, therefore, a risk exists that in reaching for Darwin's sanction, one overreaches, or at least distorts, his influence over time. One strategy that has produced that effect has been to reduce "Darwinism" to a sort of simplified and abstract explanatory entity valid throughout all historical periods, at the cost of neglecting the full complexities of evolutionary explanations (Richards 2004: 26; Delisle 2017: 147, 158).

In the continued search for the real Darwin, the specificity of the theses presented in the following pages lies in the belief that Darwin is not only very much a mid-nineteenth-century scholar but also one that was deeply committed toward the static worldview that accompanied the Scientific Revolution in the seventeenth and eighteenth centuries. This put Darwin in a difficult position with respect to embracing all the implications accompanying evolutionism. The Transformist Revolution has been a long intellectual process extending from the eighteenth century to today. One field after the other has come to grips with the need to recast its object of study under an evolutionary outlook: human history, geology, biology, astronomy, etc. No single scholar managed to move entirely from a static worldview to an evolutionary one during his or her own lifetime. Charles Darwin is no exception to this rule. In this context, many versions of evolutionism were proposed during this revolution, offering all sorts of compromises between the old and the new views. As a child of the Scientific Revolution, Darwin inherited many key concepts from a movement founded largely on a static worldview. Darwin took these concepts into consideration while erecting his theory of biological evolution. Unsurprisingly, this theory was ill-equipped to embrace the full directionality, historicity, and novelty that came with a new evolutionary world. This book will analyze what the *Origins of Species* owes to a static and stable conception of the world.[4] As such, it is an exegesis of the pages many times interpreted yet always in need of further consideration.

[4]Of the six editions of the *Origin of Species* which appeared between 1859 and 1872 [1876], Darwin presented his basic ideas more clearly in the first. Indeed, by responding to his criticisms in the other editions, he himself contributed to the muddling of his views by introducing ever more additions. Here, I shall follow the tradition established in the last few decades in the Darwin studies which consists in returning to Darwin's original exposition in 1859 [differences between the various editions are compiled in Charles Darwin (1959), *The Origin of Species: A variorum text*, edited by M. Peckham, University of Pennsylvania Press, Philadelphia]. However, this practice will not prevent us from often referring to the last edition of the *Origin*. It should be noted that although we will occasionally insist on differences between the first and the sixth editions, the main thrust of this book is concerned with identifying the *Origin*'s explanatory core by searching for fundamental similarities between the first and last editions. Analyses of some of the differences found across the various editions of the *Origin* can be found in Vorzimmer (1970), Liepman (1981), and Hoquet (2013).

For instance, after having read through various and sundry biological details, the reader of the *Origin* suddenly encounters in the very last sentence of the book this rather intriguing combination of astrophysics, biology, and theology:

> "There is grandeur in this view of life, with its several powers, having been originally breathed into a few forms or into one; and that, *whilst this planet has gone cycling on according to the fixed law of gravity*, from so simple a beginning endless forms most beautiful and most wonderful have been, and are being, evolved." (Darwin 1859: 490)[5] [my emphasis]

Surely, this apparently out of context remark placed at the very end of a long biological treatise could only have been a matter of style or rhetoric, echoing, as will be seen, similar astro-biologico-theological comments of the seventeenth and eighteenth centuries. Or is it? More than two decades prior to the publication of the *Origin of Species*, it caught the attention of the scientist-philosopher William Whewell (1837: vol. 3, pp. 615–620) that direct comparison of celestial and terrestrial phenomena were, after all, perhaps not entirely innocent or without consequence. As expressed by Whewell (1837: vol. 3, p. 618) himself: "The analysis of the heavenly motions, it has been said, supplies no trace of a beginning, no promise of an end." Indeed, in the context of a Newtonian view of our solar system eternally cycling on but static, the building of an analogy binding terrestrial phenomena to celestial ones may have been the way of transferring this stability from one area to another. In fact, the ideas contained in the preceding quote were sufficiently important to Darwin to appear repeatedly in the *Sketch* (1842), the *Essay* (1844), and the six editions of the *Origin of Species* (1859–1972). The formulation of 1842 is even more interesting, for it contains a reference to a steady-state geology, echoing a steady-state astronomy:

> There is a simple grandeur in the view of life with its powers of growth, assimilation and reproduction, being originally breathed into matter under one or a few forms, and that whilst this our planet has gone circling on according to fixed laws, and land and water, in a cycle of change, have gone on replacing each other, that from so simple an origin, through the process of gradual selection of infinitesimal changes, endless forms most beautiful and most wonderful have been evolved.[6]

As will be seen, Darwin's steady-state geology derived directly from Charles Lyell's *Principles of Geology* (1830–1833) to which he wholeheartedly subscribed. It now remains to be investigated whether Darwin also applied a steady-state view to biology and if so, the extent to which it penetrated deeply his biology.

Darwin studies have progressed far enough to reveal that Darwin was anything but an unconditional evolutionist: life was not, for him, bestowed with the natural

[5]This phrase is reproduced as such in the last edition with the important insertion of "by the Creator," an addition that has appeared since the second edition of 1860: "originally breathed by the Creator into a few forms...". See Darwin, C. (1876), *The Origin of Species*, sixth edition, with additions and corrections to 1872 (London: John Murray), p. 429.

[6]Darwin, F. (1909), *The Foundation of the Origin of Species: Two Essays Written in 1842 and 1844 by Charles Darwin* (Cambridge: Cambridge University Press), p. 52.

ability to evolve at all cost, as postulated, for instance, in Jean-Baptiste Lamarck's *Philosophie zoologique* (1809). On the contrary, Darwin's evolutionism was of a different kind, owing to his conception of the nature of the relationship between life forms and their environment: life forms are entities stable enough in need of real places in the economy nature. This is implicitly expressed in the *Origin* in the following terms:

There would be no evolution if the milieu remained stable, living matter being inherently inert.

The evolutionary process constitutes no internal drive; only an external force like natural selection can put it in motion.

Evolutionary change is a trick to keep something fundamental unchanged, adaptation being merely a way to reestablish a new stability.

The centrality of the underlying theme of stability has caught the attention of scholars. For instance, Stephen Jay Gould sees in Darwin's view based on steady-statism (its lack of biological direction and increasing complexity in structure and number of species) a kind of timelessness, with biological progress having no goal, since it is simply a by-product of the adaptive process itself. Darwin's theory of natural selection, Gould (1977a: 5, 13–14) maintains, is founded on adaptive explanations linked only to local conditions. Similarly, Camille Limoges has noted the lack of historical dimension in Darwin's theory, which seeks explanations for ecological adaptations only as they are documented from the present time.[7] These suggestions seem intriguing enough to justify further investigation.

1.2 Sketch of an Intellectual Grid

Before doing so, it will be useful to remember the context in which Darwin pursued his reflections on evolutionism. We will first consider two complementary aspects. First, the search for a single and fully coherent picture of Darwin is inherently distorting. During his lifetime, so many questions pertaining to evolutionism remained open and too few tools were available to tackle them. It is no surprise, then, that partial or inadequate answers were proposed. We know that Darwin's biological theorizing was excessively broad, ranging over many issues, including geology, embryology, biogeography, biological variation, reproduction-generation, the tree of life, the origin of life, morphology, physiology, taxonomy, the mechanism of evolution, and the place of humankind (Hodge 1983, 1985, 1989). When tackling so broad a subject, the overlapping topics and approaches of Darwin's main works published during his maturity—for instance, the *Origin of Species* (1859), the *Variation of Animals and Plants Under Domestication* (1868), and the *Descent of*

[7]Limoges, C. (1972), "Introduction", in C. Linné, *L'équilibre de la nature* (Paris: J. Vrin), pp. 7–22, see footnote p. 12.

Man (1871)—cannot conceal a lack of a unified or fully coherent view. In fact, this lack of cohesiveness even reveals itself within a single book like the *Origin*.

Second, as already alluded to, Darwin conducted his research program in the midst of a profound and fast-moving intellectual revolution which began in the eighteenth century and which continues to this day: the Transformist Revolution (Bowler 2003). A scholar's attempt at fixing his or her ideas against such a changing intellectual canvas comes with its own challenges for the unity of a single view. The Transformist Revolution impacted various research areas at a varying pace, areas that were far from entirely sealed from one another. Scholars of the time ran behind a rapidly moving revolution, proposing a wide range of hybrid positions between a fully static view and a fully evolutionary one.

Enough has been said to see, at least in a preliminary way, why Darwin's work contains both a multitude of sources of inspiration and numerous and unstable internal tensions. As John C. Greene (1986: 210) has perceptively noticed:

> The tension was apparent in Darwin's oscillation between rejecting 'necessary progression' in evolution and asserting natural improvement as a general consequence of natural selection; between discoursing about higher and lower organisms (including human races) and swearing never again to use the terms 'higher' and 'lower'; ... between asserting that everything was governed by fixed laws and acknowledging that natural selection was a statistical law; between emphasizing competitive struggle as the *sine qua non* of progress in nature and history and admiring the wonderful adaptedness and interdependence of organic beings; ... Clearly the Darwinian world view, compounded of mechanistic determinism, belief in evolution by competitive struggle and survival of the fittest, and the idea of nature-history as a single continuum undergoing progressive change in accordance with fixed laws discoverable by science, was not a seamless fabric fashioned for all time but an unstable compound of old and new ideas whose incompatibility must eventually become apparent with the progress of science and speculation.

Enough has also been said to understand why the recent historiography could come up with as different a view as can possibly be imagined, as will shortly be seen in considering the positions of John C. Greene, Michael Ruse, and Robert J. Richards. In order to make any sense at all of these widely divergent interpretations, and to put the theses of this book in context, let us sketch the "intellectual grid"[8] against which Darwin's views will eventually be positioned.[9] Considering that the roots of Darwin's ideas are historically deep and intellectually dispersed, it will be necessary to cast the net of our framework as widely as possible. The framework proposed here constitutes a complex network of ideas, something approaching a seamless fabric extending into ancient times. For convenience of presentation, information will be organized along two main axes: first, a vertical cut, the enumeration of separate *fundamental ideas* carried through history more or less

---

[8]The reader in a hurry or familiar with the presented notions may want to skip this section of the chapter.

[9]The very broad outline of this grid was freely inspired by the following sources: Collingwood (1945 [1960]), Greene (1981, 1986), Hodge and Radick (2009), and Hodge (2014).

in parallel; second, a horizontal cut, the description of *dominant systems of thought* at specific historical periods.

With respect to *fundamental ideas* relevant to nineteenth-century evolutionary biology, one must go all the way back to the original pluralism already encountered among the Ancient Greeks: (1) atomism, (2) the organizing principle, (3) evolutionary monism, (4) teleology, and (5) the chain of being. Fundamental ideas have persisted until today because they can be, and have been, conceived in a very abstract manner by scholars throughout history. Indeed, scholars have exploited them by using their abstract or neutral core only to clothe them with specific interpretations. It would be a mistake to look for filiations of fundamental ideas already clothed in specific interpretations, since under a specific form, connections may either be weak or appear to be lacking. With respect to the question of identity, ideas are not as constrained as biological species or lineages. The mind can transcend apparent differences between specific ideas and gather concealed similarities at a more fundamental level. The same thought process is involved when fundamental ideas are combined in various ways; combinations which may seem incompatible to scholars committed to different ideas. A caveat is therefore in order here: it is not because fundamental ideas can be traced back to Ancient Greeks that they should be taken to be particularly Hellenistic in character. Again, fundamental ideas are too neutral to be culturally or epistemologically associated with their original proponents and interpretations, whoever and whatever they may be.

Atomism (Epigenesis) As presented by Leucippus and Democritus, the entire diversity of the world is ultimately reducible to various combinations of eternal and indestructible tiny invisible particles (atoms) of various configurations organized by pure chance under random motion.[10] The only permanent entities of the universe are atoms, although more complex entities may emerge from atomic combinations. Largely abandoned throughout the Middle Ages, this fundamental idea was revived in the sixteenth and seventeenth centuries, for instance, by Galileo and Pierre Gassendi, and exploited by Robert Boyle, René Descartes, and Newton. With Descartes, this fundamental idea took the specific interpretation of a world conceived as a law-bound system of matter in motion. This is a view with obvious evolutionary implications, since the world can then be seen as a series of steps in the reorganization of its ultimate corpuscular basis: "the idea of a self-contained system of matter in motion seemed to imply that transiency of all particular structures produced by the combinations and permutations of the system" (Greene 1959: 13). Soon enough, in the eighteenth century, Buffon, Kant, and Laplace (in his *Exposition of the System of the World*, 1796) speculated about the origin of our solar system from a previous state of arrangement of matter.[11] In the nineteenth century, Herbert Spencer exploited this fundamental idea in full in his *Synthetic Philosophy*, by envisioning

[10]On atomism in general, see Dijksterhuis (1961), Guthrie (1965), and Barnes (1982).

[11]On the implications of a law-bound system of matter in motion for evolutionism in astronomy, biology and economics, see Greene (1959: 24–37; 1981: 128–157), Foley (1976), Depew and Weber (1995: 113–139), and Schabas (2005: 79–113).

cosmic progress as moving along the axis of matter, life, and humans under a reorganization process of basic matter from an original homogeneous state to an ever more heterogeneous one. In the eighteenth century, political economists began to conceive society as composed of competing atomized actors (individuals). They either proposed an optimistic view, as seen in Adam Smith's *The Wealth of Nations* (1776) in which it is argued that continued competition increases the wealth of nations, or a pessimistic one, as expressed, for example, in Thomas Malthus's *Essay on the Principle of Population* (1798) for whom laissez-faire policies could only lead to confrontation. The doctrine of atomism as presented here is compatible with the notion of epigenesis: a developmental process that moves from an unorganized and undifferentiated state to an organized and differentiated one, in cosmology or embryology, as in other areas.

Organizing Principle For scholars who did not subscribe to the idea of matter being organized by pure chance, a separate notion was needed to account for both the order and the diversity of entities observed in the world. The fundamental idea of an organizing principle gave matter its shape, so to speak. This abstract idea was expressed in many specific ways. In Plato's ontology (in the dialogue *The Meno*, for instance), transcendental and eternal forms existing in the perfect supralunar region permitted the existence of their imperfect and ephemeral counterparts in the sublunar region, as seen in the decaying and multiple entities on Earth. Plato's ontology bears strong affinities to the mathematicism of Pythagoreans, for whom all things are mathematical beings (numbers and proportions): different triangles (variations) having something fundamental in common (their angles add up to 180°) (Guthrie 1975: 250–252). Forms are thus the organizing principle of world's entities. In Aristotle's more immanent ontology, the diversity of entities on Earth owes its existence to as many earthly forms as there are specific living species (e.g., cats, dogs, whales)(Pichot 1993: 106–107). Under the name of "post-Platonism" the same fundamental idea persists in the modern era. Reading Buffon's *Histoire naturelle* and *Époques de la nature* (1749–1788), one gathers that distinct families of animals appear in different geological times through separate instances of spontaneous generation under as many distinct *moules intérieurs* as they are existing families. These *moules intérieurs* are thought to possess an integrative force akin to Newton's gravity allowing them to give organic molecules their shape (Roger 1989: 152–207, 528–558; Hodge 1992: 241–254). Similarly, several scholars inspired by *Naturphilosophie*[12] subscribed to the notion of archetypes, which are blueprints to living forms, like variations on a common theme. In Johann Herder's *Ideen zur Philosophie der Geschichte der Menschheit* (1784–1785), for instance, all living

[12]*Naturphilosophie* (sometimes known as "romantic biology") has often been seen as an intellectual movement replete with mystical notions such as *microcosm-macrocosm* and *introspection-intuition,* notions used as a means to inquire about the world and its supposed harmonious unity. In reality, *Naturphilosphie* constitutes a complex movement with some strands being closer to materialism, mechanism, and empiricism. For a glimpse of this multifaceted movement, consult Lenoir (1982) and Renault (1999).

forms are generated out of a unique type: "...Nature, amid the infinite variety which she loves, seems to have fashioned all the living things upon our earth after a single original model... of organization" (Lovejoy 1959: 210). For his part and simply enough, William Paley held in *Natural Theology* (1802) that species were all independently created in accordance to the Creator's plan. Looking back to Descartes, it had been noted that his system of matter in motion was "law-bound," that is, structured or organized by a limited number of laws. Ontologies giving organizational precedence to laws over matter, as seen in Descartes, also embrace the fundamental idea of an organizing principle. Of course, the role of law-like principles varies, whether laws are believed to take matter down the path of rearrangement (atomism) as implied by Descartes's view or thought to be part of a perfectly stable and contrived universe, as in Newton's. Also compatible with the idea of an organizing principle is the concept of places in the economy of nature (John Ray, Linnaeus), and by extension, of the market place with its law of supply and demand in human economic affairs. These, too, create order by organizing or channelling entities. Whether transcendent or immanent, whether called *eidos*, *archai*, laws, forces, *moules intérieurs*, archetypes, or places in the economy of nature, and whether applied at the large scale of entire kingdoms (plants, animals) or at the very narrow scale of individual species (cats, dogs, whales), these many specific expressions can all be subsumed under the abstract idea of an organizing principle.

Evolutionary Monism (Preformationism) In the ontology of Presocratics like Thales, Anaximenes, and Anaximander, the entire world is contained in a primordial materialistic and immanent principle (*arché*) which constitutes, all at once, a single and identical substance (monism) and the deterministic causes that govern its deployment in time, which generate all the world's entities (Kahn 1960; Guthrie 1962: 1–145; Cherniss 1970; Vlastos 1970; Algra 1999). In some interpretations, the world is predetermined to deploy itself in full (a directional process) and remain in such a state; in other interpretations the primordial principle is predetermined to deploy itself in full before collapsing on itself entirely, thus returning to its original state before a new deployment becomes possible (a cyclical process). While according to the atomistic doctrine everything in the world is fundamentally reducible to its smallest constituent parts, and according to the notion of organizing principle matter at all levels is structured by the ontological precedence of a principle over matter itself, one finds in the idea of evolutionary monism a fused compound of matter, form, and materialistic causes. From this primordial principle, simple matter eventually generated more complex ones like the sun, the planets, the Earth, the oceans, life, and humans. To reverse the perspective, the entire diversity of cosmic entities is believed to already be contained in the primordial principle. This is a closed process, since the sequence is entirely predetermined. Abandoned during the Middle Ages, the fundamental idea of evolutionary monism was revived in modern times under different guises. At the cosmic scale, some interpretations of the Big Bang theory today might be thought to be consistent with such an idea, whereas at the more modest scale of seventeenth- and eighteenth-century biology, the preformationist school concerned with reproduction and generation claimed that everything of the adult form already existed at conception, but in a miniature state only (Roger 1993[1963]: 325–384). Also consistent with evolutionary monism is the widespread notion among *Naturphilosophen* that nature as a whole constitutes a single and unified complex which deploys itself in time in order to reveal its manifold entities, biological or otherwise. Johann Herder, Friedrich von Schelling, and Goethe, at one time or another, presented ideas akin to this notion. Even Ernst Haeckel's view, which is closer to linear and progressive developmentalism than open and contingent evolutionism in the second half of the nineteenth century, retained something fundamental of evolutionary monism (Bowler 1988: 82–90; Richards 2008; Levit and Hossfeld 2017).

Teleology In the purest tradition of atomism, form, matter, and entities are largely generated unconstrained by internal or external conditions since matter is organized by pure chance. Nonetheless, it might be argued that living organisms are integrated wholes whose parts must be sufficiently harmonious to allow survival (internal conditions). Or, it may well be contended that living organisms have to be adapted or designed in such ways as to be able to face the environmental settings in which they live. The idea of an organizing principle discussed above may or may not entail the fundamental idea of teleology. If a benevolent and omnipotent God is believed to have created living forms to fit into specific places in the economy of nature, as argued by John Ray, Linnaeus, and William Paley, then an organizing principle and teleology are united. If, on the other hand, immanent and materialistic archetypes (unity of types) in the tradition of *Naturphilosophie* are thought to have precedence over internal or external conditions—thus placing structure before function—then an organizing principle and teleology are largely uncoupled.[13] That is why teleology constitutes a fundamental idea of its own. Generally speaking, all views assuming the existence of some kind of Creator or Designer with intentions embrace teleology. This releases us from reviewing all possible cases. More interesting is Aristotle's immanent and somewhat naturalistic notion of "final causes," which stipulate that features of organisms are organized in such a way as to ensure their well-being (Ruse 2003: 17–19). Under this view, it becomes legitimate to investigate in their own right questions pertaining to function, purpose, or design in world's entities without necessarily resorting to the ultimate ontological source of their existence. Teleology thus becomes naturalized. The fortune of that fundamental idea was manifold, notably in natural theology during the seventeenth, eighteenth, and nineteenth centuries. Newton's search for the fundamental laws (especially gravity) explaining the order, stability, and equilibrium of our solar system in his *Mathematical Principles of Natural Philosophy* (1687) was part of a quest for a planned and designed universe (Burtt 2003[1954]: 231–233, 283–299), a stability also often recognized in Laplace's *Celestial Mechanics* (1799–1825), a conclusion not necessarily shared by its author, however (Hodge 2014: 173). So were John Ray's observations in *The*

[13] The implications of the opposition between "structure" and "function" are most clearly seen in the debate between George Cuvier and Étienne Geoffroy Saint-Hilaire in 1830. See Appel (1987) and Corsi (1988, Chap. 8).

Wisdom of God Manifested in the Works of Creation (1691), in which living creatures are thought to have been designed for the collective benefit of the whole, and Georges Cuvier's laws of correlation of parts and conditions of existence that allow organisms to be both internally integrated and adapted to their milieu (Brooke 1991, Chap. 6; Grene and Depew 2004, Chap. 5). Denying naturalism all together, as noted above, Paley embraced in his *Natural Theology* (1802) creationism pure and simple: species were created to fit perfectly in their designated places in nature.

The Chain of Being (*Scala Naturae*) While atomism and evolutionary monism may not be entirely irrelevant for embracing a kind of linear arrangement of beings (Anaximander and Spencer, for instance), the formal motivation for this fundamental idea comes from the chain of being. Again, it is an abstract idea that has been expressed in a multitude of specific ways. In the *Timaeus* of Plato, an all-powerful, benevolent, and eternal Creator could only assume its plenitude by existing as a series of beings of all possible kinds, thus generating a scale of beings descending from the more perfect (Itself) to the less perfect (Lovejoy 1936: 58–59). Moving away from this purely ontological grounding, in *On the Generation of Animals*, Aristotle formulated the idea of a single graded *Scala Naturae* for all animals, one which ascends toward perfection, with zoophytes at the bottom and humans at the top (Lovejoy 1936: 46–58). In the Middle Ages, Thomas Aquinas introduced a Christianized chain of being that envisions a creation of forms organized along a hierarchy extending upward from material beings to beings provided with ever greater mental faculties and, ultimately at the pinnacle, pure spirit (Gilson 1989 [1964]: 193–254). Whether descending the scale toward exclusive materiality or ascending toward exclusive spirit, the *Scala Naturae* proceeds from basic elements such as minerals, plants, animals, humans, and angels. In the seventeenth and eighteenth centuries, several specific interpretations of the chain of being were proposed. In his *Considérations philosophiques de la gradation des formes de l'être* (1768), Jean-Baptiste Robinet temporalized the scale by imagining that while God had created all the beings simultaneously, these were only revealed in a temporal sequence beginning with the simplest living forms before moving to the most complex ones (Roger 1995). A simplified version of the chain of being even existed in a scholar like Carl von Linnaeus, who organized life forms along a trophic scale of dependence in his *Politia Naturae* (1760). Indeed, in order for the equilibrium of nature to be maintained among life forms whose potential reproduction and expansion threatens this very equilibrium, herbivorous animals must restrain the expansion of plants and carnivorous animals the expansion of herbivorous animals (Limoges 1972: 13–15). Similarly, yet reversing the logic of the argument, James Hutton in his *Theory of the Earth* (1795) held that the perpetual destruction and dissolution of land masses were essential for generating soil to grow plants, plants for providing food for animals, and plants and animals for serving the needs of humans (Greene 1959: 79–80). In the nineteenth century, yet another version of the chain of being was held by Robert Chambers, who "argued" that the main animal forms (including invertebrates, fish, reptiles, mammals, humans) evolved independently and in parallel but climbed the scale only as far as their separate intrinsic level of complexity would allow them to (Bowler 2003: 134–140). Evolutionary views that embraced the notion of "progress" all inherited something of the idea of a Great Chain of Being.

Although the persistence of *fundamental ideas* is useful for sketching out an intellectual grid for understanding nineteenth-century evolutionary biology, it would remain awfully incomplete if a place were not provided to accommodate for *dominant systems of thought* that have emerged from time to time, systems that may persist in time under a more or less marginal state. Let us consider three such systems: (1) organicism, (2) mechanism, and (3) universal evolutionism. Like fundamental ideas, systems of thought (whether dominant or marginal) have persisted to the present day because they have been conceived throughout history in a thoroughly abstract fashion. Using modern terminology, it may be legitimate to consider such systems as worldviews (Greene), paradigms (Kuhn), or *epistémè* (Foucault), provided that it is recognized that these research entities are never as self-enclosed and as all-encompassing as sometimes argued.

Organicism Among the Ancient Greeks, the organicist system of thought was more dominant than any other (especially when compared to materialistic atomism) and encompassed all cosmic entities. The entire cosmos was conceived as a living organism permeated by mind (Collingwood 1960[1945]: 3–4). Whereas all motion becomes explainable in a universe believed to be alive, its inherent order and regularity, on the other hand, are dependent upon its organizing mind, which saturates it. During this time period, organicism took the form of "animism": the world is a macrocosm, conceived on analogy with the microcosm of a rational human being whose structuring (organizing principle) actions generate order and attain desired goals (teleology) (Lenoble 1969: 219–232). Underlying the various physical or external clothing of the numerous cosmic entities exists a common and single internal condition fusing life and mind. One brand of organicism is seen in Plato's transcendent ontology of perfect, eternal, and celestial forms, upon which the decaying and earthly entities depend. Even Aristotle's somewhat immanent and more naturalistic science ultimately rests upon it. Another brand of organism existed among the Stoics, a truly immanent doctrine which sees the world as a unified, dynamic, and self-creative whole permeated and organized by a unique divine principle (*logos*), but materialistically expressed in a single principle (such as *Fire*) responsible for the activity of the world's components (Brunschwig 1998: 534–548). Organicism persisted under different guises in the Christian and Hermetic traditions. In the latter "occult" tradition of the Renaissance, in Paracelsus, for instance, a stoic-like approach to nature was embraced, to which were added nearly magical virtues of a hylozoic world whose numerous animated manifestations (world's entities) were believed to express passions (such as love, hate, pleasure, pain) (Collingwood 1960 [1945]: 94–96). A no longer dominant organicism takes a more naturalistic turn after the seventeenth century, often confining itself exclusively to biology. In the field of physiology, for instance, specific biological principles (vitalism, holism) necessary for the integration and functioning of organisms are defended against a dominant mechanistic and reductionistic tradition (Duchesneau 1982, 1998). In the field of morphology, the search for analogical structures across a wide range of living forms (variations of an archetype) echoes the ancient quest for "correspondences" between entities in the world, a legacy of the microcosm-macrocosm idea revived in some versions of idealism with the notion of a thinking (and feeling) being as the epistemological center of the world, with the various entities assumed to be part of a single integrated organic whole (Schmitt 2004: 41–42). This can be seen in Étienne Geoffroy Saint-Hilaire's attempt in his *Principes de philosophie zoologique* (1830) to unify under a single type the vertebrates and the invertebrates (Le Guyader 1998).[14] By partly relying on facts coming from both the animal series and the embryological series, St-Hilaire joined in what Johann Meckel had already fully expanded upon by considering the two factual series as perfect mirror images of each other in their development (the recapitulation theory), to which Louis Agassiz later added a third series with the data coming from the fossil record, thus offering a threefold parallelism or correspondence (Gould 1977b: 33–68). These developments in morphology, especially as expressed by Johann Herder in the late eighteenth century, were sometimes assumed to be compatible with panpsychism, thus reintroducing the notion of mind permeating the whole world (Skrbina 2005: 112–115). This being said, the fact remains that a more naturalistic organicism gradually imposed itself as the nineteenth century progressed.

Mechanism A declining organicism was replaced by a new dominant system of thought during the Scientific Revolution in the sixteenth, seventeenth, and eighteenth centuries: the mechanistic view. Whereas mechanism had existed previously—traces of it can be found in the atomism of Epicurus and Lucretius (Lenoble 1969: 98–101, 120–121)—this older manifestation cannot be compared with the depth and pervasiveness of its revival in Galileo, Mersenne, Descartes, and Bacon, especially in light of the desire of these latter scholars to manipulate nature rather than to approach it passively (Lenoble 1943; Dijksterhuis 1961[1950]; Westfall 1977[1971]; Rossi 1996[1962]). Like the original organicism, mechanism had universal pretentions of explaining everything. Abandoning the analogy of a living and minded universe, the entire world was described as a lifeless and mindless machine, that is, a clockwork universe under the purview of a Supreme Engineer. The implications of this new analogy were numerous. A quantitative strain of neo-Pythagoreanism ("Christianized Platonism") conceived the book of this mechanical universe to be "written in mathematical language," to quote Galileo's in *The Assayer* (1623) (Rupert Hall 1981[1963], Chap. 3). Furthermore, because machines are constituted of different parts, their functioning can only be comprehended by breaking them apart, hence the reductionistic tendency accompanying this view of atomism. Unlike the organicist doctrine that stipulates that wholes are by definition self-regulated entities, mechanism implies that the bodily

[14]By that time, St-Hilaire had largely given up on the notion of humankind as the reference model for all living forms. This was not the case for all the scholars between the late eighteenth century and first half of the nineteenth century who sought a unity of types, for instance, Johann Herder, Lorenz Oken, and Étienne Serres.

components of machines had to be designed to perform the task for which they were conceived. Indeed, machines cannot be a collection of randomly assembled parts, as each of them is specifically designed to perform a task in a functional and contrived whole (teleology). This, in part, explains the close ties between mechanism and natural theology during the Scientific Revolution, as expressed in the views of Robert Boyle and Isaac Newton (Westfall 1958). By extension, and following logically, since a lifeless machine cannot create itself, an intelligent Designer outside itself is required (Brooke 1991, Chap. 4). While this conceptual framework is sometimes referred to as "Christianized Aristotelianism," this expression falls short of accounting for all its dimensions. From a mechanistic perspective of contrived and mutually dependent parts, one easily embraces the idea of attributed roles and places in the economy of nature (Worster 1994, Chap. 2). In the mechanistic doctrine, order exists either as a transcendent design in the Creator's plan or as an immanent ontological reality of an eternal world always identical to itself, with no past and no future different from its current state. Because machines cannot be modified without breaking down, motion is allowed in a mechanistic world so long as balance or equilibrium is maintained, as can be seen in Isaac Newton's cyclical solar system of resolving planets, James Hutton's never-ending three-stage geological cycle (erosion, stratification, uplift), Linnaeus's contempt for varieties as deviations from the norm, and Buffon's degenerating families under unfavorable environmental conditions. The mechanistic world constitutes a perfectly stable and static one in the sense that change in one direction is compensated for by change in the opposite direction. Motion or change can only be superficial, as it conceals something more fundamental: a permanent and complete or fully constituted world.

Universal Evolutionism Although the spirit of mechanism has persisted to the present day, it gradually started to lose ground in the eighteenth century as part of a view applicable to all the world's entities, confining itself more and more to fields such as physics and physiology. In the meantime, a new and dominant system of thought, a *universal evolutionism* encompassing all entities, slowly imposed itself. Undoubtedly, the idea of "evolution" had appeared under many guises in the past. At this time, however, it acquired a universality. One scientific area after another turned to it, including geology, astronomy, biology, human history (Bowler 2003). In the field of natural history, the static, ordered, and contrived view of Ray and Linnaeus was challenged by a dynamic, causal, non-teleological, and time-oriented approach, as seen, for instance, in Buffon's *Histoire naturelle* (1749–1789) and Jean-Baptiste Lamarck's *Philosophie zoologique* (1809). Of an even wider scope are Robert Chambers's *Vestiges of the Natural History of Creation* (1844) and Herbert Spencer's Synthetic Philosophy as exposed in *The Principles of Psychology* (1855), *First Principles* (1862), the *Principles of Biology* (1864–1867), the *Principles of Sociology* (1876–1896), and the *Principles of Ethics* (1892–1893). Whereas Chambers's view combined with the fundamental idea of the chain of being, Spencer's was fused with atomism. Other such expressions are also found in C. Lloyd Morgan's *Emergent Evolution* (1923), Jan Smuts's *Holism and Evolution* (1926), and Pierre Teilhard de Chardin's *Le Phenomène humain* (1955), among others. The theory of an expanding universe is perhaps today the strongest illustration of the universality of this evolutionism.

Philosophically speaking, if we were to proceed to an ultimate synthesis or simplification of our five *fundamental ideas* and three *dominant systems of thought,* it could be said that the quest to understand the world is confined to two aspects only: (1) to explain how the world's entities come about or what kind of entities are being generated and (2) to explain the nature of such entities either by looking at their inherent structure or at how they are organized externally. More often than not, the eight ideas/thought systems reviewed here contributed to both aspects. Of course, the specificity of any given view stems from giving these two aspects meaning or content, as exemplified in a number of cases above. At this point, one may want to raise the following question about our intellectual grid: how does it relate to Darwin? Let us insist on two important points. First, Darwin's theory of biological evolution appeals to entities (traits, organisms, varieties, species) that are transformed under the filtering and channelling action of an external organizing agent, that is, the law of natural selection as applied in the context of places in the economy of nature. Darwin's theory is, therefore, about how various kinds of entities come about and how they become organized. Second, the abstraction process that was required in exposing fundamental ideas and dominant systems of thought reveals the full ontological, conceptual, and theoretical breadth that is necessary to embrace a formula as apparently simple as "the transformation of species under the action of natural selection." There is much more to understanding Darwin than this simple biological formula would suggest. Unpacking it will be our main task in the following chapters.

The intellectual grid sketched here is comprehensive enough and, at the same time, simple enough to allow us to keep sight of the key strands composing its network. It would have been possible to multiply the number of specific interpretations given to these *fundamental ideas* and *dominant systems of thought.* Doing so would merely have made this intellectual network denser still; approaching what we have already suggested was a seamless fabric of ideas since ancient times. It is the density of this fabric that permits Darwin scholars to move across its network and give meaning to his work.

1.3 The Multiple Faces of Darwin: John C. Greene, Michael Ruse, and Robert J. Richards

Equipped with this intellectual grid, we are now in a better position to review where Charles Darwin has recently been positioned by John C. Greene, Michael Ruse, and Robert J. Richards. It is useful to pay attention to the views of these three prominent Darwin scholars for the following reason: each has clearly delineated quite different inspirations underlying Darwin's ideas. Collectively, they thus have covered a significant portion of our intellectual grid.

Considering that we are talking about evolution or transformation over time, it seems quite intuitive to assume that Darwin owed something to the doctrine of atomism, as has been discussed above. Indeed, this claim is explicitly defended by Greene, who holds that Darwin should be seen as part of the larger tradition that sees the universe as a law-bound system of matter in motion, a system that moved from a mechanistic context in the seventeenth century to the one provided by universal evolutionism in the nineteenth century.[15] According to Greene, the revival of atomism in the seventeenth century came under two possible interpretations, one incorporating it into a dominant static worldview of designed and contrived entities (as in Newton), the other in a less dominant dynamic worldview of a self-contained system of matter in motion (as in Descartes). This latter worldview was eventually to become dominant by the nineteenth century, at the end of a long road that stretched from Robert Boyle's *Origin of Forms and Qualities* (1666) to Charles Darwin's *Origin of Species* (1859), passing major signposts like Descartes, Buffon, Kant, Lamarck, and Lyell along its way. In the following quote, Greene explains how a static view is put in evolutionary motion by breaking nature down with corpuscularism (into traits, varieties, species) and organized or channelled under competition:

> "The habit of regarding the physical environment as subservient to the sentient creation then lost its hold, and the adaptation of living organisms was seen to be a matter of dire necessity rather than a wise contrivance. Chance and struggle, the antitheses of pre-established harmony and providence, claimed a share in the process of creation as biologists sought to understand the production of new varieties and species". (Greene 1959: 13)

Whereas John Ray's and Carl von Linnaeus's natural history was entirely turned toward the static, designed, and permanent taxonomic structure of life forms, the new natural history of Buffon, for instance, was dynamic, causal, non-teleological, and time-oriented. It sought to explain how the motion of organic molecules organized itself under both the uniformity of laws and biological realities such as generation, inheritance, and variation. Once environmental change is acknowledged as a fact, Greene goes on, the issue of how life forms maintain themselves can no longer be evaded. For Lamarck, evolution became the universal means of avoiding extinction, as life forms were believed to be endowed with innate capacities to change. For Lyell, a mix of piecemeal extinctions of species and special creation became the explanation, as he could not subscribe to the idea of an unlimited capacity for variation, which would imply transformation. The guiding principle of Lyell's approach was elimination, not evolution. According to Greene, Darwin sought a mechanism of evolution based on natural selection by accepting from Lamarck the universality of evolution while keeping Lyell's elimination process in mind. Far from believing that Darwin was alone in his understanding of things, Greene argues that Alfred Wallace, Thomas Henry Huxley, and Herbert Spencer were all committed to the same view:

[15]For the view of John C. Greene, consult especially Greene (1957: 60, 63–64; 1959: 11–13, 25–29, 36; 1971: 6–9, 11–13, 16–21; 1981: 128–157; 1986: 204–210).

> "...Spencer, Darwin, Huxley, and Wallace reached similar views concerning nature, man, God, history, society, and science about 1860, converging on a world view that I propose to designate as Darwinism, although Spencer could rightfully demand that it be called Spencerianism... Each of these four men, in his own way, had fused several streams of Western thought into a unitary view of nature-history as a continuum undergoing progressive change in accordance with fixed laws discoverable by science. From the seventeenth-century revolution in physics and cosmology they derived their conception of nature as a wisely ordained, law-bound system of matter in motion. From Descartes, Kant, Laplace, and others they inherited the idea of deriving the present structures of the physical universe from a previous, more homogeneous state of the system of matter in motion through the operation of natural laws. From Hutton and Lyell they learned to view the surface of the Earth as a material system producing perpetual changes in geological structures and constantly modifying the conditions of existence for living organisms. From Erasmus Darwin and Lamarck they got the idea that the wisely ordained system of nature was capable not merely of producing changes in the physical world and in the conditions of life but also of generating simple organisms capable of undergoing progressive complication in structure and faculties in response to the demands of the changing environment, evolving man from a monad or living filament. From social theorists like Comte they acquired the vision of a social science of historical development that would discover the necessary laws and stages of human progress. From the British school of political economy and from British culture generally they derived their faith in the beneficent effects of competitive struggle". (Greene 1981: 148–149)

Atomistic motion of biological entities channelled along a progressive line under external pressure is what Greene sees mainly at work in Darwinism. Indeed, as he himself summarized: "To a considerable extent Spencer, Darwin, and Huxley thought of themselves as simply extending to the world of life and history the conception of nature... as a law-bound system of matter in motion." (Greene 1986: 206)[16]

Whereas Greene's thesis seems readily compatible with the idea of biological evolution, Michael Ruse's view is less intuitive especially when considering that Darwin's ideas are said to be largely founded on a mechanistic approach to nature with its constitutive living parts being under the organizing principle of teleology (adaptation).[17] Greene stressed the verticality of change: corpuscular matter is projected forward, ascending in evolutionary complexity. Ruse, on the other hand,

---

[16]As we have seen at the beginning of this chapter, Greene recognizes a number of internal tensions in Darwin's arguments. Greene (1986: 209–210) explains why this is the case in the same paper: "Thus, the triumph of the Cartesian program of deriving the present structures of nature from previous states of the law-bound system of matter in motion produced a worldview containing ideas inconsistent with the mechanical cosmology that had helped to give it birth. These new ideas—progress and the concomitant idea of levels of being; mind as a part of nature; the functional unity and interdependence of organisms; struggle, chance, and individuality as real factors in nature—rested in uneasy tension alongside conceptions of nature and natural science derived from seventeenth-century physical science." While Greene is certainly right to point out a number of such internal tensions in Darwin's views, this book will propose a different reason for their occurrence.

[17]The ideas of design, teleology, and natural theology in science have long been recognized in the historiography, as seen especially in Great Britain between the seventeeth and nineteenth centuries. Among a long list of possible references, consult Gillispie (1951), Westfall (1958), Cannon (1960),

focuses on the horizontality of adaptation: each living form is adapted and designed to fit into a particular place in nature.[18] Ruse positions Darwin within what he himself calls the "traditional interpretation" (Ruse 2004: 4), that is, within the British context of natural history inherited from natural theology permeated with notions such as design, places in the economy of nature, division of labor, and competition (struggle for existence). Biological evolution is thus the outcome of life forms competing to occupy places in the economy of nature: the main evolutionary drive being selective pressure working on organisms and species to eliminate the unfit and select the better fit. More explicitly, among key explanatory components of Darwin's theory, according to Ruse, are the following.

First, in line with Newton's physics, Darwin sought a causal explanation for biological evolution based on the notion of "force," natural selection being analogically conceived as such.

Second, following Newton's method, Darwin structured his arguments in the *Origin of Species* to be consistent with the ideal of *vera causa*: in order to be accepted, a cause must be conclusively shown to be real and not conjectural. Darwin did this in two ways: by demonstrating the factuality of natural selection through artificial selection and by showing that the cause of natural selection is a unifying one (consilience of inductions) since it can explain a series of independent facts in fields like biogeography, embryology, and morphology.

Third, artificial selection itself relates to the British context of the agricultural and industrial revolution in two interconnected ways: manipulations allow the improvement of crops and cattle; the same manipulations show that organisms are not just parts thrown together but are rather designed and adapted functional wholes organized in accordance to their well-being (teleology). This, Darwin learned from the natural theology of William Paley and the functional morphology of Georges Cuvier (conditions of existence).

Fourth, from the English cleric and scholar Thomas Malthus, Darwin took the idea of overpopulation as a means of generating selective pressure, a key condition for evolutionary change.

Fifth, from the Belgian-born Henri Milne Edwards (but ultimately going back to the Scottish political economist Adam Smith's notion of division of labor), Darwin drew inspiration for the idea that more places in the economy of nature would be available for specialized organisms than for generalized ones.

Most central to Ruse's thesis is Darwin's commitment to the analogy of mechanism in the living world. In the following quote, Ruse goes beyond the *Origin* by referring to a book on orchids published by Darwin in 1862:

> "[T]here is the question of mechanisms and machines... In this book, Darwin was laying out evolutionary biology as he hoped it would be done. 'I think this little book will do good to

---

Tuan (1968), Yeo (1979), Ospovat (1981), Cornell (1983), Gillespie (1987, 1990), Deacon (1986), Brooke (1991), and Gould (1998).

[18]For the view of Michael Ruse, especially consult Ruse (1996: 136–177; 1979[1999]; 2003: 91–149; 2004). See also Richards and Ruse (2016).

the *Origin*, as it will show that I have worked hard at details, and it will perhaps, serve [to] illustrate how natural History may worked under the belief of the modification of species' (Letter to his publisher, John Murray, September 24, 1861). It is teleological in the machine sense throughout. The very title flags you to this fact: *On the Various Contrivances by which British and Foreign Orchids are Fertilized by Insects*... Darwin was looking at the organic world as if it were an object of design by one of his grandfather's manufacturing friends: he was taking organized end-directed complexity as the absolutely crucial key to unlocking the secrets of the living world and its attributes. Contrivances are human-made objects, which are created with an end in view... This was Darwin's perspective on the living world, just as it had been for Paley". (Ruse 2004: 16)

Finally, as much as Darwin accepted the notion of biological progress, he did so not by embracing a blind progressionism or vital tendency but rather as a simple consequence of a selection process that generated new, improved forms (as compared to other existing ones) in the struggle for existence, a fundamentally British biology, according to Ruse.

Whereas Greene sees Darwin's theory as based on matter in motion projected forward (*atomism-mechanism*) and Ruse on the adaptation of forms (*teleology-mechanism*), Robert J. Richards's holds that Darwin was greatly influenced by the German Romantic movement and that, consequently, his theory contains strong strains of *Naturphilosophie*.[19] To use the terminology developed in the intellectual grid sketched above, Richards places Darwin within *evolutionary monism-organicism*.[20] At the general level, *Naturphilosophen* view nature as a single, unified, dynamic, and self-creating compound internally driven toward humans and purposively organized into several components (individual organisms, archetypes, nature as a whole). At the more specific level, continues Richards, nature is founded on the following ideas.

First, under the influence of scholars like Schelling and Goethe, and through the more immediate contact of Joseph Henry Green and Richard Owen, Darwin embraced the idea that life forms are regrouped under a number of archetypes out of which biological diversity is generated. What Darwin added to this idea is the incarnation of the archetype in a common ancestor: "Darwin thus suggested that the generalized archetype of the vertebrates did not lie hidden away as an idea in the mind of God; rather, it was the form of a creature that walked the earth many generations ago" (Richards 1999: 129).

Second, crucial to *Naturphilosophie* and its inherent organicism of a monistic brand are analogies or correspondences in nature across its several parts, such as between the zoological series, embryological development, and the fossil record. This program was filled out by Darwin in his own original way: "Darwin continued to reiterate this principle of the embryo recapitulating the adult stages of the ancestors through all editions of the *Origin*... But he actually reversed the

---

[19]To a certain extent, Phillip R. Sloan (2001, 2005) also acknowledges a connection between Darwin and *Naturphilosophie*.

[20]For the view of Robert J. Richards, consult Richards (1999, 2000, 2002, 2004, 2005). See also Richards and Ruse (2016).

methodological situation: he suggested that recapitulation might serve to confirm the progressive character of the paleontological record" (Richards 2000: 77–78).

Third, in the monistic doctrine of *Naturphilosophie*, matter and mind constitute two features of the same original complex as such, a kind of absolute mind is assumed to permeate nature since the beginning of time, although it is only clearly revealed phenomenally with the rise of complex life forms and humans. "Natural selection, in Darwin's image-driven language, displays patently the attribute... of intelligence... [Nature operates] like a subtle and refined mind that can direct development in an altruistic and progressive way." (Richards 2004: 33)

Fourth, evolutionary progress is assumed to be a natural outcome of a creative drive whose dynamics are visible in the various forms generated out of a single archetype, a drive, or causal force not incompatible with mundane physical forces, as expressed in Darwin's theory of natural selection.

Fifth, as expressed by Alexander von Humboldt, nature is a harmoniously unified network of integrated and functioning parts, a notion Darwin applied in the context of the "entangled bank":

> "The Nature that Darwin experienced with the aid of Humboldt was not a machine, a contrivance of fixed parts grinding out its products with dispassionate consequence. The nature that Darwin experienced was a cosmos, in which organic patterns of land, climate, vegetation, animals, and man were woven into a vast web pulsating with life... [t]he God of Paley had begun to give way to the God of nature, soon to be completely transmogrified into the romantics's *Deus sive Natura*". (Richards 1999: 124–125)

Sixth, like other *Naturphilosophen*, Darwin viewed nature through a moral lens. This value-laden conception can be seen in the way he conceives natural selection as a goal-directed and active agent: "Darwin never referred to or conceived natural selection as a mechanistic principle, and the nature to which selection gave rise was perceived in its parts and in the whole as a teleologically self-organizing structure" (Richards 1999: 130). Natural selection proceeds by altruistically selecting the good variations, casting out the bad ones, as it works for the good of each being and toward progress and perfection (with death generally coming promptly).

Finally, with *Naturphilosophie* comes the idea of a full experience of the world which goes beyond the mere scientific dimension. The unified whole is also grasped through a sort of poetic prose and sensibility often expressed by Darwin in numerous metaphors and images about nature, according to Richards.

1.4 A Static Worldview: The Main Theses of This Book

At first glance, it is difficult not to be bewildered by the wide disparity of theses defended by Darwin scholars in recent decades. Why is that? One can think of several reasons. Darwin's view is a multifaceted one: it has several intellectual components, whether we elect to call them *strains* and *layers* or establish a distinction between fundamental and superficial aspects. This explains why John C. Greene, Michael Ruse, and Robert J. Richards openly recognize that beyond the main source of inspiration each identifies as instrumental to Darwin's work, others influenced him (Greene 1971: 9–23; Ruse 2004: 9; Richards 2004: 34). We are not, therefore, confronted with three largely independent historiographical schools—*atomism-mechanism* for Greene, *teleology-mechanism* for Ruse, and *evolutionary monism-organicism* for Richards—but are instead faced with a complex network of ideas approaching a seamless fabric. The goal of placing all of Darwin's work within a single intellectual tradition is long gone. The rule guiding the debate over Darwin's soul, so to speak, consists in evaluating the relative importance of these inspirations for Darwin's various projects. Since the whole issue is a simple matter of the relative impact of each of these *strains* or *layers*, it is understandable why scholars might disagree over shades of difference, no matter how significant these might be.

However, the intellectual tools of historians and philosophers have rarely served them well in this pursuit. Gross labels such as "atomism," "mechanism," and "organicism" blend together to a degree greater than we usually care to recognize. Surely, to speak of biological evolution as a law-bound system of matter in motion is revealing, but only up to a certain point when talking about biological entities. Similarly, to compare nature to a machine has its obvious limitations, for, ultimately, how "mechanical" can living organisms and populations actually be? And to claim that a monistic and organic nature is permeated by analogies or correspondences can only mask some phenomenal realities and differences, unless one is prepared to follow Parmenides all the way in asserting that nature is one, timeless, uniform, and unchanging. My comment here is not meant to be critical; on the contrary, these are clearly images and analogies necessary for scholarly work. Yet, their inherent imprecision can only fuel differences of interpretation. In addition, one must also recognize that notions such as "biological progress," "evolution by competition," "close connection between embryology and the animal series," among many others, are neutral enough to be easily incorporated in the three views reviewed above. The apparent disparity of the theses promoted by Greene, Ruse, and Richards, we would argue, is not as wide as it may seem, especially once the distorting effect of our imprecise intellectual tools is accounted for.

This being said, their views rendered a great service to the Darwin industry by demonstrating that a wide intellectual net is necessary in order to fully understand Darwin. If we were to characterize their collective effort, we would say they presented us with an intellectual grid composed of three largely overlapping research programs running in parallel between the seventeenth and nineteenth centuries, as seen in *atomism-mechanism*, *teleology-mechanism*, and *evolutionary monism-organicism*, respectively. The vertical axis of our intellectual grid has therefore been fairly well covered, a sure sign of maturity in the Darwin studies. Nonetheless, the jury is still out on such studies, since among the historiographical pieces still missing, a key but neglected one concerns the horizontal axis: the transition from a static worldview in the seventeenth century to a universal evolutionism in the twentieth century. This question reintroduces into Darwin studies the connection between two successively dominant worldviews or systems of thought.[21] We argue in this book that Darwin's work rests significantly on the older worldview, one that is largely implicit and which subtly yet profoundly insinuates itself into many of his explanatory components and upon which Darwin himself never really entertained us, if only because metaphysical or ontological assumptions are rarely openly addressed.

It is all the more difficult today to detect Darwin's hidden worldview, considering that he is believed to be one of the key figures (if not *the* key figure) in charting the new territory of biological evolution. For most of us, a large part of our motivation in reading Darwin consists in seeking explanations about evolution. Unsurprisingly, this is precisely what we find. The motto "I found what I was looking for" is never more true than in Darwin's case. At this juncture, a liberating question can be raised in order to shake off the weight of the historiographical tradition: Was Darwin really an evolutionist? We typically assume that he was, especially in light of the public splash that surrounded the reception of the *Origin of Species* in 1859. But was this turmoil truly about the idea of evolution itself, an idea that had been slowing making its way among scientists during the pre-1859 decades (see Bowler 1988)?

Interestingly, it is more difficult than it might seem to explain the reasons for this reception among scholars and laymen alike. Was it because Darwin proposed an evolutionary mechanism which, for the first time, really seemed credible? Was it because Darwin's work had implications for human beings and ethics in general? Was it because Darwin was a respected member of the conservative British scientific establishment, which may have lent credence to his view? Was it because Darwin's work had materialistic implications in an age of weakening theological convictions? Was it because Darwin's publication had just the right timing, thus igniting a scientific and theological atmosphere ripe for such a conflagration?

The search for the reasons behind this turmoil is most intriguing, considering that Darwin's main mechanism of evolution—natural selection—was never largely accepted in the decades following its original publication (Hull 1973). There seems to be at the same time much more and much less to this publication than the idea of evolution and its accompanying mechanism. Once this is recognized, one can retreat from these scientific and public controversies and think again about the real Darwin. Thanks to Darwin's rhetorical skills, we tend to read more into the *Origin* than is really there. The economy of Darwin's thoughts in the *Origin* requires that we engage it with a judicious weighting, one that consists in clearly distinguishing between the levels of arguments.

Asking the apparently preposterous question "was Darwin an evolutionist?" has the merit of pointing at the numerous ways that have been used to explain change or transformation. Human beings have long acknowledged that they live in a world characterized by change: entities around them are changing, including themselves, as part of the aging process. "Change" or "transformation," therefore, is not the issue.

[21] We say "reintroduces" rather than "introduces" because the 1960s and the 1970s saw considerable discussion of Kuhn's paradigms and Foucault's *épistémè*. See Kuhn (1970) and Foucault (1966).

Rather, the issue is how to account for that change: Is it real, or merely apparent and superficial, leaving intact the essence of things? After all, an aging human being remains the same in some fundamental ways. It is beyond the scope of this book to review the full range of theoretical possibilities and subtleties in the intellectual spectrum, ranging from the recognition of a total lack of change to a comprehensive, complete, and total change.[22] For Darwin, as will be seen in the *Origin*, little room is provided to account for *genuine* evolutionary novelties. The worldview to which Darwin subscribed was elaborated in the seventeenth and eighteenth centuries, and change or motion, although accepted, was explained away as a recurring process, as was seen in Isaac Newton's cyclical solar system of resolving planets, James Hutton's never-ending three-stage geological cycle (erosion, stratification, uplift), and Linnaeus's contempt for varieties as deviations from the norm. To quote John Greene: "Change was recognized as a real aspect of nature, but a superficial aspect. It contributed variety to nature's panorama, but it could not alter her fundamental structures" (Greene 1957: 61). Understood in this way, motion or change conceals something more fundamental: a permanent, complete, or fully constituted world. It will be seen in the following chapters how Darwin stretched the conceptual envelope of this worldview to accommodate his own understanding of evolutionary motion. So, whereas Darwin is usually thought to have squarely embraced the worldview of "universal evolutionism," we hold that he never quite fully and successfully completed the transition from the previous dominant static worldview.

Let us be more explicit about the main theses presented in the following chapters. It should first be mentioned that this book is not a direct extension of any of the three views just described (those of Greene, Ruse, Richards). The arguments to be presented here are meant to fill a void in the intellectual grid already worked out in the Darwin studies. This being said, the theses developed here are not without relations to the *teleology-mechanism* of Michael Ruse and the *atomism-mechanism* of John C. Greene.[23] It will be argued, however, that what Darwin owes to mechanism, atomism, and teleology is the phenomenal (superficial) manifestations of a more profound ontological commitment toward a static worldview. Our analysis will take us to the junction of the mechanistic view of nature, natural theology, and the Scientific Revolution.

The Scientific Revolution stands as the major intellectual event of the centuries prior to Darwin's *Origin*. It is true that historians of science are now more reserved when talking about an event as broad as the "Scientific Revolution" since it tends to dissolve on closer analysis, as is expected for any complex and changing intellectual movements (Schuster 1990; Cohen 1994; Shapin 1996; Osler 2000; Hellyer 2003). This being said, few of them would go so far as to hold that nothing significant has

[22]Any decent introductory book on Ancient Greek thinkers will offer a range of theoretical possibilities and subtle discussion of this question.

[23]This is not to claim that Robert J. Richards's thesis must not find a place in a complete picture of Darwin. An attempt will be made below to consider some of his views in our own narrative.

happened between 1500 and 1800. The historiographical problem can easily be bypassed by recognizing that this intellectual event was composed of several intellectual strains, some of which served as the foundation for a mechanistic view of nature (Burtt 2003[1954]; Koyré 1939; Butterfield 1949; Hall 1954; Dijksterhuis 1961; Kuhn 1976; Westfall 1977). Among the intellectual threads contributing to this view are the following: (1) the quantification or mathematization of nature, the book of nature is assumed to be written in the language of mathematics; (2) the mechanization of nature, not unlike actual machines, the world is seen as organized like a clockwork mechanism in motion; (3) the corpuscularian view of nature, natural bodies are viewed as consisting of inert particles of matter in motion and nothing else, the whole being nothing more than the sum of its constituent parts; and (4) a new attitude toward knowledge, nature's secrets must be actively pursued through observation and experimentation. Lists of names attached to these intellectual currents vary in the historiography, but scholars often associated with them include Francis Bacon, Robert Boyle, René Descartes, Galileo Galilei, Pierre Gassendi, William Gilbert, William Harvey, Christian Huygens, Johannes Kepler, Pierre-Simon Laplace, and Isaac Newton.

The part of the Scientific Revolution which is of special interest for our thesis is the one that fused with the tradition of natural theology, the inquiry into God's plan for the creation (Brooke 1991, Chap. 4). Their combination allowed for the erection of a lasting worldview depicting it as clockwork, designed, stable, and static. The ramifications stemming from this worldview are profound with respect to our conception of the universe (Collingwood 1945: 5, 14; Greene 1981: 11–14, 33, 44): a machine is in need of an external creator for its existence; it is lifeless and is designed for a specific purpose; it requires an external force to put it in motion; it constitutes a finished, ordered, stable, closed, balanced, static, harmonious, and permanent system; it could not have come into existence in piecemeal fashion since it would not be functional and thus would have to be perfect and fully constituted; it is built of distinct yet complementary parts contrived together; it cannot be modified without breaking down, depriving it of a past or future different from the currently observed state. Our solar system was, at the time, the paradigmatic model of a perfectly ordered and stable mechanics in perpetual motion.

Epistemologically speaking, this world was not a revealed one. God had not given us the plan for His creation. Rather, it was for us to use our ingenuity to wonder and investigate the Creator's accomplishment in order to find out and understand the plan of the creation. If the combination of empiricism and inductivism became so central a doctrine for knowledge acquisition during those centuries (irrespective of whether or not it was strictly followed), it is precisely because the world was believed to be fully constituted or finished. Indeed, our curiosity could not be deceived in a world "open as a book," created as it was for us to read into its completeness.

Was Charles Darwin exposed to these ideas in the nineteenth century? He was, through several channels. Jonathan Hodge is certainly right to stress that no historical narrative can support the notion of a Darwin directly and massively engaging with the great figures of the Scientific Revolution, in readings or otherwise. Rather, one must resort to a weaker, more diffuse narrative in which Darwin received indirect influences through the intermediary of scholars of the second half of the eighteenth century and the first half of the nineteenth century (Hodge 2014: 177–178). This being said, the vitality and strength of these indirect influences during Darwin's own pre-*Origin* years should not be underestimated, especially when incarnated in the central and iconic figure of Isaac Newton (Bellon 2014). Intellectual bridges connecting Darwin to major ideas of the Scientific Revolution were numerous.[24] For the moment, let us insist on four such bridges only. First, by reading William Paley's *Natural Theology* (1802), Darwin encountered the design argument as applied to life in its most traditional presentation (Ruse 2003: 110). Second, with his personal acquaintance with Charles Lyell and his now-classic *Principles of Geology* (1830–1833), Darwin could see the application of the uniformity principle in the context of a steady-state world (Hodge 1982). Third, Darwin came across the crucial notion of a perfectly balanced economy of nature founded both on the mutual dependence and encroachment of living forms by reading the English translation of Carl von Linnaeus's treatises, *Oeconomia Naturae* (1749) and *Politia Naturae* (1760) (Pearce 2010: 495–496). Finally, the empirico-inductive tradition linking Darwin to Newton, through Thomas Reid, was open to him with the personal acquaintance of John Herschel and his *A Preliminary Discourse on the Study of Natural Philosophy* (1831) (Ruse 1979: 61). These channels were anything but impermeable to each other and were all rooted in ideas of the previous centuries.

Darwin absorbed ideas from the Scientific Revolution. He did not apply them passively but adapted them in such a way as to come up with a theory of biological evolution. The result was a very peculiar theory of biological evolution in the *Origin of Species* that owed much to conceptual tools inherited from the context of a world believed to be static. Indeed, what was so central to Darwin's explanatory core was the idea of "completeness," understood here to be the synthesis of mechanism, atomism, teleology, and natural theology: if the world today is believed to be complete, the evolutionary past has little relevance for explaining its current state; if the world is assumed to be complete, the evolutionary past will always be less than the present; if the world is acknowledged to be complete, its investigation will be most fruitfully conducted through neontology (present) rather than paleontology (past); if the world is held to be complete, empirico-inductivism constitutes a tailor-made approach for studying a world already exposed to us in full. In other words, we are claiming that what profoundly organized Darwin's commitments toward mechanism, atomism, teleology, and natural theology is the notion of *completeness*: a steady-state ontology that underlies and binds the other explanatory components together.

Our analysis of the *Origin of Species* will proceed as follows. In Part 1, it will be argued that Darwin built an evolutionary theory on the notion of a fully constituted world. This is seen is his utilization of concepts such as stability, completeness, timelessness, unity, permanence, and uniformity. Unsurprisingly, Darwin will be at

[24]See especially the contributions contained in the special issue "Charles Darwin and Scientific Revolutions," *Endeavour*, Vol. 38, No. 3–4, 2014.

pains to promote a truly open evolutionary process generating genuine evolutionary novelties. In Part 2, we will review Darwin's profound commitment to a "horizontal" approach to evolution: to travel in geographical space (today) is to travel in geological time (past). We will do so by proceeding to a thorough analysis of the *Origin of Species*, systematically moving across his book. While embracing this approach, Darwin resorts to such concepts as order, harmony, equilibrium, symmetry, and contrivance, with few exceptions that will serve as profound revealers of his ultimate ontological-methodological choices. We will argue that far from presenting a unified theory of evolution—irrespective of his efforts—Darwin managed to produce only a series of competing pictures of evolution. In Part 3, it will be seen that for each key explanation Darwin provides in favor of directional change or progress, he also offers a counterargument. In this explanatory context, Darwin's theory is more about maintaining an evolutionary equilibrium than producing an evolutionary directionality. In Part 4, we will argue that since Darwin's view of evolution is congenial to a static and fully completed world, it made perfect sense for him to support the methodologies best suited to study that kind of a world, namely, a science based on *vera causa*, empiricism, inductivism, gradualism, and uniformity. It was Darwin's challenge and evidence of originality, to find ways to "loosen up" a bit these notions devised for a steady-state view in order to generate some kind of evolutionary change.

References

Algra K (1999) The beginnings of cosmology. In: Long A (ed) The Cambridge companion to early Greek philosophy. Cambridge University Press, Cambridge, pp 45–65

Appel T (1987) The Cuvier-Geoffroy debate: French biology in the decades before Darwin. Oxford University Press, Oxford

Barnes J (1982) The Presocratic philosophers. Routledge, London

Bellon R (2014) There is grandeur in this view of Newton: Charles Darwin, Isaac Newton and victorian conceptions of scientific virtue. Endeavour 38:222–234

Bowler PJ (1983) The eclipse of Darwinism. Johns Hopkins University Press, Baltimore

Bowler PJ (1988) The non-Darwinian revolution: reinterpreting a historical myth. Johns Hopkins University Press, Baltimore

Bowler PJ (2003) Evolution: the history of an idea, 3rd edn. University of California Press, Berkeley

Brooke JH (1991) Science and religion: some historical perspectives. Cambridge University Press, Cambridge

Brunschwig J (1998) La philosophie à l'époque hellénistique. In: Canto-Sperber M (ed) Philosophie grecque. Presses Universitaires de France, Paris, pp 457–591

Burtt E (2003[1954]) The metaphysical foundations of modern science. Dover, New York

Butterfield H (1949) The origins of modern science, 1300–1800. G. Bell & Sons, London

Cannon W (1960) The problem of miracles in the 1830s. Vic Stud 4:4–32

Cherniss H (1970) The characteristics and effects of Presocratic philosophy. In: Furley D, Allen R (eds) Studies in Presocratic philosophy, vol 1. Routledge & Kegan Paul, London

Churchill FB (1982) Darwin and the historian. Biol J Linn Soc 17:45–68

Cohen HF (1994) The scientific revolution: a historiographical inquiry. University of Chicago Press, Chicago

Collingwood RG (1960[1945]) The idea of nature. Oxford University Press, Oxford

Cornell J (1983) From creation to evolution: Sir William Dawson and the idea of design in the nineteenth century. J Hist Biol 16:137–170
Corsi P (1988) The age of Lamarck: evolutionary theories in France, 1790–1830. University of California Press, Berkeley
Darwin C (1859) On the origin of species. John Murray, London
Darwin C (1872) The origin of species, 6th edn, with additions and corrections [1876]. John Murray, London
Darwin F (1909) The foundation of the origin of species: two essays written in 1842 and 1844 by Charles Darwin. Cambridge University Press, Cambridge
Darwin C (1959) The origin of species: a variorum text, Peckham M (ed). University of Pennsylvania Press, Philadelphia
Deacon G (1986) Reformation theology and the mechanistic conception of nature. In: Lindberg D, Numbers R (eds) God and nature. University of California Press, Berkeley, pp 167–191
Delisle RG (2017) From Charles Darwin to the evolutionary synthesis: weak and diffused connections only. In: Delisle RG (ed) The Darwinian tradition in context: research programs in evolutionary biology. Springer, Cham, pp 133–168
Depew D, Weber B (1995) Darwinism evolving: systems dynamics and the genealogy of natural selection. MIT Press, Cambridge
Desmond A, Moore J (1991) Darwin: the life of a tormented evolutionist. W.W. Norton, London
Dijksterhuis E (1961) The mechanization of the world picture. Clarendon Press, Oxford
Dobzhansky T (1937) Genetics and the origin of species. Columbia University Press, New York
Dobzhansky T (1951) Genetics and the origin of species, 3rd edn. Columbia University Press, New York
Duchesneau F (1982) La physiologie des lumières: empirisme, modèles et théories. Martinus Nijhoff, La Haye
Duchesneau F (1998) Les modèles du vivant de Descartes à Leibniz. J. Vrin, Paris
Fisher RA (1930) The genetical theory of natural selection. Clarendon Press, Oxford
Foley V (1976) The social physics of Adam smith. Purdue University Press, West Lafayette
Foucault M (1966) Les mots et les choses: une archéologie des sciences humaines. Gallimard, Paris
Gillespie NC (1987) Natural history, natural theology, and social order: John ray and the 'Newtonian ideology'. J Hist Biol 20:1–49
Gillespie NC (1990) Divine design and the industrial revolution: William Paley's abortive reform of natural theology. Isis 81:214–229
Gillispie CC (1951) Genesis and geology: a study in the relations of scientific thought, natural theology, and social opinion in Great Britain, 1790–1850. Harvard University Press, Cambridge
Gilson É (1989[1964]) Le thomisme: introduction à la philosophie de Saint Thomas d'Aquin. J. Vrin, Paris
Gould SJ (1977a) Eternal metaphors of palaeontology. In: Hallam A (ed) Patterns of evolution, as illustrated by the fossil record. Elsevier, Amsterdam, pp 1–26
Gould SJ (1977b) Ontogeny and phylogeny. Belknap Press, Cambridge
Gould SJ (1998) On transmuting Boyle's law to Darwin's revolution. In: Fabian A (ed) Evolution: society, science and the universe. Cambridge University Press, Cambridge, pp 4–27
Greene JC (1957) Objectives and methods in intellectual history. Mississipi Valley Hist Rev 44:58–74
Greene JC (1959) The death of Adam: evolution and its impact on Western thought. Iowa State University Press, Iowa
Greene JC (1971) The Kuhnian paradigm and the Darwinian revolution in natural history. In: Roller D (ed) Perspectives in the history of science and technology. University of Oklahoma Press, Norman, pp 3–25
Greene JC (1975) Reflections on the progress of Darwin studies. J Hist Biol 8:243–273
Greene JC (1981) Science, ideology, and world view. University of California Press, Berkeley
Greene JC (1986) The history of ideas revisited. Rev Synth 3:201–227

Grene M, Depew D (2004) The philosophy of biology: an episodic history. Cambridge University Press, Cambridge
Guthrie W (1962) A history of Greek philosophy, vol. 1: the earliest Presocratics and the Pythagoreans. Cambridge University Press, Cambridge
Guthrie W (1965) A history of Greek philosophy, vol. 2: the Presocratic tradition from Parmenides to Democritus. Cambridge University Press, Cambridge
Guthrie W (1975) A history of Greek philosophy, vol. 4. Plato, the man and his dialogues: earlier period. Cambridge University Press, Cambridge
Haldane JBS (1932) The causes of evolution. Longmans and Green, London
Hall AR (1954) The scientific revolution, 1500–1800. Longmans, London
Hellyer M (2003) Editor's introduction: what was the scientific revolution? In: Hellyer M (ed) The scientific revolution: the essential readings. Blackwell, Oxford, pp 1–15
Hodge MJS (1982) Darwin and the laws of the animate part of the terrestrial system (1835-1837): on the Lyellian origins of his zoonomical explanatory program. Stud Hist Biol 6:1–106
Hodge MJS (1983) The development of Darwin's general biological theorizing. In: Bendall D (ed) Evolution from molecules to men. Cambridge University Press, Cambridge, pp 43–62
Hodge MJS (1985) Darwin as a lifelong generation theorist. In: Kohn D (ed) The Darwinian heritage. Princeton University Press, Princeton, NJ, pp 207–243
Hodge MJS (1989) Generation and the origin of species (1837-1937): a historiographical suggestion. Br J Hist Sci 22:267–281
Hodge MJS (1992) Two cosmogonies (theory of the earth and theory of generation) and the unity of Buffon's thought. In: Gayon J (ed) Buffon 88. J. Vrin, Paris, pp 241–254
Hodge MJS (2014) On Darwin's science and its contexts. Endeavour 38:169–178
Hodge MJS, Radick G (2009) The place of Darwin's theories in the intellectual long run. In: Hodge MJS, Radick G (eds) The Cambridge companion to Darwin, 2nd edn. Cambridge University Press, Cambridge, pp 246–273
Hoquet T (2013) The evolution of the origin. In: Ruse M (ed) The Cambridge encyclopedia of Darwin and evolutionary thought. Cambridge University Press, Cambridge, pp 158–164
Horgan J (1996) The end of science: facing the limits of knowledge in the twilight of the scientific age. Addison-Wesley, Reading
Hull DL (1973) Darwin and his critics: the reception of Darwin's theory of evolution by the scientific community. Harvard University Press, Cambridge
Huxley JS (1942) Evolution: the modern synthesis. Allen and Unwin, London
Jepsen GL, Mayr E, Simpson GG (eds) (1949) Genetics, paleontology, and evolution. Princeton University Press, Princeton
Kahn C (1960) Anaximander and the origins of Greek cosmology. Columbia University Press, New York
Kohn D (1985) Introduction: a high regard for Darwin. In: Kohn D (ed) The Darwinian heritage. Princeton University Press, Princeton, pp 1–5
Koyré A (1939) Études galiléennes. Hermann, Paris
Kuhn TS (1970) The structure of scientific revolutions, 2nd edn. University of Chicago Press, Chicago
Kuhn TS (1976) Mathematical versus experimental traditions in the development of physical science. J Interdiscip Hist 7:1–31
Largent M (2009) The so-called eclipse of Darwinism. Trans Am Philos Soc 99(1):3–21
Largent M (2013) Darwinism in the United States, 1859-1930. In: Ruse M (ed) The Cambridge encyclopedia of Darwin and evolutionary thought. Cambridge University Press, Cambridge, pp 226–234
Le Guyader H (1998) Geoffroy Saint-Hilaire: un naturaliste visionnaire. Berlin, Paris
Lenoble R (1943) Mersenne ou la naissance du mécanisme. J. Vrin, Paris
Lenoble R (1969) Histoire de l'idée de nature. Albin Michel, Paris
Lenoir T (1982) The strategy of fife: teleology and mechanics in nineteenth-century German biology. University of Chicago Press, Chicago

Levit GS, Hossfeld U (2017) Major research traditions in twentieth-century evolutionary biology: the relations of Germany's Darwinism with them. In: Delisle RG (ed) The Darwinian tradition in context: research programs in evolutionary biology. Springer, Cham, pp 169–193
Liepman HP (1981) The six editions of the "origin of species": a comparative study. Acta Biotheor 30(3):199–214
Limoges C (1972) Introduction. In: Linné C (ed) L'équilibre de la nature. J. Vrin, Paris, pp 7–22
Lovejoy AO (1936) The great chain of being: a study of the history of an idea. Harvard University Press, Cambridge
Lovejoy AO (1959) Herder: progressionism without transformism. In: Glass B, Temkin O, Straus W (eds) Forerunners of Darwin: 1745–1859. Johns Hopkins University Press, Baltimore, pp 207–221
Mayr E (1942) Systematics and the origin of species. Columbia University Press, New York
Mayr E (1963) Animal species and evolution. Belknap Press, Cambridge
Oldroyd DR (1984) How did Darwin arrive at his theory? The secondary literature to 1982. Hist Sci 22:325–374
Osler MJ (2000) The canonical imperative: rethinking the scientific revolution. In: Osler MJ (ed) Rethinking the scientific revolution. Cambridge University Press, Cambridge, pp 3–22
Ospovat D (1981) The development of Darwin's theory: natural history, natural theology, and natural selection, 1838–1859. Cambridge University Press, Cambridge
Pearce T (2010) 'A great complication of circumstances'—Darwin and the economy of nature. J Hist Biol 43:493–528
Pichot A (1993) Histoire de la notion de vie. Gallimard, Paris
Renault E (1999) Naturphilosophie. In: Lecourt D (ed) Dictionnaire d'histoire et philosophie des sciences. Presses Universitaires de France, Paris, pp 674–680
Rensch B (1960) Evolution above the species level. Columbia University Press, New York
Richards RJ (1988) The moral foundations of the idea of evolutionary progress: Darwin, Spencer, and the neo-Darwinians. In: Nitecki MH (ed) Evolutionary progress. University of Chicago Press, Chicago, pp 129–148
Richards RJ (1999) Darwin's romantic biology: the foundation of his evolutionary ethics. In: Maienchein J, Ruse M (eds) Biology and the foundation of ethics. Cambridge University Press, Cambridge, pp 113–153
Richards RJ (2000) The epistemology of historical interpretation: progressivity and recapitulation in Darwin's theory. In: Creath R, Maienschein J (eds) Biology and epistemology. Cambridge University Press, Cambridge, pp 64–88
Richards RJ (2002) The romantic conception of life: science and philosophy in the age of Goethe. University of Chicago Press, Chicago
Richards RJ (2004) Michael Ruse's design for living. J Hist Biol 37:25–38
Richards RJ (2005) Darwin's metaphysics of mind. In: Hösle V, Illies C (eds) Darwinism and philosophy. University of Notre Dame Press, Notre Dame, pp 166–180
Richards RJ (2008) The tragic sense of life: Ernst Haeckel and the struggle over evolutionary thought. University of Chicago Press, Chicago
Richards RJ, Ruse M (2016) Debating Darwin. University of Chicago Press, Chicago
Roger J (1989) Buffon: un philosophe au Jardin du Roi. Fayard, Paris
Roger J (1993 [1963]) Les sciences de la vie dans la pensée française au XVIIIe siècle. Albin Michel, Paris
Roger J (1995) L'histoire naturelle au XVIIIe siècle: de l'échelle des êtres à l'évolution. In: Pour une histoire des sciences à part entière. Albin Michel, Paris, pp 237–251
Rossi P (1996[1962]) Les philosophes et les machines, 1400–1700. Presses Universitaires de France, Paris
Rupert Hall A (1981[1963]) From Galileo to Newton. Dover, New York
Ruse M (1979) The Darwinian revolution. University of Chicago Press, Chicago
Ruse M (1996) Monad to man: the concept of progress in evolutionary biology. Harvard University Press, Cambridge

Ruse M (1999) The Darwinian revolution, 2nd edn. University of Chicago Press, Chicago
Ruse M (2003) Darwin and design: does evolution have a purpose? Harvard University Press, Cambridge
Ruse M (2004) The romantic conception of Robert J. Richards. J Hist Biol 37:3–23
Schabas M (2005) The natural origins of economics. University of Chicago Press, Chicago
Schmitt S (2004) Histoire d'une question anatomique: la répétition des parties. Publications Scientifiques du Muséum national d'Histoire naturelle, Paris
Schuster JA (1990) The scientific revolution. In: Olby R, Cantor G, Christie J, Hodge MJS (eds) Companion to the history of modern science. Routledge, London, pp 217–242
Shanahan T (2004) The evolution of Darwinism: selection, adaptation, and progress in evolutionary biology. Cambridge University Press, Cambridge
Shapin S (1996) The scientific revolution. University of Chicago Press, Chicago
Simpson GG (1944) Tempo and mode in evolution. Columbia University Press, New York
Simpson GG (1953) The major features of evolution. Columbia University Press, New York
Skrbina D (2005) Panpsychism in the west. MIT Press, Cambridge
Sloan PR (2001) 'The sense of sublimity': Darwin on nature and divinity. Osiris 16:251–269
Sloan PR (2005) It might be called reverence. In: Hösle V, Illies C (eds) Darwinism and philosophy. University of Notre Dame Press, Notre Dame, pp 143–165
Stebbins GL (1950) Variation and evolution in plants. Columbia University Press, New York
Tuan Y-F (1968) The hydrologic cycle and the wisdom of god: a theme in geoteleology. University of Toronto Press, Toronto
Vlastos G (1970) Equality and justice in early Greek cosmologies. In: Furley D, Allen R (eds) Studies in Presocratic philosophy, vol 1. Routledge & Kegan Paul, London, pp 56–91
Vorzimmer PJ (1970) Charles Darwin: the years of controversy. Temple University Press, Philadelphia
Westfall RS (1958) Science and religion in seventeenth-century England. University of Michigan Press, Ann Arbor
Westfall RS (1977 [1971]) The construction of modern science: mechanisms and mechanics. Cambridge University Press, Cambridge
Whewell W (1837) History of the inductive sciences, vol 3. John Parker, London
Worster D (1994) Nature's economy: a history of ecological ideas, 2nd edn. Cambridge University Press, Cambridge
Wright S (1931) Evolution in Mendelian populations. Genetics 16:97–159
Yeo R (1979) William Whewell, natural theology and the philosophy of science in mid-nineteenth century Britain. Ann Sci 36:493–516

Part I
Historical Shallowness

Chapter 2
Evolution in a Fully Constituted World

Abstract This chapter provides an overview of many issues to be further developed in the rest of this book. In contradiction to the common assumption that the *Origin of Species* constitutes a contribution to universal evolutionism, it is argued that one finds at the core of Darwin's magnum opus an explanatory structure congenial to a static worldview. This is reflected in several key assumptions: (1) the evolutionary past is imagined by projecting extant forms backward in geological time; (2) the unity of the tree of life is denied and replaced by isolated and segmented entities called classes or phyla, each having its own independent evolutionary history; and (3) classes and phyla are believed to be largely modern-looking as far back in time as one can probe. Given such commitments, it is no surprise that Darwin was unable to complete his transition to universal evolutionism successfully. With James Hutton, Charles Lyell, Thomas Henry Huxley, and Richard Owen, Charles Darwin assumed that the world had long been fully constituted. This view emerges more clearly if the reader is willing to go beyond Darwin's rhetoric in the *Origin*.

Keywords Static worldview · Classical age · Epistémè · Uniformitarianism · Actualism · Neontology · Preformationism · Ancestors · Archetypes · Shadows · Rhetorics

Perhaps the greatest obstacle for Darwin on his way to universal evolutionism was his implicit, yet profound, commitment to a fully constituted world. Envisioning the world on analogy with a machine, as Darwin subtly does, comes with a cluster of overlapping and mutually reinforcing concepts: stability, completeness, timelessness, unity, permanence, and uniformity.[1] Let us be more specific by articulating the various explanatory components implicated here, as they will be deployed in this chapter and elsewhere in this book:

[1]A number of the ideas exposed in this chapter were originally presented in R.G. Delisle (2014a), "Evolution in a Fully Constituted World : Charles Darwin's Debts Toward a Static World in the *Origin of Species* (1859)", *Endeavour*, 38: 197–210.

R. G. Delisle, *Charles Darwin's Incomplete Revolution*, Evolutionary Biology – New Perspectives on Its Development 1, https://doi.org/10.1007/978-3-030-17203-9_2

1. The world (inorganic and organic, celestial and terrestrial) is a single and unified system. Darwin came as close as he could to subscribing to an all-encompassing static worldview while at the same time providing just enough room to allow for some amount of biological evolution: he embraced a stable solar system, a steady-state geology, and a biology erected upon a modified version of a steady-state view.
2. This world is composed of such parts as needed to function properly. As such, the parts are continuous and contiguous, as would be expected in the mechanistic doctrine, irrespective of the level of matter at which it is applied. At the biological level, Darwin conceived atomized entities in motion pressed against each other under the impetus of competition: traits, organisms, species, and genera.
3. This world is permanent, with no beginning and no end beyond its known current state. The imperative functionality of a machine guarantees that it remains the same and stable, with no past and no future. This actualist notion accompanies Darwin's commitment to a complete and permanent world, as he maintained the essence of this idea by taking extant forms to be the incarnation of a complete world. Past and extinct forms are seen as diminished representations of a state seen in full only today.
4. This single and unified world is permeated with, or held together by, the principle of uniformity, broadly construed (including the notion of gradualism-gradation). As such, it is in a state of being uniform, unvarying, and regular, bounded through laws and processes now in operation that have been the same everywhere and for all eternity. In its full expression, the principle of uniformity extends to cover identical and homogeneous entities lacking in diversity or variation. This latter portion of the definition is of interest to us, since Darwin's approach to evolution is not based on a truly open and contingent process of evolution generating evolutionary novelties but rather on a recycling process of variation observed today and projected backward in geological time.

2.1 The Completeness of the World

For scholars not committed to evolutionism, the question of the beginning of the world was a nonissue, for the obvious reason that they assumed that the world presented itself exactly as seen today. In his *Opticks* (1730), for instance, Newton declared inquiry into origins to be "unphilosophical," taking for granted that the world did not arise out of a changing process involving a previously chaotic state of freely combining matter in motion (Greene 1981: 27, 33, 130, 204). Newton's finished world was modeled on a static celestial mechanics in perpetual motion.[2] The impact of Newton's work on the imagination of scholars in all scientific areas

[2]For more details and some qualifications about Newton's support of a static worldview, see Chap. 8 of this book.

cannot be stressed enough. This model was explicitly imported, for instance, from astronomy into terrestrial geology, explaining why James Hutton's *Theory of the Earth* (1795) subscribed to an "Earth's machine" whose parts are both continually wearing out and being renewed in a never-ending cycle of three stages: (1) rivers and ocean waves erode the surface of the Earth; (2) eroded debris is deposited in stratification whose accumulation generates heat in the lower strata under weight pressure; and (3) new surface lands are uplifted under the impetus of expanded melted sediments and magmas (Gould 1987, Chap. 3). Hutton (1788: 304) coined the famous motto: "we find no vestige of a beginning,—no prospect of an end." Assuming that the Earth had a divinely created beginning, Hutton knew perfectly well that it might eventually suffer destruction at its creator's hands. The epistemological point Hutton wanted to make was elsewhere, being concerned with the acquisition of scientific knowledge: wherever we look on our planet, we find no evidence of a world different from ours. The geological cycles point in no particular direction, the last one being not fundamentally different from the cycle currently unfolding. The view offered by Hutton is truly that of a steady-state world. In a world believed to be fully constituted as envisioned by Newton and Hutton, science has no business venturing into the metaphysical quest of beginnings and ends, a question more properly part of the search for first cause or God's plan for the creation.

An inspirational figure as important for Darwin as Charles Lyell did not conceal in his first volume of the *Principles of Geology* (1830: 61–65, vol. 1) his sympathy for Hutton, commending the latter for promoting a geology based on fixed principles, as Newton had done for astronomy, thus excluding questions of origins. Under Lyell's uniformitarianism (Rudwick 1970: 7, 8), the world is exclusively explained by currently observed processes: "The characteristic feature of the Huttonian theory was... the exclusion of all causes not supposed to belong to the present order of nature" (Lyell 1830: 63, vol. 1). Lyell erected his own neo-Huttonian steady-state geological system, arguing that while aqueous factors erode the surface of the Earth, igneous factors act as a counterbalance to restore it (Hodge 1982: 7; Rudwick 1970: 12–18). This process generates change in the limited sense that eroded and restored features are not replaced by identical ones—seas, lakes, rivers, islands, and continents—although this remodeling of Earth's surface features has a balancing effect when viewed as a whole and over time. Lyell's cyclical geology thus makes but a very small opening to a genuinely historical process responsible for introducing new features (Gould 1987: 150–155). Interestingly, Hutton was defended by Lyell when it came to confronting opponents who held that the former promoted an eternal world without a beginning (Lyell 1830: 64, 65, vol. 1). As expected, Lyell made the distinction between the beginning of the world and science's inability to find any evidence of it (Rudwick 1970: 8–10). Lyell's position is perfectly consistent for someone committed to a fully constituted world. In the very last phrases of the third volume of the *Principles of Geology* (1833), Lyell joined the list of scholars who justified their commitment to this unified and permanent world by combining celestial and terrestrial references:

> . . .although it may strengthen our conviction, that the present [geological] system of change has not gone on from eternity, it cannot warrant us in presuming that we shall be permitted to behold the signs of the earth's origin, or the evidences of the first introduction into it of organic beings. In vain do we aspire to assign limits to the works of creation in *space*, whether we examine the starry heavens, or that world of minute animalcules which are revealed to us by the microscope. We are prepared, therefore, to find that in *time* also, the confines of the universe lie beyond the reach of mortal ken. (Lyell 1833: 384, vol. 3) [italics original]

Can it be entirely coincidental, then, that Darwin (1859: 490, 1872: 429) also concluded the *Origin* with the words "whilst this planet has gone cycling on according to the fixed law of gravity, from so simple a beginning endless forms. . . are being, evolved"? Apparently not, if we follow Jonathan Hodge and Gregory Radick (2009: 262):

> And in rejecting and replacing Lyell's view of the organic world, [Darwin] explicitly upheld Lyell's view. . . of the physical world. That view of the physical world included not merely Lyell's neo-Huttonian balanced, stable system of igneous and aqueous causation for the earth's surface, but Lyell's uncontroversial acceptance of the old (eighteenth-century) Laplace and Lagrange stability thesis for the planets' orbits around the sun, and his controversial rejection, sanctioned by John Herschel, of any nebular hypotheses about the origin of this stable solar system. As with the solar system and the earth, so, for Darwin, with life and its origin: he took it as given; even implying, at the end of the book. . . that it was God's work not nature's.

Whereas Hodge (2014: 173) holds that Darwin applied his dynamic and progressive evolutionary biology against this stable astrogeological background, we will go one step further and argue that the notion of stability (and accompanying concepts) penetrated deeply into Darwin's *biology* as well. To better grasp the issue, it may be useful to reverse the perspective: if Darwin had truly supported an unconstrained biological evolutionism, would it not have been easier for him to embrace also the nebular hypothesis in astronomy, thus transposing the evolutionary impetus of one area (astronomy) to another (biology) in a single and universal view? Instead, Darwin subscribed to a static and cyclical universe while adopting a steady-state geology rather than one based on directionalism. Why, then, stop at the threshold of biology? Something in Darwin's thought system resisted full and open evolutionism.

Newton, Hutton, Lyell, and Darwin embraced the same approach to knowledge: the investigation of the world must exclusively be done through an analysis of its current configuration. "Actualism" is the name given to this approach. The present time becomes the sole horizon for studying the past (or the future). As an explanatory component included in the more inclusive principle of uniformity, actualism extends to the notion that the same processes and laws permeate the world throughout time and space, being homogeneous and unvarying. The *methodology* of actualism is never as legitimate as when it is applied against the *ontological* belief that the world is fully constituted, with the past and future not being significantly different from the present and with no novelties hiding from us in these other time dimensions.[3] In fact, since everything to be known about the world is already there before our eyes, it is almost an abuse of language to speak of other time dimensions. In a real sense, the past and the future have no real existence in a world assumed to be self-identical throughout.

At this juncture, the rise of the fields of geology and paleontology in the eighteenth and nineteenth centuries posed a potential threat to the uniformity principle by potentially revealing geological processes and biological forms with no modern-day equivalents. Apparently, this left only two options open: (1) to accept the irruption of novelties in the world, thus shattering the idea of a fully constituted world, or (2) to preserve this latter notion by denying the reality of so-called novelties by finding a way to include them among currently known entities. These options opened up a dichotomy between a largely ahistorical view and a truly historical one. For their part, Georges Cuvier and Alexandre Brongniart's *Description géologique des environs de Paris* (1825, 2nd ed.) and William Buckland's *Geology and Mineralogy Considered With Reference to Natural Theology* (1836) opened up the intellectual road leading to the idea of novelties. Their works were instrumental in suggesting an Earth's history characterized by a series of largely independent and successive creations of faunas, each creation being separated by a major episode of extinction and having no obvious analogies to extant counterparts, especially when the most ancient forms were concerned (Bowler 2003: 108–120).[4]

The notion of a fully constituted world faced the major challenge of the extinction of life forms. Indeed, how could life forms now be absent from the world through extinction if the world is permanent and complete? It is instructive to follow the retreating steps taken to save this notion in the face of a growing challenge. First, it was denied that individual species went extinct, as these were believed to be merely waiting to be discovered in unexplored regions of the Earth, still well and alive. This view was adopted by John Ray, Carl von Linnaeus, James Hutton, and Thomas Jefferson (Greene 1959, Chap. 4; Gould 1987: 87, 88). In the first third of the nineteenth century, when species extinction became too obvious to be denied, Lamarck even resorted to calling the idea of transmutation itself to the rescue. After all, should transformed species really be considered extinct? Although Lamarck's system leaves the reader with the impression of a vertical and progressive evolutionary drive, at a more fundamental level, order and stability are maintained: while changing (lost) species climb the scale of complexity, their vacated places are continually reoccupied by new and lower ones, taking up exactly the same places they have left open (Limoges 1970: 36).

---

[3]Chapter 8 of this book will analyze in some depth the relation between ontology and methodology in Isaac Newton, Thomas Reid, Charles Lyell, and Charles Darwin.

[4]Although Buckland argued that the distinct successions of faunas were divinely created, Cuvier was more inclined to hold that extinctions were localized, implying that such life forms existed elsewhere. It should be noted that Cuvier died in 1832. Those carrying on Cuvier's work were less scrupulous than Cuvier himself in squarely embracing clean and definitive clear-cut breaks in the history of life.

For an anti-evolutionist like Lyell, another way existed to save the notion of a fully constituted world.[5] Although Lyell acknowledged the numerous extinctions of individual species faced with ever-changing physical conditions, he refused to recognize that the main biological types seen today (such as mosses, mammals, conifers, and mollusks) had not always existed (Rudwick 1970: 14, 19–21; Bowler 1976: 70–76; Hodge 1982: 6, 7, 92). Taking into consideration that the annals of life were known to be quite incomplete in the 1830s, Lyell's position was anything but unreasonable. He therefore recognized the reality of the extinction of groups at very low taxonomic levels but denied it at higher taxonomic levels (Bartholomew 1979: 285). In a real sense, Lyell's system left no place for novelties in the history of life, promoting instead cyclical fluctuations in depletion and renewal of existing types in time. Unfortunately for Lyell, his system implied a gradual depletion of biomass over time for which no explanation of replenishment was provided. As explained by Martin Rudwick (1971: 225), Lyell's view faced a challenge:

> Throughout the 1830s, in fact, while Lyell saw his actualistic policy used increasingly in geological research, his steady-state system gradually seemed less and less tenable in the light of that research. Edouard Lartet's discovery of the first fossil primates within the Tertiary epoch, Louis Agassiz's delineation of an age of fish preceding the age of reptiles, Roderick Murchison's description of a vast Silurian system largely pre-dating even the age of fish—these and many other discoveries seemed to confirm still more clearly the essential validity of the directionalist view of the history of life, rather than Lyell's steady-state view.

Nevertheless, it will be seen that Darwin retained Lyell's logic by applying it to higher taxonomic levels still, a logic to which Thomas Henry Huxley and Richard Owen also subscribed (as will be seen below). Whereas Darwin accepted the extinction of groups at fairly high taxonomic levels (species, genera, families, and orders), he managed to save the notion of a fully constituted world by applying it to the highest taxonomic groups (classes and phyla) by denying their extinction. In this way, the idea of a permanent, complete, and unified world was recast within an "evolutionary" context. From that angle, Darwin is perhaps better seen as being among the last proponents of a steady-state view than among the very first genuinely modern evolutionists.

To many in the eighteenth century and early nineteenth century, the world was assumed to be stable enough to allow for the sustainability of life at all times: inorganic and organic worlds were held to be perfectly synchronous, since they were designed and contrived for each other. On this view, no realm took precedence over the other in the sequence of the creation (Greene 1981: 42). For believers of a genuinely changing world, however, it might be argued that the physical conditions in the early periods of Earth were too harsh to sustain life, which, it was thought, required a precise sequence of events. This is the view of Robert Chambers in his *Vestiges of the Natural History of Creation* (1844: 89, 90, 146–150), when he held, for instance, that plants had first to modify the air's composition before animals

[5]In the 1820s, 1830s, and 1840s, Lyell was clearly an anti-evolutionist. By the 1860s, he had made an opening to the idea of evolution. See Bartholomew (1973).

could appear while reptiles came before mammals in light of their lower need for oxygen (Egerton 1973: 340). Of course, proponents of a fully constituted world could not share this analysis. For James Hutton, the Earth always remained a stable abode for life throughout continuous and cycling changes (Gould 1987: 74, 81, 85; Laudan 1987: 115, 116). How could it be otherwise in a finished world? But Hutton's view was under stress, as Greene (1959: 81, 82) explains:

> The strategy of granting the recent origin of man in order to gain acceptance for the indefinite antiquity of the earth was a shortsighted one. So long as the earth and man were assumed to have been created at approximately the same time, it was easy and natural to regard the earth and its flora and fauna as designed for the use and instruction of man. The steadily accumulating evidence pointing to a progress from lower to ever higher forms of life in the geologic series could not shake the traditional view of the relation of the inorganic world to the organic and rational worlds so long as the progression was thought to have taken place in a few thousand years. But if Hutton was right and the earth millions of years old, whole economies of vegetable and animal life must have flourished and expired long before man made his appearance. How, then, one might logically ask, could the maintenance of human life be regarded as the final cause of these vast revolutions on the globe? Were not the means out of proportion to the end?... [A] passage of [Hutton's] *Theory of the Earth* suggests that he would rather grant the antiquity of man than face the consequences of the opposite view... The study of the earth's crust, Hutton seemed to suggest, would eventually disclose a history of human races and empires comparable in duration and vicissitudes with the globe's turbulent history... Hutton assumed man's antiquity in order to save belief in the wise contrivance of nature.

Whereas Hutton elected to at least contemplate the possibility of the existence of human beings in very ancient times, Lyell found a way to save the idea of the habitability of the world at all times while keeping humans within recent geological horizons. As Michael Bartholomew (1973: 274, 275) argues, one way Lyell avoided the possible implications of an evolutionary perspective for humans was to put them in a late geological period while simultaneously placing all the other mammals in the earliest geological period:

> Lyell had found a 'scientific' way of dividing man, who, axiomatically, was regarded as a recent creation, from the higher mammalia, which Lyell now pushed back as far down into the past as the earliest rocks... [T]ransmutationists could be blocked if it could be shown that, far from there having been a gradual complexification of organic forms through time, the most complex classes existed with all other classes, in the earliest times, alongside the most simple, and that when creative additions were made to floras and faunas, the addition represented no advance in complexity... The implication of Lyell's new position was that man's origination must have been special, as he was, according to Lyell, preceded by no orderly sequence of progressively complexifying mammal species, from which he could have been genealogically derived.

By putting humans at the top of geological creation, however, was not Lyell undermining his own position, at least in the minds of the most committed directionalists or progressionists? Realizing this, Lyell (1830: 155, vol. 1) formulated a preemptive strike:

> Assuming, then, that man is, comparatively speaking, of modern [recent] origin, can his introduction be considered as one step in a progressive system by which, as some suppose, the organic world advanced slowly from a more simple to a more perfect state? To this

> question we may reply, that *the superiority of man depends not on those faculties and attributes which he shares in common with the inferior animals, but on his reason by which he is distinguished from them.* [my emphasis]

Martin Rudwick (1970: 14) concludes that Lyell extricated himself from the situation by distinguishing between the novelty offered by the human mind—an entirely separate ontological category from the rest of nature—and the lack of novelty seen in his physical conformation: "…on the physical level his [man's] appearance denotes no progress in the organic world, while on the 'moral' level his appearance is a phenomenon of an entirely different kind." Thus the argument for a fully constituted world was saved. Although the traditional historiography stresses the differences and tensions between Lyell's and Darwin's positions—as evidenced by Darwin's letters to Lyell, pressuring him to recognize both his evolutionary theory and humankind's place among the animals (see Bartholomew 1973: 293–302)—it has been somewhat overlooked that the cogitations of Hutton and Lyell paved an intellectual road that offered Darwin the possibility of salvaging the argument for a fully constituted word by recasting it. From Darwin's perspective, the fact that the human body shares many similarities with other representatives of its class (Mammalia) and phylum (vertebrates) means one can probe the annals of life by tracing the commonalities of these taxonomic groups as far back in time as one can. Apparently, Darwin had found a way to synthesize their views: while Lyell recognized the bodily similarities between humans and other animals, Hutton contemplated a deep history for humans.

2.2 Neontology Versus Paleontology: The Double Epistemological Standard

Let us now engage with the specifics of Darwin's view. Although our argumentation will require the span of the entire book to be fully deployed, the nature of Darwin's commitment to a fully constituted world can be investigated, at this early stage, under three headings:

1. An overestimation of the virtues of neontology as an approach to reconstructing the past. It is assumed that the present time is the key to unlocking the past.
2. A depreciation of the epistemological value of paleontology on the presumption that we know too little about the evolutionary past.
3. The restriction of the history of life on Earth to a mere deployment of forms already known today; past variations are shoehorned to fit within today's.

The field of paleontology will provide, by the 1840s and 1850s, the beginning of a rough approximation of life's history. According to Peter Bowler (1996: 41), by "the middle of the nineteenth century an outline of the history of life had become available from the fossil record," while Rudwick (2008: 546) states that by "the 1840s, it was becoming almost impossible to doubt that the fossil record was a broadly reliable trace of the true history of life." Quantitatively speaking, 4000 fossil species were known to Defrance in 1824, 27,000 to Bronn in 1849, and 34,500 to d'Omalius d'Halloy in 1868 (Laurent 1987: 398, 399). Qualitatively speaking, some fossil groups began to appear in their geological settings. For instance, in their *Description géologique des environs de Paris* (1825, 2nd ed.), Cuvier and Brongniart presented a series of forms increasingly more different from those present today, as we go further back in time: close to living elephants was the mastodon from America (*Mastodon americanus*) and the woolly mammoth from Siberia (*Elephas primigenius*), both found in fairly recent deposits; in the older strata of the Tertiary, an unknown mammal genera akin to *Palaeotherium* was recovered, a form presenting vague affinities with living tapirs, rhinoceros, and pigs; in the still older Secondary deposits, a form with no living counterpart was found: *Mosasaurus*, a giant marine lizard from Germany (Bowler 2003: 113, 114). In Buckland's *Geology and Mineralogy Considered With Reference to Natural Theology* (1836), very old Paleozoic extinct marine arthropods (trilobites) are depicted and presented to a wide audience (Rudwick 2008: 423–436). Similarly, Chambers reviewed fossil groups in the editions of his *Vestiges of the Natural History of Creation* (1844; 10th ed. in 1853) such as very ancient Silurian fish (*Holoptychius*, *Coccosteus*, *Cephalaspis*, and *Pterichthys*), forms said to be "entirely different from any fishes now existing," "some of which are of most extraordinary and surprising forms" (Chambers 1844: 67, 68). There, Chambers disseminated knowledge extracted from specialized works like Louis Agassiz's *Recherches sur les poissons fossiles* (1833–1843) to a wider audience.

In the mid-nineteenth century, the annals of life revealed a number of possible facts: first, that there is no life in the Precambrian or Primary, the oldest known geological deposits; second, that aquatic animals and plants came before their terrestrial counterparts; third, that while fish and invertebrates were abundant in the older Paleozoic era and the reptiles well represented in the older Mesozoic era (Secondary), the mammals became more numerous in the more recent Cenozoic era (Tertiary); and fourth, that some preliminary evidence may be suggestive of an evolutionary complexification, extending from simple invertebrates to the complex vertebrates or, among the vertebrates, from fish to reptiles to mammals.[6] This being said, it should not be overlooked that in the early 1860s, Thomas Henry Huxley (1862) was still anything but impressed by the value of the fossil record at supporting a number of assumed features of the evolutionary history of life. Such empirical uncertainties played against a directionalist view of life.

Be that as it may, Darwin addressed the issue of the annals of life in Chapters 9 and 10 of the *Origin of Species* (1859),[7] as one would expect from his topic (Herbert and Norman 2009). Revealingly, Darwin did not proceed to a systematic survey of the fossil record, carefully avoiding directly raising the major evolutionary

---

[6]See table of geological periods in Bowler (1976: 5).

[7]In the sixth edition of the *Origins of Species* (1872), the two chapters devoted to paleontology are Chaps. 10 and 11.

questions of the period, such as the rise of life from nonlife. Rather, Darwin's approach was more partial and selective. Let us postpone answering the question "why did Darwin opt for such a strategy" for a little while until more contextual remarks are provided.

It is true that during the late 1820s and the 1830s, the young Darwin was interested in the rise of the simplest life forms under spontaneous generation—infusorian monads—at the multiple junction of nonlife and life, asexual and sexual reproduction, and plants and animals (Gruber 1974: 129–149; Kohn 1980: 109–113; Hodge 1982: 66–80; Sloan 1985, 1986). Darwin was at that early period willing to investigate a whole spectrum of evolutionary issues, including those concerned with origins. Yet, despite its title, the *Origin of Species* does not provide the reader with a consideration of any such issues, many of them being beyond the reach of the biological science of the day. In a real sense, Darwin's science resembled that of Newton, Hutton, and Lyell in carefully avoiding the issue of origins. Living in an age of rising evolutionism, however, Darwin could not ground his view on the same suppositions as Newton and Hutton. One epistemological constraint (among others to be reviewed) faced by Darwin concerned the canons of scientific investigation—the proper methodological rules of inductive science—heuristic rules that prompted Darwin to publicly give up on the issue of origins (Farley 1977: 80, 81; Strick 2000: 37–40, 72, 73, 92, 93). This is not to say that Darwin had entirely given up on implicitly embracing in the *Origin* the idea that an early phase of life may provide evidence for a unified evolutionary past. This is seen in instances such as (1) the limited number of postulated progenitors for plants and animals, no more than four or five each,[8] or perhaps a single ancestor for all of life,[9] and (2) in the systematic comparison of plant and animal conditions, on the assumption that both kingdoms shared many features in common, whether in terms of reproduction, generation, or physiology.

Once this is acknowledged, the fact remains that the issue of origins has little place in the *Origin of Species*. As expressed by Darwin (1859: 207) himself when studying instincts, evolution is embraced from the perspective of an object already largely constituted: "I must premise, that I have nothing to do with the origin of the primary mental powers, any more than I have with that of life itself. We are concerned only with the diversities of instinct and of the other mental qualities of animals within the same class" (see also Darwin 1872: 205). It is no surprise, then, that the reader of the paleontological chapters of the *Origin* encounters a world quite similar to ours. More often than not in these chapters, Darwin elects to refer to ancient forms by general designations only—speaking of ancient mollusks, crustaceans, shells, cephalopods, mammals, birds, or reptiles—reading our modern categories into ancient times.[10] Darwin refers only rarely to precise fossil groups; this is

---

[8]Darwin (1859: 484, 488, 490, 1872: 425, 429).

[9]Darwin (1859: 488, 490, 1872: 429).

[10]Darwin (1859: 288, 289, 297, 298, 303–306, 313, 316, 318, 319, 321, 323, 324, 330, 331, 334, 339, 341).

the case, for instance, for mastodons, *Megatherium*, *Toxodon*, trilobites, *Lingula*, *Nautilus*, and ganoid fish.

Significantly, in 1859, Darwin recognizes that the Silurian constitutes a temporal barrier (which becomes a Cambrian barrier in 1872) with the following characteristics: after the barrier, life is largely organized around well-defined groups as known today; before the barrier, nothing is known of life; and around the barrier itself, life seems to spring full-fledged, suddenly, out of nothing. For that reason, Darwin has no other choice when speaking of remote ancestors but to postulate entirely *hypothetical* progenitors. On his theory, such ancient progenitors must not only exist, but they must also be different and intermediate in features, even though empirical support is entirely missing:

> . . .number of species of the same group, suddenly appear in the lowest known fossiliferous rocks. Most of the arguments which have convinced me that all the existing species of the same group have descended from one progenitor, apply with nearly equal force to the earliest known species. For instance, I cannot doubt that all the Silurian trilobites have descended from some one crustacean, which must have lived long before the Silurian age, and which probably differed greatly from any known animal. Some of the most ancient Silurian animals, as the Nautilus, Lingula, &c., do not differ much from living species; and it cannot on my theory be supposed, that these old species were the progenitors of all the species of the orders to which they belong, for they do not present characters in any degree intermediate between them. (Darwin 1859: 306)[11]

Darwin (1859: 302–310) was forced to confront the apparent sudden rise of whole groups of animals, fully constituted (see also Darwin 1872: 282–289). Here, we detect in Darwin's arguments a tension between the kind of biology he actually builds and the kind of biology he intends to. Close attention to this *rhetorical* dimension in the *Origin of Species* is essential to distinguishing between the levels of arguments presented. This is key to detecting the influence of the static worldview on Darwin's ideas. Before continuing our analysis, it is important to note that the expression "rhetorical Darwin" will be used throughout this book in different, yet sometimes overlapping, ways (see Table 2.1):

Sense 1: When Darwin assumes that his theory (a) fully adopts an evolutionism wherein all forms are ultimately derived from a limited number of common ancestors, if not a single one, having gone through profound changes accompanied by the rise of genuine evolutionary novelties and/or (b) provided a structure organized around a causal core (natural selection) giving explanatory unity to a series of independent sets of descriptive phenomena found in a host of disciplines and/or (c) fully embraced evolutionism in all its manifestations and openness in accordance to the reality of historical contingency.

Sense 2: To this rhetorical Darwin, we will oppose another Darwin holding to a rigid pan-divergent view negating the evolutionary contingencies, complexities, and novelties, a view organized around fixed independent classes/phyla looked at

[11]See Darwin (1859: 463, 465, 489) on the Silurian barrier. See also Darwin (1872: 285, 286). In the sixth edition, the Silurian barrier becomes the even older Cambrian barrier.

Table 2.1 Argumentative layers in the *Origin of Species*

Layer 1 (rhetoric sense 1) A somewhat superficial reading exposes a theory of biological evolution organized around a causal explanatory core (natural selection + divergence + economy of nature) giving meaning and unity to apparently unrelated disciplines (such as paleontology, systematics, and embryology). Such a theory gives the impression of being dedicated to explaining a unified tree of life fully open to all evolutionary manifestations under the conceptual tool of the strong contingency thesis. Layer 1 offers the illusion of a truly modern evolutionary view
Layer 2 (rhetoric sense 2) A more critical reading shows a theory designed to confine evolution to a pan-divergent strait-jacket, with the dual principle of divergence-gradation as its explanatory core. This core is protected from refutation through the adjustments of flexible and auxiliary explanatory variables such as the geographical distance, the geological timeframe, the taxonomic level, and the amount of selective pressure. This argumentative layer jettisoned any pretention of explaining universal evolutionism in all its manifestations with the help of the strong contingency thesis. Rather, the evolutionary process is simplified to fit a segmented tree of life organized around nearly permanent, stable, and isolated classes and phyla, and for which variability observed among extant forms is believed to be more than enough when it comes to embracing the evolutionary past. Layer 2 reveals a Darwin strongly indebted to the tradition of natural theology and the ontology accompanying a static worldview. Darwin's sole Diagram in the *Origin* appears in all its rigidity and limitations at this explanatory layer
Layer 3 A more careful analysis sees Darwin's profound commitment to empirico-inductivism overwhelming him from all sides with a wide array of evolutionary phenomena that could not possibly be synthesized under the purview of his narrow theory. Darwin failed to contain threatening complexities such as reproductive networks, sustained directional and progressive trends, the rise of analogical forms, reticulate evolution, random evolutionary walks, ancestors without apparent descendants, and the causes behind the exhaustion of the evolutionary drive. Layer 3 is most revealing of Darwin's incomplete transition to modern and universal evolutionism, caught as he was between two revolutions (the Scientific Revolution and the Transformist Revolution), two worldviews (the static and the evolutionary), and two epistemologies to make sense of them
To summarize, what originally seemed a relatively inclusive interpretative framework (layer 1) proves on closer analysis to be too restrictive to deal with evolution's complexities (layer 2). The more Darwin tightened his grip on evolutionary phenomena the more they slipped through his fingers, revealing rhetorical dimensions that concealed Darwin's incomplete transition to evolutionism (layer 3)

Darwin's *magnum opus* is a challenging piece of work in light of the several argumentative layers it contains, all confusingly fused together. A close analysis reveals three main layers

from the present time only, in addition to having failed at proposing a theory of evolution even remotely unified. This other Darwin ultimately based his view on a commitment toward a steady-state view of biology, wherein evolutionary motion (change) constitutes a superficial feature of a world fundamentally static.

Sense 1 and Sense 2 are said to be rhetorical in this book because Darwin will, ultimately, be unable to cope with what emerges from his own application of the empirico-inductivistic approach (see Chap. 8 of this book), one that overwhelms him from all sides with a range of evolutionary phenomena incompatible with his theory, whether he tries to contain them with the theoretical structure already described in Sense 1 or the theoretical structure organized around a rigid pan-divergent view as presented in Sense 2. The rest of this book will attempt to provide meaning and content to the Darwin that emerges beyond such rhetorical dimensions.

Reading Chapters 9 and 10 of the *Origin* (1859) leaves the reader with the strange impression of a history of life largely at a standstill. Darwin promotes a view of the past that presents itself as a unified, single, and continuous network of forms fusing the present and the past: "As all the living forms of life are the lineal descendants of those which lived long before the Silurian epoch, we may feel certain that the ordinary succession by generation has never once been broken, and that no cataclysm has desolated the whole world" (Darwin 1859: 489; see also 1872: 428). Darwin's view is opposed to the discontinuous view that eventually emerged from the works of Cuvier, Brongniart, Buckland, and others. Darwin's (1859: 303–306) approach is clearly revealed in a passage in Chapter 9.[12] In it, he relates more and more of the new fossil discoveries to living counterparts:

1. Although Cuvier once declared that monkeys only lived in recent times (Quaternary era), we know them now from the oldest portion of the Tertiary.
2. Whereas mammals we once restricted to the Tertiary and Quaternary eras, they can now be extended throughout almost the entire Secondary.
3. Just as whales were apparently dated back to the Tertiary only, they are now found in the later part of the Secondary.
4. Once sessile cirripedes (crustaceans) were confined to the Tertiary; they are now not only found in the Secondary but these also look quite closely to their extant counterparts.
5. Finally, teleostean fish of the pre-Secondary era (Paleozoic) are mentioned, a group of forms that includes the large majority of existing species.

Interestingly, what Darwin does in this particular passage of the *Origin* was quite similar to what Charles Lyell had done throughout the 1830s and 1840s: to assimilate extinct forms with extant ones. To quote Michael Bartholomew (1973: 289–291) on Lyell:

> Lyell claimed that during the earliest periods of which we have geological records, conditions were no more 'primitive' than are certain areas of the globe today, but were capable of supporting the most complex forms of life… Lyell said that wherever the geological and palaeontological evidence enables us to reconstruct, imaginatively, a past environment capable of supporting mammals, there mammals may reasonably be expected to have lived… [D]uring the 1830s and 1840s, Lyell was able to produce evidence of classes of organisms being found in earlier and earlier formations, and moreover, he was able to assemble a good case to show that within classes there had not been an invariable development through time: some Silurian fish were as 'developed' as fish swimming in a modern era… In the twenty years that had elapsed since the first volume of *Principles* was published,

[12] An extensive quote of this passage is provided in Chap. 5 of this book. Darwin introduced some minor changes and additions in the sixth edition, but the spirit of the passage remains the same. See Darwin (1872: 283–285).

> the work of Sedgwick and Murchison had extended the stratigraphical column... down through the Silurian and Cambrian Systems. Consequently Lyell had a greater problem than in 1830 in denying that mammals were absent from the earliest formations, and in his [1851] address he judiciously refrains from pushing his theory of selective deposition to its very limits: he never clearly says that mammals abounded during the Cambrian period. But his position is substantially the same as in 1830. He still maintains that selectivity of deposition will account for the absence of fossils of mammals, birds, insects, and flowering plants in the earliest strata, and he maintains that when examples of these classes are found, they are often as developed as their modern counterparts.

Darwin's biology is much closer to Lyell's than is usually assumed in the historiography. Far from being struck by the strangeness of some fossil groups, their common bias consisted in adopting an *epistemology of assimilation*, with the principle of uniformity being applied in its full force by including homogeneous entities believed to be lacking in diversity or variation. Of course, Darwin would not entirely overlook the differences separating some fossil groups and their counterparts whether extinct or extant,[13] but it should be noted how, for him, past variability is organized by its manifestation among extant groups. In Chapter 10 of the *Origin* (1859), one reads this remarkable statement:

> Let us now look to the mutual affinities of extinct and living species. They all fall into one grand natural system; and this fact is at once explained on the principle of descent. The more ancient any form is, the more, as a general rule, it differs from living forms. *But... all fossils can be classed either in still existing groups, or between them.* That the extinct forms of life help to fill up the wide intervals between existing genera, families, and orders, cannot be disputed. For if we confine our attention either to the living or to the extinct alone, the series is far less perfect than if we combine both into one general system. With respect to the Vertebrata, whole pages could be filled with striking illustrations... showing how extinct animals fall in between existing groups... (Darwin 1859: 329).[14] [my emphasis]

Darwin applies his theory based on the principle of divergence literally or strictly: the variability of past forms falls between the one of extant forms as expected when the time dimension is reversed on the divergence process, hence convergence backward in time. Apparently, genuine evolutionary novelties find no theoretical place in Darwin's theory, since nothing goes beyond the biological diversity of extant forms. In fact, the paleontological chapters are less devoted to studying real and precise fossil groups, as to exposing Darwin's theory, as seen in long and purely abstract developments having little grounding in empirical reality and much in entirely hypothetical scenarios.[15] That argumentative strategy is part of the rhetorical clothing found throughout the *Origin of Species*, in which an empirical part is fused with hypothetical scenarios in such an intricate and subtle fashion that the reader does not always notice that Darwin has moved from one to the other. Darwin's

---

[13]See Darwin (1859: 306, 313, 323, 429, 1872: 286, 287, 290, 291, 297, 298, 378)

[14]This quote is reproduced in full in Chap. 5 of this book. In the sixth edition, information is added to bolster the case. See Darwin (1872: 301, 302).

[15]See, for instance, the long development in which Darwin lays out the way the fossil record is supposed to comply with his view of evolution, as presented in the Diagram of the *Origin of Species*. See Darwin (1859: 331–336, 1872: 303–307).

strategy caught the attention of a number of his contemporaries. As summarized by Alvar Ellegard (1990[1958]: 189): "A host of press voices accused Darwin of mingling together 'facts and assumptions, probabilities and speculations… in most illogical confusion'… and of using his imagination to supply the lack of factual observation."

In order to grasp Darwin's way of approaching paleontology, one should compare it to the main paleontological treatises of his time written, for instance, by Louis Agassiz (1833–1843, 1844–1845), Joachim Barrande (1846, 1852–1881), Georges Cuvier (1825, 1830), Hugh Falconer (1868; Falconer and Cautley 1845–1849), Alcide d'Orbigny (1842–1860), Richard Owen (1849–1884, 1860), and Jules Pictet (1853–1857), names mentioned by Darwin himself. Irrespective of the fact that these paleontologists had views about theoretical issues (Bowler 1976, 1996; Laurent 1987; Rudwick 2008), their treatises were both empirically oriented and dense. Compared to them, Darwin's paleontology in the *Origin of Species* is theoretically driven and assertive in an age of empirical uncertainties. Among theoretical components promoted in these two paleontological chapters, the principle of gradualism-gradation occupies a central place:

> The main cause, however, of innumerable intermediate links not now occurring everywhere throughout nature depends on the very process of natural selection, through which new varieties continually take the places of and exterminate their parent-forms. But just in proportion as this process of extermination has acted on an enormous scale, so must the number of intermediate varieties, which have formerly existed on the earth, be truly enormous. Why then is not every geological formation and every stratum full of such intermediate links? Geology assuredly does not reveal any such finely graduated organic chain; and this, perhaps, is the most obvious and gravest objection which can be urged against my theory. The explanation lies, as I believe, in the extreme imperfection of the geological record. (1859: 279, 280; see also 1872: 264, 265)

Darwin will then go on to expand on the numerous reasons why such intermediate forms are not found in the fossil record. In Darwin's theory, the principle of gradation comes with a precise meaning, consisting in viewing genealogical connections as documenting the slow pattern-process of divergence as one or more forms move away from the ancestor's state. For Darwin, the links are intermediate in the full sense of the word. Or to reverse the perspective, as already alluded to—when "divergence" is read backward in time, it should be seen as "convergence":

> By the theory of natural selection all living species have been connected with the parent-species of each genus, by the differences not greater than we see between the varieties of the same species at the present day; and these parent-species, now generally extinct, have in their turn been similarly connected with more ancient species; and so on backwards, *always converging* to the common ancestor of each great class. So that the number of intermediate and transitional links, between all living and extinct species, must have been inconceivably great. (Darwin 1859: 281, 282)[16] [my emphasis]

[16]See also Darwin (1859: 301, 330–334,1872: 266, 282, 303–305).

This comes as no surprise, given that the theory is founded on the dual core of the principle of divergence and the principle of gradualism-gradation.[17] Of course, this raises an obvious question: should one really assume that the evolutionary past conforms to Darwin's theoretical expectation? Clearly, Darwin evades issues such as evolutionary parallelism, genuine evolutionary gaps, and the non-representativeness of today's variability. Yet, the principle of gradation was so central to Darwin's theory that the full complexity of evolution was levelled to comply with it. This epistemological choice had the advantage of keeping all of the past within the purview of the present but with the disadvantage of displaying an evolutionary history largely at a standstill, that is, deprived of significant novelties when compared to extant forms. Indeed, if most of the past echoes in today's variability, nothing significantly new could have happened during that past.

Given his theory, Darwin uneasily exploited the field of paleontology to support a view of life deprived of discontinuities, offering us a long list of reasons justifying its limitations (1859: 280, 287, 289, 293–301, 306). The assumed inconveniences were of two main orders. First are the biological limitations: species transformation is too slow to be properly recorded in geological formations; species migration leaves gaps in the geological record; fossil species are too often known from a single specimen only; soft body parts cannot be preserved; and the continuity of evolutionary lineages is obscured by taxonomic debates. Then, geological reasons are given: the lack of decent knowledge beyond the United States and Europe; the imperfect correlation between geological formations; missing strata not being recorded; a lack of fossilization when sediments are not accumulating or when seabeds are rising (rather than subsiding) above sea level; and a discrepancy in the paces of geological sedimentation and biological evolution. Darwin reiterates again and again the limitations of paleontology for recording the actual pattern-process of evolution.[18] Borrowing a metaphor from Lyell, he writes:

> ...I look at the natural geological record, as a history of the world imperfectly kept, and written in a changing dialect; of this history we possess the last volume alone, relating only to two or three countries. Of this volume, only here and there a short chapter has been preserved; and of each page, only here and there a few lines. Each word of the slow-changing language, in which the history is supposed to be written, being more or less different in the interrupted succession of chapters, may represent the apparently abruptly changed forms of life, entombed in our consecutive, but widely separated, formations. (Darwin 1859: 310, 311; see also 1872, 289)

More specifically, Darwin's theory makes two complementary demands on the field of paleontology[19]: first, to expose the small and numerous connections binding together transforming lineages at low taxonomic levels (varieties, species) and second, to trace back common ancestors of evolutionary groups climbing the

---

[17]The epistemological implications accompanying Darwin's dual core (divergence-gradation) will become clearer as we pursue our analysis in the following chapters of this book, but see especially our Chapters 5 and 7.

[18]Darwin (1859: 310, 311, 333, 345, 464, 465, 475).

[19]Darwin (1859: 279–282, 299, 301–303, 306, 307).

taxonomic scale (species, genera, families, and orders). Unfortunately, paleontology could not provide that kind of support for Darwin's theory. To answer the question we raised earlier, we now understand better why Darwin refused to engage in a thorough review of the annals of life: the fossil record did not meet his theoretical expectations. In order to mitigate the risk of seeing the annals of life undermining his theory, Darwin delivered a preemptive strike, by arguing that future research in that area is unlikely to resolve the limitations he noted:

> We shall, perhaps, best perceive the improbability of our being enabled to connect species by numerous, fine, intermediate, fossil links, by asking ourselves whether, for instance, geologists at some future period will be able to prove, that our different breeds of cattle, sheep, horses, and dogs have descended from a single stock or from several aboriginal stocks; or, again, whether certain sea-shells inhabiting the shores of North America, which are ranked by some conchologists as distinct species from their European representatives, and by other conchologists as only varieties, are really varieties or are, as it is called, specifically distinct. *This could be effected only by the future geologist discovering in a fossil state numerous intermediate gradations; and such success seems to me improbable in the highest degree.* (Darwin 1859: 298, 299; see also 1872: 279, 280) [my emphasis]

One can legitimately summarize Darwin's strategy in the two paleontological chapters of the *Origin* with the expression "double epistemological standard." When some paleontological facts seemed to support Darwin's theory, he was only too happy to accept them; when other facts seemed to go against his theory, he simply discarded them on the grounds that they were merely artifacts of an imperfect fossil record.

Having summarily disposed of the field of paleontology, Darwin moved beyond this unpromising area. He could now subscribe in full to an area more congenial to his epistemology—neontology—and from there impose his research agenda by reading the entire history of life backward, that is, by projecting today's variability into the past. For Darwin, then, it is only natural that taxonomy applied among extant forms is assumed to capture the essence of the past. Although some traces of descent are sometimes somewhat obscured by the action of natural selection—two forms may look alike because either they are phylogenetically related or responded similarly to selective pressures (Darwin 1859: 414–416)—Darwin is nonetheless confident that life forms are characterized by non-vital structures (embryological structures, rudimentary organs) allowing for the genealogical traceability of forms, with only rare exceptions.[20] Armed with this theoretical commitment, Darwin will go on to explain in his sole Diagram in the *Origin* the evolution of several related lineages part of a single class (see Fig. 2.1).[21] It is worth quoting him in full:

> The reader will best understand what is meant, if he will take the trouble of referring to the diagram... We will suppose the letters A to L to represent allied genera, which lived during the Silurian epoch, and these have descended from a species which existed at an unknown anterior period. Species of three of these genera (A, F, and I) have transmitted modified descendants to the present day, represented by the fifteen genera (a^{14} to z^{14}) on the

[20]Darwin (1859: 416–421, 425, 426, 434, 436, 439, 440, 452, 455).

[21]The Diagram in the *Origin* is found in the middle of Chapter 4 in all editions.

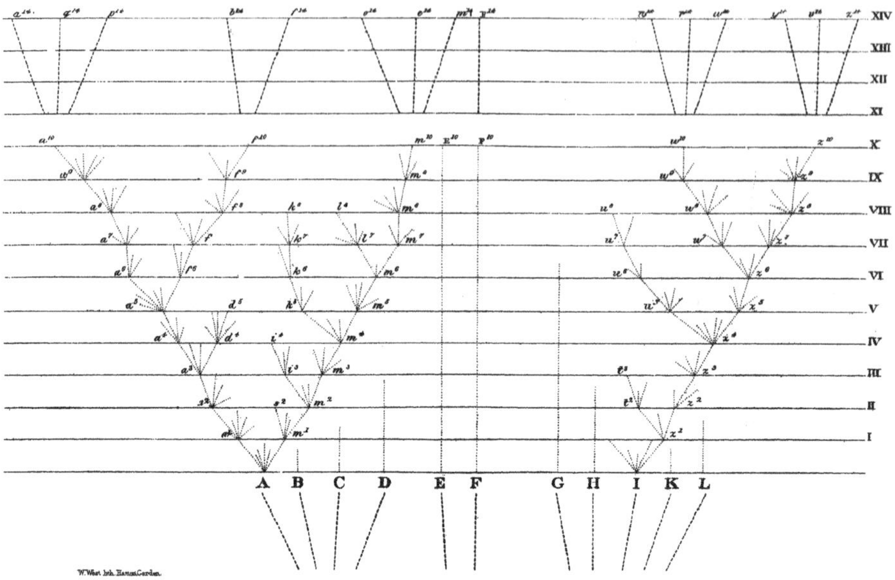

Fig. 2.1 Darwin's Diagram. This sole illustration in the *Origin of Species* is a genealogy of extant forms and not a phylogenetic tree

> uppermost horizontal line. Now all these modified descendants from a single species, are represented as related in blood or descent to the same degree... yet they differ widely and in different degrees from each other. The forms descended from A, now broken up into two or three families, constitute a distinct order from those descended from I, also broken up into two families. Nor can the existing species, descended from A, be ranked in the same genus with the parent A; or those from I, with the parent I. But the existing genus F^{14} may be supposed to have been but slightly modified; and it will then rank with the parent-genus F; just as some few still living organic beings belong to Silurian genera. So that the amount or value of the differences between organic beings all related to each other in the same degree of blood, has come to be widely different. *Nevertheless their genealogical arrangement remains strictly true, not only at the present time, but at each successive period of descent.* All the modified descendants from A will have inherited something in common from their common parent, as will all the descendants from I... All the descendants of the genus F... are supposed to have been but little modified, and they yet form a single genus. But this genus, though much isolated, will still occupy its proper intermediate position; for F originally was intermediate in character between A and I, and the several genera descended from these two genera [A and I] will have inherited to a certain extent their characters. (Darwin 1859: 420–422; see also 1872: 369, 370) [my emphasis]

This key passage contains an array of issues that can profitably be unpacked under two complementary headings:

1. The entire length of geological time since the Silurian (or Cambrian) was required for forms within a single class to diversify to their actual state (a^{14} to z^{14}), going from distinct genera in the Silurian, to distinct families, to distinct orders today. According to Darwin, evolution is an extremely slow process. Let us also keep in mind that, for Darwin, variability observed among extant forms serves as a gauge

of past forms: "all fossils can be classed either in still existing groups, or between them" (Darwin 1859: 329; see also 1872: 301). By combining the two notions just mentioned, one can deduce that Darwin holds that no high taxonomic groups (classes, phyla)[22] ever went extinct in the history of life, in contradiction to current views.[23] Knowing as we do that Darwin thinks that the time period separating the Silurian (or Cambrian) from today constitutes about half the length of time for the whole history of life (Darwin 1859: 307; 1872: 286), we can conclude that the original species at the root of the class in the Diagram needed huge amount of time to merely reach the genus level by the Silurian (or Cambrian) period.

2. The so-called arrangement or separation characterizing the three main progenitors (A, F, and I) in the Silurian or Cambrian was preserved up to today. The evolutionary process has not significantly blurred the phylogenetic relationships since that ancient time. Nothing happened in the past (i.e., the rise of novelties) of enough significance to erase from today's taxonomy past traces.

The Diagram is a window into Darwin's neontological preconceptions, in the sense that it reveals his theoretical expectations. Without any significant recourse to the field of paleontology—in light of his own dismissal—Darwin's view of the history of life is largely a neontology projected backward in time, a sort of perpetual neontology.

While Darwin was committed to a fully constituted world, he could not claim that *all* past events are registered in taxonomy. On the road to evolutionary diversification, numerous forms became extinct within a class. It is the very essence of Darwin's theory to postulate an evolutionary dynamics based on competition, with extinction being its by-product.[24] To get the view of a full world, Darwin had to collapse or superimpose all the extinct forms that ever existed onto the ones currently living. This he did in a thought experiment: "Extinction has only separated groups: it has by no means made them; for if every form which has ever lived on this earth were suddenly to reappear, though it would be quite impossible to give definitions by which each group could be distinguished from other groups, as all would blend together by steps as fine as those between the finest existing varieties, nevertheless a natural classification, or at least a natural *arrangement*, would be possible" (Darwin

[22]It should be noted that taxonomy was not yet stabilized in Darwin's time, hence his somewhat loose or ambiguous utilization of the "class" level to sometimes designate, for instance, either the vertebrates as a whole (fish, reptiles, birds, and mammals) or these last four groups separately. See Darwin (1859: 338, 434, 439, 1872: 310, 382).

[23]Scholars today recognize the extinction, for instance, of invertebrate classes like *Trilobita* and *Blastoidea*, of vertebrate classes like *Placodermi* and *Acanthodii*, of an animal phylum like *Vetulicolia*, and of other possible phyla part of the Ediacaran and Burgess Shale faunas. It is true that Darwin recognized the extinction of trilobites and ammonites, but for him, these groups were either of the family or order level only. See Darwin (1859: 321, 1872: 297). Also relevant for the topic here is our discussion few pages below in this chapter of the interpretation given to the Burgess Shale fauna.

[24]Chapter 7 of this book analyzes the nature of the evolutionary dynamics proposed by Darwin in greater detail.

1859: 432)[my emphasis].[25] Recognizing that the present cannot *entirely* register all there is to know about the past, Darwin nonetheless saves the argument by telling us that the past is *largely* represented in the present. In the worst-case scenario, extinct forms will be directly aligned with extant ones, as seen in the near identity of their appearances. In the best possible case, extinct forms will fall between two extant groups, as predicted by a theory founded on the notion of common ancestry. Although the centrality of the concept of "divergence" in Darwin's theory has rightly caught the attention of Darwin scholars (Limoges 1968; Sulloway 1979; Browne 1980; Schweber 1980; Ospovat 1981; Kohn 1985, 2009; Tammone 1995; Richards 2012), it has not been sufficiently realized that such a concept is, for him, implicitly tied to the notion of a fully constituted world.

In light of Darwin's unique approach to the past, it was to be expected that his postulated progenitors or common ancestors would have a strangely familiar look. For instance, reflecting upon the progenitors of various groups, Darwin writes:

> In the vertebrata, we see a series of internal vertebrae bearing certain processes and appendages; in the articulata, we see the body divided into a series of segments, bearing external appendages; and in the flowering plants, we see a series of successive spiral whorls of leaves. An *indefinite repetition* of the same part or organ is the common characteristic... of all low or little-modified forms; therefore we may readily believe that the unknown progenitor of the vertebrata possessed many vertebrae; the unknown progenitor of the articulata, many segments; and the unknown progenitor of flowering plants, many spiral whorls of leaves. (Darwin 1859: 437; see also 1872: 384, 385) [my emphasis]

Not unlike "miniatures," Darwin's hypothetical ancestors look like downsized forms waiting to express themselves in their full-grown state of today, in a process akin to preformationism. There is a cyclical twist to Darwin's way of thinking about evolution, with the same variations being recycled within separate classes or phyla. In contradistinction to Darwin's conception, today's biologists recognize that classes and phyla of the past included some features significantly beyond today's variability, thus providing room for genuine evolutionary novelties. Taking the vertebrates as an example, it is now recognized that the vertebrae (bone or cartilage) which compose the vertebral column of vertebrate animals find their phylogenetic origin in a more primitive structure called a notochord (a long axial rod composed of a fibrous connective tissue wall around cells and/or a fluid-filled space), a structure characteristic of chordates possibly derived from an even more primitive proto-notochord state (cephalochordates), which may ultimately be traceable to a mere dorsal nerve cord (hemichordates) (Kardong 2012, Chaps. 2 and 3; Saxena and Saxena 2008, Chap. 1; Gee 1996). One wonders, then, how Darwin's nearly full-blown progenitors could possibly have originated from genuinely simple and primitive conditions, especially for someone so profoundly committed to gradualism, as Darwin was? Indeed, although Darwin (1859: 306, 307, 1872: 285, 286) provided for extensive amount of time during the pre-Silurian or pre-Cambrian period to allow for the rise

[25]In the sixth edition, the phrase is slightly compressed, but the message remains the same. See Darwin (1872: 380).

of such simple beings, the fact remains that his theory is powerless to conceive truly primitive beings. Darwin's commitments to "neontology" and "gradualism" were, in this particular case, condemned to remain in insoluble tension.

2.3 Historicity and the Unknowable Past

We know that it was easy for Newton, John Ray, and Linnaeus to embrace a world fully constituted. Things were more challenging for those in the nineteenth century living at a historical moment when the world was increasingly seen as changing. It is fascinating to see how Darwin and others kept that tradition alive.[26] As the ground beneath his feet opened up, revealing an evolutionary past, Darwin rushed to assimilate it to extant forms. By so doing, he exposed himself to the epistemological criticism already formulated by William Whewell against Lyell in 1832 concerning historical object of study:

> It seems to us somewhat rash to suppose, as the uniformitarian does, that the information which we at present possess concerning the course of physical occurrences, affecting the earth and its inhabitants, is sufficient to enable us to construct classifications, which shall include all that is past under the categories of the present. Limited as our knowledge is in time, in space, in kind, it would be very wonderful if it should have suggested to us all the laws and causes by which the natural history of the globe... is influenced—it would be strange, if it should not even have left us ignorant of some of the most important of the agents which, since the beginning of time, have been in action... (Whewell 1832: 126)

The word "classification" in the quote above should not be equated with "taxonomy." Here, Whewell uses the word to denote the proper ordering of phenomena into different branches of knowledge in accordance to their distinct nature, a classification performed by Whewell himself in his *History of the Inductive Sciences* (1837). Let us follow Marjorie Grene and David Depew (2004: 171) in their understanding of Whewell's warning: "In light of mass extinctions spread across quite different geological eras, Whewell thought it decidedly unscientific for Lyell to shoehorn the history of life into categories cobbled up to fit contemporary life forms." In our view, however, the issue is less a matter of "scientificity" than of confronting two distinct epistemologies design to make sense of the past. Whereas Whewell anticipated that the present would be blind to novelties of the past, Darwin refused to seriously contemplate that possibility. Whewell's *History of the Inductive Sciences* (1837: 618–620, vol. 3) is instructive, precisely because it grasps the tipping point of two temporally successive yet overlapping epistemologies accompanying, respectively, the static worldview and the evolutionary worldview. Whewell saw that the areas of geology and history of life could be approached from either a static-cyclical view of the universe (as in Newton and Lyell) or an

---

[26]As will be seen in several places in this book, Darwin was not alone in maintaining the tradition of belief in a fully constituted world. He will be joined by Charles Lyell, Thomas Henry Huxley, and Richard Owen.

evolutionary-cosmological view (as in Chambers and Spencer). To again quote Hodge's (1982: 95, 96) account of the young Darwin:

> [Darwin] has nothing to say about the ultimate constituents and operations of everything from here to infinity; and leaves the stars and the solar system, even—indeed especially—the earth, arranged exactly as it was before; an oblate terraqueous spheroid, the earth is left orbiting steadily and stably in a slowly and slightly varying elliptical path around the sun, exactly in fact as it is found in Lyell and in John Herschel's and Mary Somerville's digest of Lagrange and Laplace... [That] leaves him with the solar system going about its balanced Lagrangean-Laplacean business as before... When Darwin, Lyell, and Herschel talk of 'astronomical causes' of climate changes, they mean long-run reversible orbit changes rather than... 'cosmological' or 'cosmogonical' causes of climate change, that is causes tracing to a putative origin of the planet in, say, a hot molten fluid chaos or in an incandescent nebula.

It will be seen in the following chapters how the biology of Darwin was erected in the context of a major theoretical tension, uncomfortably sitting as he was between two scientific revolutions (the Scientific Revolution and the Transformist Revolution), two worldviews, and two epistemologies designed to tackle them (Delisle 2014b). To better understand these two alternatives, it is useful to momentarily project ourselves into the twentieth century. Stephen Jay Gould (1989: 99, 100) has argued that today's biological variability (which he calls "disparity") cannot accommodate all that is seen in the fossil record. As an example, consider the animal kingdom as observed today. It is recognized that it is divided in 20 or 30 different basic body plans, also called *phylum* (plural: phyla), such as chordates (vertebrates), arthropods (insects), mollusks (clams), echinoderms (starfishes), sponges, corals, and annelids (earthworms). Gould holds that one finds in the Cambrian period of British Columbia (Canada) 15–20 extinct body plans (phyla) whose architectures are entirely unlike anything now existing (see Fig. 2.2). Gould's detractors aside,[27] the fact remains that many scholars recognize that the diversity or disparity observed at the Precambrian-Cambrian transition (the Ediacaran and Burgess Shale faunas) is not easily assimilated to extant forms (Gould 1991; Seilacher 1992; Gee 1996: 295–299; Conway Morris 1998: 12–14, 27–30, 169–218). This is even more evident when considering that disparity within the same class or phylum might have been greater in earlier evolutionary phases than during later ones. Darwin's neontological approach could not fail but be blind to significant evolutionary manifestations of the history of life (see Fig. 2.3). Darwin's Diagram, therefore, is less a portion of tree of life than a genealogy of extant forms.

[27]For a clear and concise presentation of the main issues involved in this debate, see Sterelny (2007, Chap. 10).

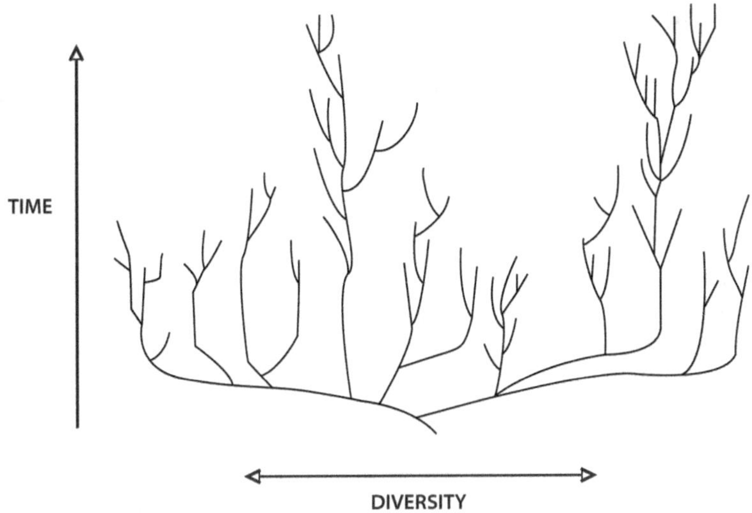

Fig. 2.2 Decimation Model. This evolutionary model is based on a rapid multiplication of types followed by a significant amount of decimation (extinction). Many body plans that existed in the past have left no traces today

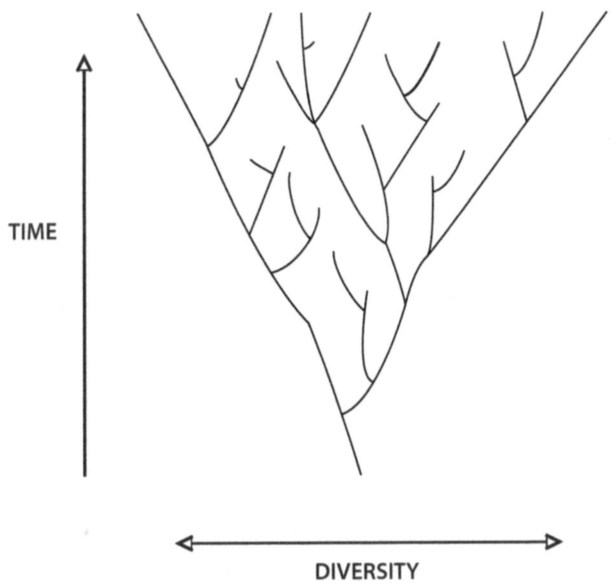

Fig. 2.3 Cone of Increasing Diversity. This evolutionary model assumes that variability observed among extant forms entirely captures that of the past, since all fossils are either aligned with or fall in-between extant forms. This is Charles Darwin's view, which he based on the notion of a slow deployment of increasing diversity within a class

2.4 A Closed, Permanent, and Segmented Tree of Life

It may be tempting for a modern reader to assume that Darwin's view of the history of life represents a single "cone of increasing diversity", to use Gould's (1989: 38–47) terminology, which binds all life together under a common ancestor. In reality, Darwin regrouped extant forms under a number of separate cones of increasing diversity. His overall picture of life, therefore, is characterized by a series of phylogenetic groups evolving in parallel that independently and suddenly sprung, full-fledged, out of the Silurian or Cambrian barrier. It is striking that his view does not present life as an evolutionary process projecting itself vertically in time, as seen especially in Robert Chambers' *Vestiges of the Natural History of Creation* (Hodge 1972; Secord 1989) and Herbert Spencer's encyclopedic *A System of Synthetic Philosophy* (Spencer 1857; La Vergata 1995). In opposition to verticality, Darwin embraces horizontality: today's main living groups have always independently existed, at least as far back as we can probe in the geological annals.

We have seen that this logic was originally applied by the anti-evolutionist Lyell in his *Principles of Geology* (1930–1833) and was recast in an evolutionary context for separate and independent classes and phyla. Darwin scholars have not sufficiently stressed the extent to which he organized the history of life within a rigid structure of independent and parallel classes (fish, reptiles, birds, mammals) and phyla (such as vertebrates or mollusks), each understood by reference to their living counterparts. It is useful here to introduce Darwin's contemporaries into the picture. Reviewing what can be extracted from the fossil record, Thomas Henry Huxley had this to say in 1862:

> [E]nough has been said to justify the statement that, in view of the immense diversity of known animal and vegetable forms, and the enormous lapse of time indicated by the accumulation of fossiliferous strata, the only circumstance to be wondered at is, not that changes of life, as exhibited by positive evidence, have been so great, but that they have been so small. (Huxley 1862: 1)

Huxley was struck by the fact that extant forms at high taxonomic levels are similar to those of the past. Speaking of orders among plants, he writes: "There are two hundred known orders of plants; of these not one is certainly known to exist exclusively in the fossil state. The whole lapse of geological time has as yet yielded not a single new ordinal type of vegetable structure" (Huxley 1862: xlviii). Richard Owen applied the same reasoning to classes among the invertebrates. He writes in 1860: "But as to the larger groups of… true invertebrate animals, it may be affirmed that every known fossil belongs to some one or other of the existing classes, and that the organic remains of the most ancient fossiliferous strata do not indicate or suggest that any earlier and different class of being remains to be discovered, or has been irretrievably lost" (Owen 1860: 18). It seems, therefore, that several scholars in the 1860s worked within the tradition of a world believed to be fully constituted, and most of them were by then committed to one brand of evolutionism or another.

For Darwin, today's separate classes and phyla are an expression of the world's fullness; before that, one finds them in a presumed diminished state as one travels back in time. Of course, Darwin could not have held that the world had always been as full as today. Nonetheless, he saved the argument by implicitly embracing two complementary ideas. First, the past was full but only under a preformationist state, with only time being required for classes and phyla to express themselves in full. Second, since fullness is only seen among extant classes and phyla, it is epistemologically justified to study it only through extant forms. Once this is recognized, the full meaning of the following statement can be appreciated:

> It may be asked how far I extend the doctrine of the modification of species... All the members of whole classes can be connected together by chains of affinities, and all can be classified on the same principle, in groups subordinate to groups. Fossil remains sometimes tend to fill up very wide intervals between existing orders... Throughout whole classes various structures are formed on the same pattern, and at an embryonic age the species closely resemble each other. *Therefore I cannot doubt that the theory of descent with modification embraces all the members of the same class.* (Darwin 1859: 483, 484)[28] [my emphasis]

Darwin confines his analysis of transmutation to isolated classes and phyla, giving up on a *universal evolutionism* transgressing the frontiers of classes and phyla, except in entirely hypothetical reasoning. Darwin may have thought that evolutionism defended at this fairly high taxonomic level was ambitious and illustrative enough for his time, and in a sense it was. Yet, it has often been overlooked that Darwin's segmented evolutionism also responded to the epistemological imperative of a world believed to be fully constituted. No surprise, then, that his transformism could only have been characterized by a process recycling today's variability by projecting it into the past, leaving the reader of the *Origin* with the impression that evolution within a class is largely at a standstill.

If the fullness of the world had to wait for the complete deployment of the classes and phyla over time to express itself, it should be noted that Darwin recognized that it had, factually speaking, already been largely attained in very ancient times—in the Silurian or Cambrian period, perhaps—judging, for instance, from the separate and independent state of the mammal, reptile, and fish classes at such an early period. Moving beyond the mere theoretical notions of his Diagram, Darwin writes:

> Thus the embryo comes to be left as a sort of picture... of the ancient and less modified condition of each animal. This view may be true, and yet it may never be capable of full proof. Seeing, for instance, that the oldest known mammals, reptiles, and fish strictly belong to their own proper classes, though some of these old forms are in a slight degree less distinct from each other than are the typical members of the same groups at the present day, it would be vain to look for animals having the common embryological character of the Vertebrata, until beds far beneath the lowest Silurian strata are discovered—a discovery of which the chance is very small. (Darwin 1859: 338; see also 1872: 310)

[28]In the sixth edition, the passage was reproduced with minor changes only, although it is worth noting the following addition : "Therefore I cannot doubt that the theory of descent with modification embraces all the members of the same great class *or kingdom*" [my emphasis]. See Darwin (1872: 424). In this last edition, Darwin's rhetoric concerning hypothetical ancestors becomes more ambitious, since he claims that it should be possible to seek common ancestors of kingdoms rather than simply of classes or phyla.

This quote is instructive on two counts. First, it establishes an epistemological distinction between the hypothetical notion of an unknown common ancestor and the factual reality of a world already constituted. Second, it puts significant explanatory stress on the pre-Silurian or pre-Cambrian period: if half the time Darwin (1859: 307, 1872: 286) allots for the entire history of life is situated before the Silurian or Cambrian, then nearly the entire deployment of classes and phyla would have had to occur during that first half. Although he wanted to impose his neontological agenda, the evanescent picture of an assumed common vertebrate ancestor dimly enlightened by embryology did not loom large in the face of the "factuality," so to speak, of very ancient and full-blown mammals, reptiles, and fish.

Darwin's double epistemological standard with respect to paleontological facts should here be recalled: if they fit within his theory, he accepts them; if they don't, he dismisses them. In the case that concerns us here, however, Darwin does not contest the value of paleontological facts, as he accepts the lack of proof for a common ancestor between mammals, reptiles, and fish. This allows us to catch a glimpse of Darwin's argumentation in the *Origin*:

1. Extant and independent classes and phyla are the permanent and ultimate units to be employed in thinking about the evolutionary past (Darwin 1859: 483, 484; 1872: 424).
2. Paleontology shows that classes such as mammals, reptiles, and fish have been clearly distinct from one another for a very long time (Darwin 1859: 338; 1872: 310).
3. From the morphology of living forms, we gather that the unknown progenitor of the vertebrate phylum, for instance, already possessed many vertebrae, just as observed in its living counterparts (Darwin 1859: 437; 1872: 385).
4. Far from being explicit about the progenitor's picture that emerges from the field of embryology, Darwin nonetheless thinks that he can use it to probe more deeply in time than with any other fields. This being said, whatever exactly he extracts from embryology, Darwin remains within the confines of extant classes and phyla: "Embryology rises greatly in interest, when we thus look at the embryo as a picture, more or less obscured, of the common parent-form of each great *class* of animals" (Darwin 1859: 450; 1872: 396) [my emphasis].

   The first four steps in Darwin's argumentation fall within the purview and explanatory power of a world believed to be fully constituted.
5. Only when Darwin moves beyond extant classes and phyla to evoke an animal kingdom that arose out of a limited number of progenitors—no more than four or five or perhaps fewer (Darwin 1859: 484, 488, 490; 1872: 424, 429)—does he tread into entirely hypothetical, abstract, and uncharted territory with no living counterparts. There, however, Darwin's theory is powerless and does not venture, except under rhetorical arguments only.

2.5 Archetypes, Ancestors, or Shadows?

During the pre-1859 decades, scholars increasingly began to recognize that life forms seem to fall naturally into a number of separate biological types—such as articulata, radiata, mollusca, and vertebrata—united by their morphological similarities. Schelling, Goethe, Oken, Cuvier, Geoffroy St-Hilaire, and Owen, among others, all subscribed to this notion in one form or another. Some did it in the name of a unified and monistic view of nature, others under the principles of the laws of correlation of parts and conditions of existence, and still others under empirical realism. Knowing as we do that Darwin also recognized the unity of types, Darwin scholars have recently debated the ultimate underpinnings of his position. Robert J. Richards (1999: 129) argues that this is best explained by the impact of German romantic biology:

> Darwin thus suggested that the generalized archetype of the vertebrates did not lie hidden away as an idea in the mind of God; rather, it was the form of a creature that walked the earth many generations ago... As an instance of the molding force exercised by archetype theory on the deep structure of the *Origin*, one might point to Darwin's chronic presumption that transformation of species occurred without common descent... In the *Origin*, Darwin advanced the same conviction that 'animals have descended from at most only four or five progenitors, and plants from an equal or lesser number'. This hypothesis, he maintained, was sufficient for his general theory.

A more traditional interpretation consists in reversing the perspective: Darwin did not recognize the unity of biological types because he had earlier embraced the very idea of archetype (as claimed by Richards) but rather because he realized that homologies would be a natural outcome of a divergent evolutionary process which proceeds out of a common ancestor. Jonathan Hodge (2014: 176, 177) explains:

> In German comparative anatomy, invocation of structural archetypes are prominent. But such invocations in such terms are absent from Darwin's notebooks in the 1830s and only made in the *Origin* in order to trump them with invocations of ancestors. This explanatory trumping is a trumping by historical interpretations. Furthermore, Darwin insists that ancestral structures have arisen from prior adaptive changes, so that histories and functions are reinforcing each other here in trumping structures. German romantics typically stood for the privileging of structural over historical and functional interpretations of nature and of art. Darwin's privilegings of history and function are directly opposed to such views.

Getting straight to the point, Michael Ruse (2004: 17) reinforces Hodge's claim: "[F]or Darwin homology was always a consequence of evolution rather than something put right up as the fundamental starting point of discussion." The choice, then, seems to be between *Archetypes* or *Ancestors*.

In light of the material covered in this chapter, we would like to propose a third option: *Shadows*. Richards is right to stress the segmented nature of Darwin's tree of life organized around distinct biological types. As we have seen, Darwin's theory was very actively constructed around classes and phyla. This important fact has been overlooked by many Darwin scholars, most likely because they have assumed that Darwin's types were not isolated from each other, since he vaguely allude to some ultimate and unique common ancestor (Darwin 1859: 484; 1872: 425). It is precisely at this explanatory juncture that one must distinguish between the active explanatory components that do real work in Darwin's theory (evolution within classes) and the vague, hypothetical, and rhetorical components that merely fill otherwise unbridgeable theoretical gaps (ultimate progenitors). This being said, there is an alternative to following Richards down the path of romantic biology. As a proponent of a world believed to be fully constituted, Darwin worked his way backward in time from the variability (homologies) observed today. In so doing, Darwin's so-called ancestors were deprived of their existence; Darwin's progenitors or ancestors are ultimately more like "shadows," which are generated from images cobbled together from extant classes or phyla and abstractly projected into geological times. The various biological types recognized by Darwin are today, under his ontologico-methodological commitments, secure and distinct from one another for all geological eternity. That is why Darwin's rhetoric about the ultimate common ancestors of kingdoms or life as a whole is meaningless by his own lights. For similar reasons, the assessments of Hodge and Ruse about Darwin's ancestors seem to have missed the essential point. Darwin's rhetoric would suggest that extant types are united in homologies because they had common ancestors, whereas in reality he reversed the perspective: it is because they shared extant homologies that Darwin imagined they had common ancestors. No matter how much one projects extant homologies backward into geological time, one will always end up with biological types being both separated from one another and strangely similar to their living counterparts.

2.6 Conclusion of Part I

This chapter is meant as a preliminary overview of issues and topics in need of further investigation. These will be provided in the following chapters. Enough has been said thus far to at least shed some dim light on Darwin's own special brand of evolutionism in the *Origin of Species*. Darwin inherited from the previous centuries a number of concepts that guided him in his thinking: stability, completeness, timelessness (eternity), unity (within classes), permanence, and uniformity. Scholars of the Darwin industry have long insisted on his commitment to the doctrine of uniformitarianism, usually focusing on its *methodological* aspect. We argue that this kind of analysis should also be extended to cover its *ontological* aspect, both of which are fused into a coherent ontologico-methodological epistemology strongly congenial to the static worldview. It was part of Darwin's originality to "loosen up" this array of concepts somewhat by inserting them in a peculiar theory of evolution. This he did through two complementary tasks.

First, he reaffirmed the prerogatives of the present by reducing the significance of the past. The world is seen in its completeness only in the present time, so the argument goes. This explains a number of choices made by Darwin: paleontology is instrumentalized for the profit of neontology; fossils are aligned with or classed in-between living counterparts; the world is assumed to be continuous and unified by being deprived of major evolutionary gaps as far back in time as we can empirically probe (the Cambrian); and a segmented evolutionary history of extant classes and phyla was reconstructed by merely projecting onto the past the variation observed today (shadows), giving them a sense of permanence and near eternity. Second, Darwin had to introduce some kind of evolutionary motion into this fully constituted world. He did so through the recycling of today's variations by projecting them backward in time, in addition to allowing for the full deployment of classes and phyla in time from their original potential states (miniatures; preformationism).

Irrespective of what we often think today, Darwin's *Origin* has a built-in ahistorical explanatory structure that creates conceptual obstacles on the road to universal and modern evolutionism. To quote Michael Ruse (2003: 119): "in a way the science of the *Origin* already had a rather old-fashioned look at the time of its publication in 1859." Ruse is suggesting that Darwin was more concerned with issues of the 1830s than those of the 1860s when he published his *magnum opus*. The truth of Ruse's claim, it seems, extends even beyond his original intentions: Darwin owed much to the previous two centuries. Whereas scholars like Jean-Baptiste Lamarck, Georges Cuvier, William Buckland, and Robert Chambers, in their own separate ways, managed to convey us a certain impression of temporal verticality—biological differences in geological horizons—Darwin's bold and counterintuitive approach to the question was to take us for "a walk across his garden," so to speak, through a biogeographical exploration of currently living forms. Darwin's Earth is flat, being deprived of a genuine temporal dimension. This has been perceptively noted by Camille Limoges. Taking advantage of Michel Foucault's analysis in *The Order of Things*,[29] Limoges writes:

> As always during the 18th century, recourse to the notion of origin serves the purpose of thinking in all clearness about what would still exist today and under the very same mode but for a different state, more opaque to our representation... The *Origin of Species* was in no way a treatise about origins: never features of the economy of nature were founded on any such question; on the contrary, transmutation of species finds its rules in contemporaneity, or rather in atemporality, that is, the synchronic ecological operation of natural selection.[30]

To refresh the reader's memory on what Foucault (2002: 358) had to say about the Classical Age of the eighteenth century, it is useful to read again his own words:

> In the eighteenth century, to return to the origin was to place oneself once more as near as possible to the mere duplication of representation... The order of nature was conceived,

[29]Foucault, M. (1966), *Les mots et les choses: une archéologie des sciences humaines* (Paris: Gallimard), pp. 339–346. Translated as *The Order of Things: An Archaeology of the Human Sciences* (London: Routledge Classics, 2002), pp. 358–365

[30]Translation mine. See Limoges, C. (1972), "Introduction," in C. Linné, *L'équilibre de la nature* (Paris : J. Vrin), pp. 7–22, quote on p. 12, and footnote #19 of the same page. The original text reads as: "Comme toujours au dix-huitième siècle le recours à l'origine sert à penser en toute limpidité ce qui existerait encore aujourd'hui et selon un même régime mais en un autre état, plus opaque à la représentation... l'*Origine des espèces* ne sera en rien un traité sur l'origine ; jamais les traits propres à l'économie de la nature ne se fonderont sur quelque origine : tout au contraire, la transformation des espèces y trouve ses règles dans la contemporanéité, ou plutôt dans l'intemporel, le synchronique du fonctionnement écologique de la sélection naturelle ."

> prior to any catastrophe, as a table in which beings followed one another in so tightly knit an order, and upon so continuous a fabric, that in going from one point of this succession to another one would have moved within a quasi-identity, and in going from one extremity of it to the other one would have been led by the smooth expanse of 'likeness'.

While Foucault (Gutting 1989: 190–193) and many others today believe that Darwin went beyond this classical representation of things by having moved well into the modern representation, we argue, on the contrary, that Darwin was very much an heir of the Classical Age. To borrow Foucault's concept for a moment without embracing his discontinuous view of history and epistemology, it might be said that Darwin had a strong footing in the older *epistémè*.

References

Agassiz L (1833–1843) Recherches sur les poissons fossiles, 5 vols. Imprimerie de Petitpierre, Neufchâtel

Agassiz L (1844–1845) Monographie des poissons fossiles du Vieux Grès Rouge ou Système Dévonien (Old Red Sandstone) des Iles Britanniques et de Russie, 2 vols. Aux fraix de l'Auteur, Neufchâtel

Barrande J (1846) Notice préliminaires sur le Système Silurien et les trilobites de Bohême. Leipzig

Barrande J (1852–1881) Système Silurien du centre de la Bohême, 6 vols. Prague

Bartholomew M (1973) Lyell and evolution: an account of Lyell's response to the prospect of an evolutionary ancestry for man. Br J Hist Sci 6:261–303

Bartholomew M (1979) The singularity of Lyell. Hist Sci 17:276–293

Bowler PJ (1976) Fossils and progress: paleontology and the idea of progressive evolution in the Nineteenth century. Science History Publications, New York

Bowler PJ (1996) Life's splendid drama: evolutionary biology and the reconstruction of life's ancestry, 1860-1940. University of Chicago Press, Chicago

Bowler PJ (2003) Evolution: the history of an idea, 3rd edn. University of California Press, Berkeley

Browne J (1980) Darwin's botanical arithmetic and the 'principle of divergence,' 1854–1858. J Hist Biol 13:53-89

Chambers R (1844) Vestiges of the natural history of creation. John Churchill, London

Conway Morris S (1998) The crucible of creation: the Burgess Shale and the rise of animals. Oxford University Press, Oxford

Cuvier G (1825) Recherches sur les ossements fossiles de Quadrupèdes, 3rd edn, 7 vols. GF-Flammarion, Paris

Cuvier G (1830) Discours sur les révolutions du globe, 6th edn. Bourgois, Paris

D'Orbigny A (1842–1860) Paléontologie française, 9 vols. Forgotten Books, Paris

Darwin C (1859) On the Origin of Species. John Murray, London

Darwin C (1872) The origin of species, 6th edn, with additions and corrections [1876]. John Murray, London

Delisle RG (2014a) Evolution in a fully constituted world: Charles Darwin's debts toward a static world in the Origin of Species (1859). Endeavour 38:197–210

Delisle RG (2014b) Can a revolution hide another one? Charles Darwin and the Scientific Revolution. Endeavour 38:157–158

Egerton F (1973) Changing concepts of the balance of nature. Q Rev Biol 48:322–350

Ellegard A (1990 [1958]) Darwin and the general reader: the reception of Darwin's theory of evolution in the British periodical press, 1859–1872. University of Chicago Press, Chicago

Falconer H (1868) Palaeontological Memoirs, 2 vols. R. Hardwicke, London

Falconer H, Cautley PT (1845–1849) Fauna Antiqua Sivalensis, being the fossil zoology of the Siwalik Hills, in the North of India, 2 vols. Smith, Elder, London

Farley J (1977) The spontaneous generation controversy from Descartes to Oparin. Johns Hopkins University Press, Baltimore

Foucault M (1966) Les mots et les choses: une archéologie des sciences humaines. Gallimard, Paris

Foucault M (2002) The order of things: an archaeology of the human sciences. Routledge Classics, London

Gee H (1996) Before the backbone: views on the origin of the vertebrates. Chapman and Hall, London

Gould SJ (1987) Time's arrow time's cycle: myth and metaphor in the discovery of geological time. Harvard University Press, Cambridge

Gould SJ (1989) Wonderful life: the burgess shale and the nature of history. W.W. Norton, New York

Gould SJ (1991) The disparity of the Burgess Shale arthropod fauna and the limits of cladistic analysis: why we must strive to quantify morphospace. Paleobiology 17:411–423

Greene JC (1959) The death of Adam: evolution and its impact on western thought. Iowa State University, Iowa

Greene JC (1981) Science, ideology, and word view. University of California Press, Berkeley

Grene M, Depew D (2004) The philosophy of biology: an episodic history. Cambridge University Press, Cambridge

Gruber HE (1974) Darwin on man: a psychological study of scientific creativity. E.P. Dutton, New York

Gutting G (1989) Michel Foucault's archaeology of scientific reason. Cambridge University Press, Cambridge

Herbert A, Norman D (2009) Darwin's geology and perspective on the fossil record. In: Ruse M, Richards RJ (eds) The Cambridge companion to the "Origin of Species". Cambridge University Press, Cambridge, pp 129–152

Hodge MJS (1972) The universal gestation of nature: Chambers's Vestiges and Explanations. J Hist Biol 5:136–145

Hodge MJS (1982) Darwin and the laws of the animate part of the terrestrial system (1835–1837): on the Lyellian origins of his zoonomical explanatory program. Stud Hist Biol 6:1–106

Hodge MJS (2014) On Darwin's science and its contexts. Endeavour 38:169–178

Hodge MJS, Radick G (2009) The place of Darwin's theories in the intellectual long run: In: Hodge MJS, Radick G (eds) The Cambridge companion to Darwin, 2nd edn. Cambridge University Press, Cambridge, pp 246–273

Hutton J (1788) Theory of the earth. Trans Roy Soc of Edinb 1:209–305

Huxley TH (1862) The anniversary address. Q J Geol Soc Lond 18:xl–liv

Kardong K (2012) Vertebrates: comparative anatomy, function, evolution, 6th edn. McGraw-Hill, New York

Kohn D (1980) Theories to work by: rejected theories, reproduction, and Darwin's path to natural selection. Stud Hist Biol 4:6–170

Kohn D (1985) Darwin's principle of divergence as internal dialogue. In: Kohn D (ed) The Darwinian heritage. Princeton University Press, Princeton, pp 245–257

Kohn D (2009) Darwin's keystone: the principle of divergence. In: Ruse M, Richards RJ (eds) The Cambridge companion to the 'Origin of Species'. Cambridge University Press, Cambridge, pp 87–108

La Vergata A (1995) Herbert Spencer: biology, sociology, and cosmic evolution. In: Maasen S, Mendelsohn E, Weingart P (eds) Biology as society, society as biology: metaphors. Kluwer Academic, Dordrecht, pp 193–229

Laudan R (1987) From mineralogy to geology: the foundations of a science, 1650–1830. University of Chicago Press, Chicago

Laurent G (1987) Paléontologie et évolution en France, 1800–1860. Éditions du Comité des Travaux Historiques et Scientifiques, Paris

Limoges C (1968) Darwin, Milne-Edwards, et le principe de divergence. Actes du XIIe Congrès International d'Histoire des Sciences 8:111–115
Limoges C (1970) La sélection naturelle. Presses Universitaires de France, Paris
Limoges C (1972) Introduction. In: Linné C (ed) L'équilibre de la nature. J. Vrin, Paris, pp 7–22
Lyell C (1830) Principles of geology, vol 1. John Murray, London
Lyell C (1833) Principles of geology, vol 3. John Murray, London
Ospovat D (1981) The development of Darwin's theory: natural history, natural theology, and natural selection, 1838–1859. Cambridge University Press, Paris
Owen R (1849–1884) History of British fossil reptiles, 4 vols. Cassel and Company, London
Owen R (1860) Palaeontology or a systematic summary of extinct animals and their geological relations. Adam and Charles Black, Edinburgh
Pictet F-J (1853–1857) Traité de paléontologie, 2nd edn, 4 vols. J.B. Baillière, Paris
Richards RJ (1999) Darwin's Romantic biology: the foundation of his evolutionary ethics. In: Maienchein J, Ruse M (eds) Biology and the foundation of ethics. Cambridge University Press, Cambridge, pp 113–153
Richards RJ (2012) Darwin's principles of divergence and natural selection: why Fodor was almost right. Stud Hist Phil Biol Biomed Sci 43:256–268
Rudwick MJS (1970) The strategy of Lyell's Principles of Geology. Isis 61:5–33
Rudwick MJS (1971) Uniformity and progression: reflections on the structure of geological theory in the age of Lyell. In: Roller D (ed) Perspectives in the history of science and technology. University of Oklahoma Press, Norman, pp 209–227
Rudwick MJS (2008) Worlds before Adam: the reconstruction of geohistory in the age of reform. University of Chicago Press, Chicago
Ruse M (2003) Darwin and design: does evolution have a purpose? Harvard University Press, Cambridge
Ruse M (2004) The Romantic conception of Robert J. Richards J Hist Biol 37:3–23
Saxena R, Saxena S (2008) Comparative anatomy of vertebrates. Anshan, Kent
Schweber SS (1980) Darwin and the political economists: divergence of character. J Hist Biol 13:195–289
Secord JA (1989) Behind the veil: Robert Chambers and Vestiges. In: Moore JR (ed) History, humanity and evolution. Cambridge University Press, Cambridge, pp 174–182
Seilacher A (1992) Vendobionta and Psammocorallia: lost constructions of Precambrian evolution. J Geol Soc Lond 149:607–613
Sloan PR (1985) Darwin's invertebrate program, 1826-1836: preconditions for transformism. In: Kohn D (ed) The Darwinian heritage. Princeton University Press, New Jersey, pp 71–120
Sloan PR (1986) Darwin, vital matter, and the transformism of species. J Hist Biol 19:369–445
Spencer H (1857) Progress: its law and cause. Westminster Rev 67:244–267
Sterelny K (2007) Dawkins vs. Gould: survival of the fittest, 2nd edn. Icon Books, Cambridge
Strick JE (2000) Sparks of life: Darwinism and the Victorian debates over spontaneous generation. Harvard University Press, Cambridge
Sulloway FJ (1979) Geographic isolation in Darwin's thinking: the vicissitudes of a crucial idea. Stud Hist Biol 3:23–65
Tammone W (1995) Competition, the division of labor, and Darwin's principle of divergence. J Hist Biol 28:109–131
Whewell W (1832) [Review of volume 2 of Lyell's] Principles of geology. Q Rev 47:103–132
Whewell W (1837) History of the inductive sciences, 3 vols. John Parker, London

Part II
To Travel in Geographical Space Is to Travel in Geological Time

Chapter 3
Imposing Order Upon Complexity: Divergence Forward in Time (*Origin*, Chaps. 1–5, 8)

Abstract With its twin focus on the mechanism of natural selection and the principle of divergence, Chapters 1–5 of the *Origin of Species* are often described as presenting the explanatory core of Darwin's theory of evolution. Although Darwin might have pushed for this theoretical agenda, the critical reader willing to go beyond Darwin's rhetoric encounters in them another dominant narrative. This alternative narrative sees life forms at low taxonomic levels as entangled in tight and complex reproductive networks from which it is difficult to break free considering, among other reasons, that nature does not favor inbreeding among forms that are too close and that life forms are dependent upon each other in multiple ways (symbiosis, coadaptation, predator-prey relationship, parasitism). An heir of the tradition of natural theology, Darwin's intellectual toolkit was founded on such notions as contrivance (adaptation), interconnection (competition), and interdependence-symbiosis (harmony), which put him at a conceptual disadvantage when it came to reflecting upon the process-pattern of biological evolution. One gathers from Darwin's exposition that life is less about neat irreversible divergent patterns propelled by a selective impetus than about complex genealogical strains and sub-breeds undergoing a perpetual fission-fusion process (reticulate evolution), a reality that blurs Darwin's self-proclaimed proofs in favor of divergence. These contradictions bring to light the theory-laden character of Darwin's account. In light of the solidarity of Chapter 8 of the *Origin* with Chapters 1–5, this former chapter will be included in our analysis here.

Keywords Reticulate evolution · Polyphyletism · Reproductive networks · Low taxonomic levels · Natural theology · Carl von Linnaeus · Principle of divergence · Principle of specialization · Competitive exclusion principle · Allopatric speciation · Sympatric speciation · Entangled bank · Economy of nature · Vera causa

Thus far we have mainly focused on the notion of variability observed within separate extant classes projected backward in geological time. This explanatory device was used by Darwin to remain true to the spirit of a fully constituted world. Part II of this book will continue to explore this central device in Darwin's work by

R. G. Delisle, *Charles Darwin's Incomplete Revolution*, Evolutionary Biology – New Perspectives on Its Development 1, https://doi.org/10.1007/978-3-030-17203-9_3

developing another set of corollaries: if the world is deployed in full only today, then the explanatory weight of an evolutionary theory founded on such an assumption must, of necessity, rest with a horizontal approach (geographical space) to evolution rather than a vertical one (geological time). The reader of the *Origin of Species* is given the impression by Darwin that he or she is presented with events placed in the time dimension. It is incontestably a *tour de force* on Darwin's part to have created the *impression* of an evolutionary depth out of a flat time horizon (the present). Yet, whether talking about biological variation, competition, natural selection, the economy of nature, the principles of divergence and specialization, geographical distribution, rudimentary organs, and even the fossil record (with only few exceptions), all Darwin does, factually speaking, is expose the reader to realities that can be observed among currently living organisms.

As might be expected, Darwin's attempt at generating a time depth out of the horizontality of the present will reveal a theoretical tension in his argumentation. We will seize upon this tension to probe deeper into Darwin's epistemological and methodological choices, thus exposing the main intellectual articulations contained in the *Origin*. Darwin was confronted with the impossibility of explaining the ontological reality of vertical evolution. In order to get out of this theoretical deadlock, he will have no other choice but to resort to the subterfuge of giving up, in Chapters 6 and 7 of the *Origin*, a significant portion of the explanatory toolkit he tries to put in place elsewhere in his book. The outcome of this explanatory strategy will be paid for by offering the careful reader two largely competing and incompatible evolutionary theories: one grounded in horizontality and another on verticality. In fact, we will see that this major epistemological/logical break hides a series of subordinate breaks in the *Origin*. We argue that Darwin's *Origin* offers a series of different pictures of evolution, several of them inconsistent, even contradictory.

3.1 The *Origin of Species*: The Argumentative Structure of the Book

Let us first present an overview of how Darwin structured his arguments in the *Origin*. What follows will not be a systematic summary of each chapter's content[1]: rather, we shall be concerned with the main argumentative divisions of the book.[2] A cursory look at the Table of Contents reveals at least two main divisions: (1) a group of chapters (Chaps. 1–5) concerned with *topics* relevant to the rise of variation and

---

[1]For such a systematic summary, see Ruse (1979: 188–197) and Bowler (1990:114–125).

[2]We are using here the first edition of the *Origin of Species* (1859) as a reference, although the main argumentative structure remains the same throughout all six editions. It should be noted that in the last edition of 1872, a new chapter (Chap. 7) entitled "Miscellaneous Objections to the Theory of Natural Selection" was inserted in between the others, taking the total number of chapters in this last edition to 15 as compared to 14 in all previous five editions.

the action of selective processes in the context of places in the economy of nature and (2) a group of chapters (Chaps. 9–13) surveying evolution in various *fields* such as paleontology, biogeography, systematics, morphology, and embryology. The remaining middle chapters (Chaps. 6–8) do not clearly reveal their exact nature on such a cursory inspection. Taking note of this dual division, Michael Ruse (1975, 1979, 2000, 2009) has formalized it by distinguishing, in the *Origin*, a first part devoted to showing the effectiveness of natural selection as a real cause (vera causa) or mechanism of biological evolution empirically supported and, a second part, dedicated to showing the unifying and explanatory power of this cause when applied against apparently unrelated fields belonging to distinct classes of facts, such as paleontology, morphology, and embryology (consilience of inductions). Ruse is not alone in pondering the argumentative structure of the *Origin* and has been joined in this effort by, among others, Jonathan Hodge (1977, 1989), Doren Recker (1987), and Kenneth Waters (2009). This collective endeavor soon reached a level of sophistication involving some degree of shuffling of the *Origin*'s chapters and contents. Interestingly, these studies retain the idea of a part concerned with natural selection as a real cause and one that applies this mechanism to various disciplines. More importantly, they give the *Origin*'s middle chapters (what constitutes "middle" in this context is a matter of some debate) an epistemological status: they are concerned with raising difficulties about the theory and responding to possible criticisms. We now have a threefold division of the *Origin of Species*.

While recognizing the appeal of such a threefold division, we would like to stress the fact that this depends on identifying Darwin's intentions. However, the question can be approached differently by turning to a critical investigation of the *Origin*'s argumentative structure: instead of searching for Darwin's epistemological intent, we may want to evaluate his actual achievements. As mentioned, we are not here advocating a normative approach to the *Origin;* we are simply suggesting that Darwin's own assessments be evaluated for their internal consistency. In light of this critical approach, we offer the following fivefold division of the *Origin of Species* (1859) (see Table 3.1).

Chapters 1–5 Darwin tries to establish a part of the foundation of his theory. Evolution is described as a process (natural selection) and a pattern (divergence) observable today at low taxonomic levels. This is a "flat" or "horizontal" view of evolution. Here, Darwin was influenced by natural theology and walked in the footsteps of scholars like John Ray and Linnaeus. Life as a whole is seen as caught up in tight network of complex interrelationships. In these chapters, Darwin mainly stresses the theoretical component projecting evolution forward in time through divergence. The content of these chapters of the *Origin* will be analyzed in Chap. 3 of this book. It will be seen that, irrespective of Darwin's intentions, he truly struggled to impose his theory upon the messy evidence generated by tight reproductive networks. Darwin did a poor job of defending the so-called vera causa component of his theory.

Chapters 6 and 7 Against the arguments presented in the first five chapters, Darwin offers in Chapters 6 and 7 an entirely different picture of evolution. In contradiction

Table 3.1 The *Origin of Species* (1859): a fivefold division

Chapters 1–5: Life as a whole is presented as being enmeshed in tight reproductive networks from which it is difficult to break free. Darwin truly struggled to impose his theory—centered on natural selection and divergence—upon the messy evidence that emerges from his own presentation. His defense of the so-called vera causa component of his theory was less than adequate
Chapters 6 and 7: Natural selection is depicted as a wild power to which nothing can resist: transitional forms across evolutionary grades and taxonomic levels are imagined. This vertical view of evolution is non-Darwinian, insofar as it blurs the fundamental conceptual difference so essential to Darwin's distinction between homologies and analogies
Chapter 8: A reflection on the issue of hybridization between extant forms. More than ever Darwin is caught up in the notion of tight reproductive networks
Chapters 9 and 10: Paleontological facts are largely interpreted in light of currently living forms, with all of the past falling under the purview of the present. Darwin's conceptual tool in this intellectual conquest is convergence backward in time: all past forms are claimed to be either aligned with or fall in-between extant forms
Chapters 11–13: Darwin tried to dispose of the evolutionary complexity, historicity, novelty, and genuine contingency that threatened his theory. Using fields like biogeography, systematics, morphology, embryology, and comparative anatomy, Darwin projects himself backward in time through convergence: the principle of divergence applied to the past. Darwin's attempt at a "consilience of inductions" was poorly defended and filled with internal explanatory tensions, inconsistencies, and contradictions often outlined by Darwin himself

Going beyond Darwin's rhetoric, a closer inspection reveals a fivefold division of the book

to what has been presented above, we argue that these so-called middle chapters are not mainly concerned with addressing difficulties in Darwin's theory. Rather, they are devoted to presenting an entirely different theory of evolution than the one seen in the rest of the *Origin*. In them, natural selection is depicted as a brute force to which nothing can resist: transitional forms across evolutionary grades and taxonomic levels are imagined. This is a vertical and directional view of evolution. In these two chapters, Darwin applies a thick layer of rhetoric, resorting to speculative evolutionary scenarios and blurring the fundamental conceptual difference between homologies and analogies so necessary to his theory. The content of these two chapters of the *Origin* will be analyzed in Chap. 4 of this book.

Chapter 8 Darwin returns to the foundation of his theory by reflecting upon the issue of hybridization between extant forms, which by definition can only occur at low taxonomic levels. More than ever, Darwin is entangled in tight reproductive networks, to the point of struggling to piece together a convincing demonstration that varieties could ever break free from each other and independently climb the taxonomic scale through divergence. This chapter is another contribution to the horizontal view of evolution. It is a chapter genuinely devoted to the difficulties of his theory. The content of this chapter of the *Origin* will also be analyzed in Chap. 3 of this book.

Chapters 9 and 10 Darwin tackles the fossil record by applying a double epistemological standard to paleontology: facts supporting his theory were accepted, and facts speaking against it were rejected. He incorporated all of the past under the purview of the present. This is yet another contribution to the horizontal view of evolution. Darwin's ultimate conceptual tool in this intellectual conquest of the past is convergence backward in time: all past forms are claimed to be either aligned with or fall in-between extant forms. The content of these two chapters of the *Origin* have already been analyzed in Chap. 2 of this book; it will be revisited in our Chap. 7.

Chapters 11–13 In these three final chapters, Darwin tries to close the deal by using fields like biogeography, systematics, morphology, embryology, and comparative anatomy (rudimentary organs) to capture all of the past under the present, pursuing the work begun in Chapters 9 and 10 and working more often than in the other chapters at high taxonomic levels in Chapter 13. Darwin mainly projects himself backward in time through convergence: the principle of divergence applied to the past. Thus was established the other part of the foundation of Darwin's theory (the other being in Chapters 1–5 and 8). This is yet another contribution to the horizontal view of evolution. The wild power of natural selection spoken of in Chapters 6 and 7 of the *Origin* is tamed under a rigid application of the principle of divergence. By this means, Darwin hoped to dispose of the complexity, historicity, novelty, and genuine contingency that threatened his theory. The content of these three chapters of the *Origin* will be analyzed in Chaps. 5 and 6 of this book. It will be seen that Darwin's attempt at a "consilience of inductions" in these chapters of the *Origin* was poorly defended inasmuch as it was filled with internal explanatory tensions, inconsistencies, and contradictions.

From this fivefold division in the *Origin of Species*, it is possible to simplify matters by creating a new tripartite division, thus pushing our analysis one step further (see Table 3.2).

Chapters 1–5, 8, 11–13 Part 1 is devoted to grounding the epistemological core of the theory in the horizontality of the present time. An atemporal theory of evolution rigidly structured around the principle of divergence emerges from these chapters.

Chapters 6 and 7 Part 2 is dedicated to displaying the brute and wild power of natural selection at generating directional and transitional forms across evolutionary grades and taxonomic levels. This wild power is harshly tamed by Part 1 in order to comply with the predictions of Darwin's theory.

Table 3.2 The *Origin of Species* (1859): a threefold division

Chapters 1–5, 8, 11–13: Part 1 is an attempt at grounding the epistemological core of the theory in the horizontality of the present time. An atemporal theory of evolution rigidly structured around the principle of divergence emerges from it.
Chapters 6 and 7: Part 2 displays the brute and wild power of natural selection at generating directional and transitional forms across evolutionary grades and taxonomic levels. This wild power is tamed in Part 1 in order to comply with the predictions of Darwin's theory.
Chapters 9 and 10: Part 3 handles the risk of seeing the past overflow the present with evolutionary novelties unseen among extant forms. Part 1 tries to dispose of that risk by unifying the past under the present

From the fivefold division presented in Table 3.1, it is possible to simplify it by creating a new threefold division

Table 3.3 The *Origin of Species* (1859): a twofold division

Chapters 1–5, 8, 11–13: Part A. Places where Darwin tried to expose the main explanatory components of his theory
Chapters 6, 7, 9 and 10: Part B. Threatening notions Darwin hoped to keep under the strict control of Part A

It is possible to simplify the argumentative structure of the *Origin* presented in Table 3.2 even further by resorting to a twofold division

Chapters 9 and 10 Part 3 handled the risk of seeing the past overflow into the present with evolutionary novelties unseen among extant forms. Part 1 disposed of that risk by unifying the past under the present.

It is possible to simplify the argumentative structure of the *Origin* even further, under a twofold division, bringing our analysis to completion (see Table 3.3).

Chapters 1–5, 8, 11–13 Part A. Main explanatory components of Darwin's theory.

Chapters 6, 7, 9 and 10 Part B. Threatening notions kept under the strict control of Part A.

Whether adopting the fivefold, threefold, or twofold division of the *Origin of Species*, our analysis reveals that Darwin's view of evolution contains important explanatory tensions. Because a number of notions promoted by Darwin himself threaten the integrity of his theory, he had to try to contain them. Our critical analysis will exploit these tensions in order to throw some light on Darwin's ontological and epistemological choices.

3.2 The Tradition of Natural Theology

Chapters 1–5 and 8 of the *Origin* may be best understood as stemming from the long tradition of natural theology. It is certainly not unreasonable to assume, as Darwin does, that two extant species with similarities may have shared a common ancestor in the geological past. As argued in Chap. 2 of this book, however, to project today's variability in the past comes with obvious limitations when assessing evolutionary novelties of the past. Darwin's atemporal theory of evolution, therefore, comes with its fair share of fabricated illusions. Once demystified, the temporal dimension of Darwin's theory largely reduces to a horizontal approach to evolution, the same horizontality that characterized the static worldview, for instance, of John Ray and Carl von Linnaeus in the seventeenth and eighteenth centuries. Unsurprisingly, many key concepts used by Darwin were already in common use within that tradition: order, equilibrium, harmony, symmetry, interconnection, and contrivance. Darwin revived them under new names, including adaptation (contrivance), competition (interconnection), principle of specialization (equilibrium-symmetry), and interdependence-symbiosis (harmony). This will become more obvious as our analysis proceeds. This inheritance meant these concepts could only allow limited adherence to an evolutionary theory appealing to universal evolutionism. To see why this is so, let us return to the seventeenth and eighteenth centuries.

In a famous passage of his *Opticks*, Newton (1730: 375, 376) succinctly formulated the atomistic doctrine to which he subscribed: "All these things being consider'd, it seems probable to me, that God in the beginning form'd Matter in solid, massy, hard, impenetrable, movable Particles, of such Sizes and Figures, and in such proportion to Space, as most conduced to the End for which he form'd them..." The multiplicity of world's entities are thus reducible to their atomic components, the ultimate level of matter. At this level the world was created fully constituted: building blocks were permanent, immutable, and complete; nothing else could be expected out of this full-fledged act of creation. Nevertheless, however complete the world might be at the atomic level, the fact remains that, since the macroscopic level is the product of their combinations and permutations, a theoretical option opened up in the form of a system of unstable matter in motion, generating novel states at the macroscopic level (Greene 1981: 13, 14). While Descartes rushed into the breach by speculating about the evolution of the universe—as seen in the introductory chapter of this book—scholars like Newton, Boyle,[3] and Ray sealed the breach by adding another layer of permanence and immutability to the world. As stated by Ray (1701, preface): "...the Works created by God at first, and by him conserved to this Day in the same State and Condition in which they were first made." The atomic level was believed to be definitively constituted, as were higher levels. For Newton, then, the stability of our universe in its manifold manifestations—such as, stars not colliding despite the gravitational pull, or planets maintaining the right velocities and their perfect distances in our solar system—was a clear sign of contrived activity divinely planned (Brooke 1991: 135–151).

It may not be out of place here to remember that Darwin (1859: 490, 1872: 429), like Ray, Linnaeus and Lyell,[4] was not insensitive to the connection between astronomy and biology. This being said, as much as celestial mechanics first served as the paradigmatic model for a designed, machine-like, and contrived universe, the fact remains that this model did not need to be transferred as such to the living world. As a mere *analogy*, this rigid mechanical model could offer some flexibility in its application, especially if it were to show the existence of an even more powerful

[3]For Robert Boyle's application of natural theology to the natural world, see Sargent (1995, Chap. 4).

[4]John Ray unified God's multiple creations on Earth and in the heavens under natural theology in the following words: "There is no greater, at least no more palpable and convincing argument of the Existence of a Deity, than the admirable Art and Wisdom that discovers itself in the Make and Constitution, the Order and Disposition, the Ends and uses of all the parts and members of this stately fabric of Heaven and Earth". See Ray, J. (1709), *The Wisdom of God, Manifested in the Works of Creation*, 5th edition (London : Samuel Smith), p. 33. See also: Linnaeus, C. (1744), "Oratio de Telluris habitabilis incremento", in C. Linné, *L'équilibre de la nature*, translated from Latin to French (Paris: Vrin), p. 30; and Lyell, C. (1833), *Principles of Geology*, Vol. 3 (London : John Murray), p. 384.

Creator. Indeed, it has been argued that a purely mechanical universe so well designed only showed the power of a Creator who has since deserted the world by leaving it to run as it is. Or even worse, this perfect mechanical universe might hint at a universe in need of no supervision and divine progenitor at all. To avoid the risk of sliding down this materialistic and atheistic slope, it was argued that the living world could offer an even more grandiose spectacle of wisdom, arts, and sublime craftsmanship (Olson 1987: 20–33; Amundson 1996: 18, 19). As impressive as machine-like entities may be in themselves, their inertness, rigidity, completeness, and passivity in performing tasks for which they were designed are not as impressive as, for instance, the autonomy of the multiple living beings in acting for their own independent ends; the possibility of many such beings gradually learning appropriate behaviors during their upbringing; or the expression of pain or pleasure observed among these same living creatures. In contradistinction to the automata or clockwork model, life could offer examples of a higher expertise in their complexity and flexibility. Along the mechanistic-vitalistic theoretical spectrum, the living world was often conceived by proponents of a fully constituted world and natural theology as being slightly less rigid than celestial mechanics. John Ray, Linnaeus, Darwin, and others will take advantage of this loosened up mechanistic view to allow for some kind of motion.

Natural theology, when applied to the living world, was anything but abstractly conceived. Let us follow the instructive analysis provided by Antonello La Vergata (1988). In his *Oeconomia naturae* of 1749, Linnaeus anchored natural theology in the context of the economy of nature, which he defined as the "wise arrangement and distribution which was imparted to natural things by the Almighty Founder, and by which they are fitted to accomplish their common ends and mutual uses" (La Vergata 1988: 140). What kind of arrangement is that? Far from being new, the concepts implicated by Linnaeus's definition echoed widespread notions held at the turn of the seventeenth and eighteenth centuries by Ralph Cudworth, John Ray, Robert Boyle, Samuel Clarke, and William Derham. Such concepts include: multitude, variety, connection, order, contrivance, and symmetry. This conceptual network was deployed in the following way. The greatness of the Creator is seen in the multitude or fullness of our world, explaining why Ray wondered at Anton van Leeuwenhoek's and Robert Hooke's estimate that as many as 8,280,000 microscopic animalcules could be found in a single drop of water! Without a doubt, more of the same is impressive, but not as impressive as the great multiplication of different kinds of things (variety) that display still greater skill, as argued by Boyle. That is why in his *Physico-Theology* (1713), Derham went on to review a wide variety of disparate things: handwriting, voices, kind of foods, muscles, and animals.

Of course, a created world cannot be a collection of things randomly assembled; chance can explain isolated events but not interconnected systems. A clever and beneficent connection must therefore exist between the world's entities, one that is revealing of a designed order, contrivance, and symmetry. Such an order is maintained under the balance of nature; Derham, for instance, speaks of "harmony" and "just proportion" in nature. Considering that the surface of the Earth cannot sustain an indefinite increase of biological populations, counterbalances were incorporated in the plan of creation: prey animals would devastate other species if it were not for their low fecundity, while harmless forms are allowed to reproduce abundantly; long-lived animals are prevented from exhausting the resources by their low fecundity, while short-lived ones are entitled to abundantly reproduce. In addition to these preventive checks on overpopulation, nature was also provided with a regulating mechanism whereby the numerical equilibrium in the food chain is maintained by forms preying on each other; this war among species, while cruel in its details, is beneficial as a whole. In fact, this latter idea was common among scholars like Linnaeus, Charles Bonnet, Buffon, Erasmus Darwin, Lamarck, Paley, and Charles Darwin. The *Origin of Species* closes with: "…from the war of nature, from famine to death, the most exalted object… the production of the higher animals, directly follows" (Darwin 1859: 490, 1872: 429).

Within a regulated system, the harmony and balance of nature are the products of wars and conflicts with creatures pressing against each other, constraining each other and leading to one or another's death. Lost forms, therefore, become the means by which the structure of life as a whole is preserved. To quote La Vergata (1988: 144, 145):

> And what does an individual count with regard to a species? Nothing. This is why, in Tennyson's words, nature seems to be 'so careful of the type… so careless of the single life'. Individuals are ephemeral things, species are thoughts of God: as such, they are immortal; properly speaking, the individuals perpetuate the species, not themselves. Paradoxical though it may seem, species are maintained through a continuous destruction of 'single lives'. This idea runs throughout the 18th century and finds expression not only in naturalists, but also in philosophers… *It accorded well with the taxonomical primacy given by naturalists to species and genera over varieties…* [my emphasis]

The practice of paying greater attention to higher taxonomic categories (genera and species) than to lower ones (varieties, individuals) may be called, for our purposes here, a *downward taxonomic approach.*

When Linnaeus built his taxonomic system around varieties, species, genera, orders, and classes in his *Systema naturae* (1758–1759, 10th edition), his formal investigative method consisted in climbing its levels, by regrouping similar varieties into a species and similar species into a genus and so on. Linnaeus's formal method, therefore, is based on an *upward taxonomic approach.* However, the fact remains that once the taxonomic system is largely established—with many forms already being identified and classified—it becomes possible to reverse the perspective by adopting a downward taxonomic approach by using, for instance, only very few features to regroup together several orders under the same class. From there, one can go on looking at more features to properly classify the genera and species within that same class (Bowler 2003: 68). Differently stated, in the real world of taxonomists, it is often much more practical, when classifying a form, to first identify the higher taxonomic categories to which it belongs, working its way downward to lower taxonomic categories afterwards.

The point we want to make here is that the downward taxonomic approach was preserved by proponents of a world believed to be fully constituted—such as in Lyell, Darwin, Huxley, and Owen—as it became handy for them to read the variability of the past in light of the present. Whereas the anti-evolutionist Lyell recognized the extinction of individual species but not of other, higher, taxonomic categories, the evolutionist Darwin accepted the extinction of species, genera, families, and orders but transposed the notion of a permanent and fully constituted world at the levels of classes and phyla. As revealed by the following quote from the *Origin*, Darwin knew about the downward taxonomic approach and recognized its value: "Such expressions as that famous one of Linnaeus, and which we often meet with in a more or less concealed form, that the characters do not make the genus, but that the genus gives the characters, seem to imply that something more is included in our classification, than mere resemblance" (Darwin 1859: 413, 1872: 365). That Darwin enshrined this statement right in the middle of a passage devoted to showing that Linnaeus's taxonomic system captured something important about evolutionary relationships is somewhat peculiar. The system of taxonomic boxes nested into ever more inclusive boxes (species within a genus, genera within a family, and so on) is certainly suggestive not only of descent with modification but perhaps more specifically of a Darwinian pattern-process of evolution based on forms diverging from one another while climbing the taxonomic scale. In other words, the utilization of an upward taxonomic approach could have been useful for Darwin to support his theory of evolution.

Why, then, resort to a downward taxonomic approach, as the quote above suggests, and thus follow the anti-evolutionists of the eighteenth century? To argue that higher taxonomic entities are more important than lower taxonomic ones—by being more permanent—was not necessarily the best strategy for Darwin to adopt. "So careful of the type... so careless of the single life" to again cite Tennyson's words quoted above by La Vergata. In his haste to profit from the field of taxonomy, it seems that Darwin went too far in his desire to blur the differences that exist between the practice of classifying forms (the real world of taxonomists) and that of suggesting genealogical relationships. This is perhaps to our advantage, as it seems another revealing slip by Darwin, one that hints at his way of thinking, thus indicating how he transformed the central notion of the "balance of nature" typical of natural theology, by retaining the same logic but applying it at higher taxonomic levels. Indeed, Darwin's brand of evolutionism needed to create more motion than Linnaeus had done through the turnover of individual organisms (see below). Darwin allowed for the extinction and replacement of species, genera, families, and orders as well. At the same time, Darwin stopped at the threshold of classes/phyla by treating them as unmovable, as will be seen more fully in Chap. 7 of this book. This allowed him to face the new empirical reality of extinction—a destabilizing factor in light of the notion of a balance of nature, for how can nature maintain its equilibrium if whole groups are going extinct? (Greene 1959: 89–127)—while simultaneously remaining true to the spirit of natural theology and the ideas of permanence and stability.

3.3 From the Economy of Nature (Linnaeus) to the Principle of Divergence (Darwin)

Enough has been said of the general concepts associated with natural theology for us to focus more closely on Linnaeus's own view[5] and on the intellectual changes introduced to natural theology after him. Special attention is paid here to Linnaeus, because his view is too rarely considered in full in relation to the Darwinian studies; it is usually confined to piecemeal expositions. In a real sense, there is something remarkably modern and exhaustive about Linnaeus, who provides us with a key intellectual link leading directly to Darwin.

Taking the universal flood as a starting point of his reflection in the *Oratio de Telluris habitabilis incremento* (1744),[6] Linnaeus imagined an original island populated by a single pair for each sexually reproducing species. As the water level decreased, opening up new habitable territories, species conquered them by propagating their kind in number through the natural process of generation (reproduction). Linnaeus's view is one of an expanding life through historical times (Egerton 2007: 80), wherein nothing entirely new is introduced, as the same proportions between species are maintained and regulated beforehand to avoid encroaching species driving others to extinction, such as those between herbivores and carnivores; between fish, insects, and birds; or more generally between plants and animals (Linnaeus 1744: 48, 49). A pre-established equilibrium between species is thus maintained throughout an ongoing demographic expansion (Limoges 1972: 10, 11; Egerton 1973: 336). Indeed, the "world machine"[7] would be prone to a great disaster if only one of its key components were to be missing (Linnaeus 1760: 118). Motion exists in this world, but circular motion with no sign of a beginning or ending (Linnaeus 1760: 104). Should we, then, be surprised that Linnaeus (1744: 30) also appealed to the determined trajectories of celestial bodies? According to him, nature is everywhere permeated by cyclicity: the perpetual circulation of water (evaporation/rain), the alternation of seasons, the course of a day, the birth and death of life, and the propagation-preservation-destruction cycle in the three interrelated kingdoms (Worster 1994: 34, 138).

Linnaeus subscribed to a linear scale of being, but only in very broad outlines: the mineral kingdom constitutes the physical support for all of life, the plant kingdom both covers this physical substratum and extracts from it the needed resources, and

[5]The Linnean corpus is composed of several publications, some of which were directly written by Linnaeus himself with others like *Oeconomia naturae* (1749) and *Politia naturae* (1760) written by others under his supervision. With this caveat in mind, it is standard practice in the literature today to refer to the whole corpus as Linnaeus's own contribution.

[6]This text was actually written by Linnaeus himself.

[7]The expression "world machine" is used by Linné himself. See Linnaeus (1744: 29).

the animal kingdom lives at the expense of plants.[8] Far from being a hierarchical scale ranging from the minerals up to the animals, the world's order is, for him, reversed, as the role of herbivorous animals consists in avoiding the overpopulation of plants and the role of carnivorous animals to prevent herbivore proliferation. The organizational structure and equilibrium of the whole system are preserved by a series of counterbalances or interdependent teleologies (ends) (Linnaeus 1760: 107–109). More importantly for Linnaeus, however, this balance is maintained by complex relationships between species and the environment—relations not necessarily following taxonomic categories or "kinship"—with forms living in specific milieus and feeding on specific sources, thus producing networks of interdependence (a reality extending beyond the simple scale of being).

Linnaeus saw nature as always full to capacity, with any available open space immediately filled by all the same expanding species, species pressing against each other in this ever-contiguous world (Barsanti 2005: 108–111; Egerton 2007: 82). In this "war of all against all"—an expression repeated three times in the *Politia naturae* (1760: 103)—animals mistreat plants to the point of inspiring pity, while worms, insects, fish, amphibians, birds, and mammals are busy tearing each other apart without compassion and surprising tyranny. Linnaeus describes the goals pursued by species in the following terms: eating, protecting, and reproducing. Each living form maintains itself at the price of murdering another (Linnaeus 1744: 32). While the idea of the *Scala naturae* (verticality) persisted in the minds of many scholars of the time, it is interesting to note how it gradually faded in the background in favor of a foreground composed of complex relationships among species living in contemporaneity (horizontality). Peter Bowler (2003: 66, 67) nicely captures this shift:

> The chain of being assumed that similarities between species naturally would define a linear pattern [A being closer to B, B close to C, etc.]. In practice this view was already losing support as exploration revealed an ever wider range that had to be fitted in. There were too many different degrees of relationship involved. By the end of the seventeenth century, naturalists such as John Ray had sought to understand the relationships between species by ignoring the chain of being and creating a more flexible system of biological taxonomy (classification). Linnaeus built on those foundations to create the basis for modern taxonomy... [He] assumed that God had created an array of distinct species which perpetuated themselves unchanged to the present.

Even when Linnaeus contemplates the possibility of species transformation—a local variety becoming distinct enough by natural means to be classified as a new species or a new species being produced by hybridization between distinct species—nothing fundamentally new is allowed, as these instances are merely recombinations of traits already present in the parental stock (Bowler 2003: 69, 70). Many thinkers of the time, then, including Linnaeus, embraced what could be called a double horizontality of the world: (1) the idea of evolution itself could not be

[8]Linnaeus, C. (1749), "Oeconomia naturae", in C. Linné, *L'équilibre de la nature*, translated from Latin to French (Paris: Vrin), pp. 57–101, see p. 59. This text was not actually written by Linnaeus but by I.J. Biberg.

contemplated, since the world had been divinely created full-fledged; (2) the historical dimension of the plan of creation (Adam and Eve, Noah's flood, repopulation of Earth) was recognized but largely neglected in favor of ever-existing complex relationships between currently living forms. In short, the past was assumed to be more of the same, except perhaps under a preformationist and hypothetical state, as seen in our earlier discussion of Michel Foucault and Camille Limoges (see Chap. 2 of this book).

Linnaeus's view of nature articulated several key overlapping explanatory components found later in Darwin (some under a modified aspect). First, nature is structured around the idea of the economy of nature, a principle that regulates the relationship between life forms dependent on each other for their survival. Second, these interactions are arranged in accordance with finalized or teleological ends: the needs of each individual species. Third, species are at war against each other to maintain their position in this economy and preserve the general harmony of the whole system. Fourth, each species occupies a specific physical or geographical place in this economy. On this last count, Linnaeus listed various conditions under which species could live: oceans, lakes, swamps, rocks, islands, lands, mountains, hills, valleys, and plains; earth, sand, clay, and limestone; and warm climates in the Indies, cold climates at the poles, and temperate climates in between (Linnaeus 1760: 105). Against this physical background, each individual species is assigned a natural place or station within which it is easier for it to survive, and outside which it is at a disadvantage to both fill its specific needs and face other species (Linnaeus 1760: 105–107).

Darwin's intellectual connections to natural theology and Linnaeus have long been recognized in the historiography (Stauffer 1960; Cannon 1961; Ospovat 1981; Ruse 2003; Pearce 2010). In order to generate some kind of evolutionary motion out of this static framework, Darwin had to "loosen things up" by making a number of conceptual reinterpretations and innovative additions. On this intellectual road toward greater motion, Darwin encountered scholars like Alexander von Humboldt, Augustin-Pyramus de Candolle, and especially Charles Lyell. They helped him think about the nature of the conceptual connections between the biological characteristics of distinct species, the array of physical conditions encountered in nature, and the way species relate to their biotic and abiotic environment (Egerton 1968; Rudwick 1970: 19–21; Hodge 1982; Worster 1994: 137–144; Pearce 2010: 504–506). Linnaeus had locked up the world's system through a fairly tight connection between a given species and a single physical place in nature and a balance of nature conceived as a device maintaining the same proportions between life forms. To claim, however, that Linnaeus's view constitutes a static representation of the world is not quite accurate. His view is, indeed, a contribution to the static worldview but one that does acknowledge the migration of forms, the demographic fluctuation in populations, the necessity of forms to relate to their environment and other forms, and the perpetual war between encroaching species. This is genuine *motion* but against a backdrop that remains stable and immutable by evening out changes through a recalibration process in historical times. In order to create more motion, Darwin had two main tasks:

1. To dissolve the one-to-one correspondence between one particular species and one specific physical place in nature. This Darwin accomplished by recognizing that species are usually more flexible or adaptable, and were not necessarily confined to a single physical space. This flexibility was further enhanced by allowing for the introduction of new combinations of variation within biological species.
2. To release the balance of nature from its internal constraints, which were originally founded on pre-established proportions between species. Instead, Darwin re-founded the balance of nature on the principle that the equilibrium is not pre-established but is rather the outcome of the various forces confronting each other (this being said, we will see in other chapters of this book how Darwin failed to free entirely the balance of nature from pre-established constraints).

Lyell was instrumental in paving Darwin's intellectual pathway by illustrating how species in geological times were constantly being pushed around by a perpetually changing environment, leading to occasional extinctions of species. Whereas Lyell was implicitly thinking along the lines of a nature which gradually depletes itself of its life forms, Darwin, on the contrary, saw that the removal of extinct forms opened up opportunities for remaining forms. Through such conceptual innovations, one intellectual path among several possible others lay open to evolutionists. It is at this precise epistemological juncture that Darwin took another decision that would give his evolutionism its special and unique flavor:

3. He elected to organize evolution around distinct yet genealogically related forms bound in common ancestry. Darwin tried to rigidly impose a view of evolution structured around the principle of divergence (in the following chapters we will see why Darwin had to pay a high price for this imposition).

   Let us suspend for a moment our critical analysis and resume our exposition of what is usually considered the hallmark of Darwin's evolutionary theory in the traditional historiography. These developments are only too well known to be repeated here,[9] except for a very brief summary. No attempt is made here at systematically tracing the actual historical sequence of Darwin's ideas up to their manifestation in the *Origin of Species*. Instead, we will present these ideas in thematic outline:

Relative Adaptation Although at first Darwin embraced the notion of perfect adaptation for each species, as seen in Linnaeus's work and Paley's *Natural Theology* (1802), among others, he eventually gave this up in favor of relative adaptation: some species are better adapted than others (Ospovat 1978, 1981; Amundson 1996). On this new view, species are no longer confined to pre-established stations in the economy of nature but are instead free to conquer spaces in accordance to their natural abilities. Teleology is also implied in the notion of adaptation, as species need to be sufficiently adapted to meet their survival needs.

[9]For overviews, consult Hodge (1983), Bowler (1990), Ruse (1999), and Sloan (1999).

Variation Species' abilities are conditional to the rise of natural variations in individuals and populations, some being novel (Geison 1969; Bartley 1992; Winther 2000; Bulmer 2004). Darwin and scholars of his time were not able to fully understand the laws governing the rise of biological variation, despite their efforts.

Demographic Pressure and Competition It had long been recognized, by Linnaeus and others, that life's natural tendency to expand pushed species into conflict with other species. Through a timely reading of Robert Malthus's *Essay on the Principle of Population* (1798), a political economist, Darwin became receptive to the notion that under the natural tendency of species to increase demographically, significant competition would result between species to access the limited resources (Vorzimmer 1969; Herbert 1971; Bowler 1976; Ospovat 1979; Young 1985). This being said, it should be remembered that Darwin explicitly refers to Linnaeus's arithmetic in the *Origin of Species* in his discussion of the demographic explosion of species (1859: 64, 1872: 51). Expanding upon Lyell's idea of interspecific competition in the second volume of his *Principles of Geology* (1832), Darwin imagined a perpetual competition in nature against ever-changing environmental conditions and fluctuating neighboring population pressures.

Selection This idea was very much in the air when Darwin's (1872: xiii–xxi) *Origin* was written. A full-fledged conception of selection had earlier been arrived at independently of Darwin by Alfred Wallace. In Darwin's hands, this concept ultimately became the process of "natural selection" which he tried to exemplify through an analogy with "artificial selection."[10] Under permanent biotic and abiotic constraints, the argument goes, only the most fit organisms and species carrying the best features will manage to survive and reproduce, to the exclusion of others.

Extinction The idea of extinction had been making its way into the life sciences since the second half of the eighteenth century. Lyell applied it in his *Principles of Geology* by arguing that spatially constrained species may eventually be driven to extinction (Lyell 1832: vol. 2, pp. 130–140). We know that Darwin expanded upon Lyell's application by allowing extinction to apply also to genera, families, and orders.

Common Ancestry Buffon had already contemplated the idea of common ancestry in his *Histoire naturelle* (1749–1789) by reuniting what he called the degenerating "varieties" around a series of distinct "families" (cat-like, horse-like, etc.). This idea persisted under different guises, as alluded to by Darwin himself (1872: xiii–xxi), although it became for him a central theoretical piece (Mayr 1982: 426–476). Indeed, since the best adapted organisms and populations manage to persist in geological times, they are eventually conducted under differential selective pressures to diversify into separate species. As such, they are united under a common ancestor.

---

[10]See Limoges (1970), Kohn (1980), Ospovat (1980), Evans (1984), Hodge and Kohn (1985), Sterret (2002), and Largent (2009).

Divergence Although the concept of common ancestry is hardly incompatible with the notion of divergence, the fact remains that the two need not be logically equated. For Darwin, however, they were, as he fused them together under a unified and rigid pattern-process of evolution whereby diversification under selective pressure is envisioned as a gradual process of divergence of populations drifting apart from each other, climbing independently the taxonomic scale.[11]

A Self-regulated Economy of Nature Linnaeus's notion of pre-established proportions between species is replaced, in Darwin's view, by the notion of self-regulated proportions in accordance to the actual "forces" facing each other, that is, by the encounter of specific variations carried by species under various biotic and abiotic conditions (Limoges 1970: 69–85).

Irrespective of what is novel in Darwin's modified explanatory system, we should keep in mind that it was elaborated within the same tradition as Linnaeus's own system (a view further developed in Chap. 7 of this book with regard to the notion of biological equilibrium). Let us put this claim in perspective. As an evolutionist, we know that Darwin largely refused to follow the paths taken by Lamarck, Chambers, and Spencer, who have managed to convey to us a sense of vertical projection or ascent of life toward complexity. Darwin was too committed to the horizontal view focused on the interrelationship of life forms to embrace this unconstrained verticality.[12] Two other options, then, were open to Darwin if he was to continue working in the spirit of natural theology. The first was to retain this horizontal view by embedding it in a geological framework that allowed life to be organized around major, distinct, and largely discontinuous episodes in the history of life, as more or less openly seen in the catastrophism of Cuvier and Buckland. However, to select that option would have implied him renouncing a unified view of the world bound in continuity and contiguity (at least within classes and phyla). As a proponent of a fully constituted world, Darwin opted for the alternative option: the present is the sole reference point for evolution; no distinct and isolated evolutionary episodes are to be recognized in the history of life.

As already alluded to, the principle of divergence became for Darwin a "keystone" for two interrelated reasons.[13] First, to vertical evolution freely projecting itself forward, he will oppose the notion of "sideways" evolution, that is,

---

[11]See Limoges (1968), Sulloway (1979), Browne (1980), Schweber (1980), Kohn (1985, 2009), and Tammone (1995).

[12]Chapters 6 and 7 of the *Origin of Species* (1859) are exceptions, as Darwin promotes in them a strong directional view of evolution. It will be seen, however, that the rest of the *Origin* was designed to negate such a directionality. See also our discussion of biological progress in Darwin in Chap. 7 of this book.

[13]"Keystone" is actually the word used by Darwin himself in a letter to J.D. Hooker dated to June 8, 1858. See F. Burkhardt and S. Smith (eds.), *The Correspondence of Charles Darwin*, Vol. 7 (Cambridge: Cambridge University Press), p. 102.

divergence.[14] Precisely because life is caught up in the complex web of interrelationships, it cannot evolve freely, since everywhere it encounters resistance from other living forms and the environment. The direct line is no option. In fact, forms coming into contact with each other constitute the precondition for evolutionary change. This resistance can only be overcome by taking indirect and convoluted evolutionary pathways. Second, for Darwin the principle of divergence represents a perfect time machine operating under the control of the present: projecting himself in the past, he assumes that forms with similarities share a common ancestor; projecting himself in the future, he presupposes that forms with similarities are in the process of moving apart from each other. Darwin's time machine (divergence-convergence), however, can only work properly if his assumed pattern-process of evolution is not mitigated by other evolutionary factors. Darwin's struggle against this major obstacle will be our privileged investigative tool for the remainder of this chapter.

3.4 Divergence Forward in Time

In the recent historiography (Ruse 1979: 188–192; Bowler 1990: 115–119), Chapters 1–5 of the *Origin* are usually described as the conceptual core of a theory of evolution founded on the mechanism of natural selection. This conceptual core is unpacked in the following ways:

1. Organisms and populations are bearers of natural variations transmitted through reproduction; variations whose ultimate source, expression, and redistribution across generations are not clearly understood.
2. Just as breeders tap into the natural reservoir of variations in domestic populations to select them with the intent of creating varieties with specific features, natural selection acting on populations in nature selects the best fitted organisms and populations to the exclusion of others, thus instituting evolutionary change. Sexual selection (mate selection) constitutes another specific mechanism part of the more general selective process. The entire process is conceived as producing adaptations.
3. The discrimination between fit and unfit organisms and populations is the outcome of the encounter of two incompatible imperatives: the tendency of life to overreproduce and the limited availability of resources.
4. The inevitable struggle for existence (survival and reproduction) happens in the context of available "places" in the economy of nature, that is, against both the environmental conditions imposing limitations on life and the competition generated by forms pressing against each other in a world largely filled to capacity.

[14]While most of the *Origin of Species* tries to impose a pan-divergent view of evolution, Chapters 6 and 7 constitute an important exception by promoting vertical evolutionary lines. This question will be treated separately in Chap. 4 of this book.

5. In such a competitive world, forms presenting variations that allow them to avoid competition or beat out competitors will increase their chance of survival. This process will favor forms moving away from each other, for it reduces competition (by defeating it or avoiding it all together), as reflected in the pattern of evolutionary divergence. The competitive exclusion principle and the principle of specialization are two specific concepts that are part of this more general argumentation.
6. When looked as a whole, life on Earth is organized around common ancestors whose descendants are fanning out (divergence) while evolving and climbing the taxonomic scale, with branches rapidly running to extinction and others persisting longer.

Because of its solidarity with Chapters 1–5, the content of Chapter 8 will be incorporated in our analysis here. The recent historiography (Ruse 1979: 192, 193; Bowler 1990: 120, 121) usually presents this chapter as being devoted to showing that varieties, species, and genera are not entities sharply separated from one another but are instead organized along a gradation and continuum ranging from perfect fertility to perfect sterility, as would be expected of a theory founded on the idea of diverging lineages moving away from their common ancestor by climbing the taxonomic scale.

The essential point Darwin was hoping to make in Chapters 1–5 and 8 of the *Origin* is the following: *evolution should be conceived as a divergence process-pattern projecting itself forward in time, as seen by looking at extant forms of low taxonomic levels (varieties, species, and genera).*

Although the recent historiography highlights the conceptual components laid out above, it has been somewhat overlooked that such chapters also contain another dominant narrative. This narrative has been neglected under the assumption that Darwin succeeded in proposing a theory based on universal and open evolutionism. This is precisely what is being contested here. The underlying assumptions in these chapters reveal the ultimate epistemological underpinning of Darwin's theory which he inherited from the tradition of natural theology, namely, the notion that life forms are organized in complex and interdependent ways within the economy of nature, as seen in instances such as coadaptations, predator-prey relationship, parasitism, symbiosis, and reproductive networks. Our main thesis in this chapter can thus be summarized:

> Just as we thought that Darwin had managed to extract himself from this intellectual tradition and cut through the conceptual obstacles by reorganizing nature along groups of forms bound in evolutionary kinships or common descent, the conceptual knot he promotes—kinship, analysis at low taxonomic levels, horizontal evolution—does not put him in a significantly better position than natural theologians when it comes to addressing life's history.

The key notion here is Darwin's new organizing principle: kinship or common descent. In contradistinction to what is usually believed about Darwin, this new organizing principle was no epistemological panacea in light of his intellectual heritage. Indeed, if forms in close contact are genealogically related, how can they become segregated from one another? If related forms are looked at from very low taxonomic levels, how can they be distinguished from each other? If the present time is the ultimate temporal horizon to think about the evolutionary past, how can one create enough time depth to clearly see segregated entities?

Darwin therefore faced a major challenge in Chapters 1–5 and 8 of the *Origin*: life forms today are so tightly intertwined in a web of complex relationships that it becomes difficult to imagine a universal evolutionary pattern-process based on the dual notion of common ancestry-divergence. Darwin's hope was that so much built-in pressure existed in the system (overpopulation) that organisms and populations carrying the best variations would somehow be "expulsed" forward along this pattern-process, leaving others behind in the cemetery of life. Precisely because Darwin refused to approach evolution from the vertical standpoint of life inherently projecting itself forward, he had to confront the horizontality of the complex web of interrelationships among low taxonomic entities. Unfortunately, to disentangle that network of extant forms ultimately proved impossible to him. This major challenge is reflected in these chapters of the *Origin* under three overlapping themes: (1) Darwin's struggle to build a convincing case that entities at low taxonomic levels (varieties, species, genera) could eventually be segregated from one another and their common reproductive networks; (2) Darwin's difficulties in presenting natural selection as a force generating only divergence, and not other blurring patterns like evolutionary analogies; and (3) Darwin's theoretical limitations in terms of conceiving an evolutionary dynamics that could sustain divergence in time all the way up to the highest taxonomic levels, without exhausting itself. These themes will be taken in turn.

3.5 How to Segregate Entities Bound in Reproductive Networks?

To be well founded, Darwin's theory had to rest on a corpuscular or atomistic ontology: organisms and populations are conceived as a collection of largely independent traits upon which breeders (under domestication) and natural selection (in nature) can act to direct evolution. In many places, Darwin writes as if he assumed that kind of ontology.[15] This is reflected in a general statement like: "[I]t is really surprising to note the endless points in structure and constitution in which the varieties and sub-varieties differ slightly from each other. The whole organization seems to have become plastic, and tends to depart in some small degree from that of the parental type" (Darwin 1859: 12, 1872: 9). But more concrete still is Darwin's (1859: 21, 22, 1872: 15–17) long list of traits among domesticated pigeons: differences in beaks; differences in skulls; development of carunculated

[15]For characteristic passages of this atomistic ontology, see Darwin (1859: 29–42, 45, 46, 80–87; 1872: 22–31, 34, 35, 62–69).

skin on the head; shape of eyelids; size of external orifices to the nostrils; size of the gap of the mouth; general differences in body size and shape; flying habits; size of feet; length of necks; size of wings; size of tails; length of legs; development of crop; reversed feathers; differences in utterance; number of tail-feathers; shape and size of the ramus of the lower jaw; number of vertebrae; number of ribs; size and shape of the apertures in the sternum; size of the two arms of the furcula; length of the tongue; development of the oil-gland; number of primary wing and caudal feathers; proportions of wing and tail to each other and to the rest of the body; proportions of legs and feet; number of scutellae on toes; development of skin between toes; age of acquirement of perfect plumage; shape and size of eggs; manner of flying; sexual differences; and yet more features.

The modern reader is likely to be struck by the vast scope of this list. But Darwin was making a crucial point: these variable traits are individually selected by breeders to produce new forms. Even better, Darwin was prepared to argue that breeders are not only able to produce distinct varieties but also distinct species and genera:

> Altogether at least a score of pigeons might be chosen, which if shown to an ornithologist, and he were told that they were wild birds, would certainly, I think, be ranked by him as well-defined species. Moreover, I do not believe that any ornithologist would place the English carrier, the short-faced tumbler, the runt, the barb, pouter, and fantail in the same genus; more especially as in each of these breeds several truly-inherited sub-breeds, or species as he might have called them, could be shown him. (Darwin 1859: 22, 23; see also 1872: 17)

This extraordinary statement, which caught the attention of some of Darwin's contemporaries,[16] was repeated in Chapter 13 of the *Origin*: "Now some of these birds, when mature, differ so extraordinarily in length and form of beak, that they would, I cannot doubt, be ranked in distinct genera, had they been natural productions" (Darwin 1859: 445). This statement is made in even stronger terms in the last edition of the *Origin*, as the last part reads like this: "they would certainly have been ranked as distinct genera if found in a state of nature" (Darwin 1872: 392). Darwin assumes that the selection process, whether in breeders' hands (artificial selection) or in nature (natural selection), was powerful enough to both institute profound evolutionary change and segregate close forms from their common reproductive network. Superimposing upon the atomistic ontology the notion of gradation—traits being contiguous and continuous—Darwin moves to the idea of a

[16]For instance, in a review of the *Origin of Species* originally published anonymously by Joseph Dalton Hooker, the latter writes: "gardeners and cattle and bird breeders have made races which are not only more dissimilar than many species in a state of nature are, but if found in a state of nature would unquestionably be ranked as species and even genera." See Hooker, J.D. (1859), [Review of the *Origin of Species*], *The Gardeners' Chronicle and Agricultural Gazette*, December 31, pp. 1051–1052, quote on p. 1052. See also Carpenter, W.B. (1860), "Darwin and the Origin of Species," *National Review* (January), 10: 188–214. Both papers are reproduced in Hull, D.L. (1973), *Darwin and His Critics: The Reception of Darwin's Theory of Evolution by the Scientific Community* (Cambridge: Harvard University Press), pp. 81–86, 88–114.

powerful selective force acting on such traits to transform individual differences imperceptively into differences belonging to varieties and eventually species:

> Hence I look at individual differences... as being the first step towards such slight varieties as are barely thought worth recording in works on natural history. And I look at varieties which are in any degree more distinct and permanent, as steps leading to more strongly marked and more permanent varieties; and at these latter, as leading to sub-species, and to species... I attribute the passage of a variety, from a state in which it differs very slightly from its parent to one in which it differs more, to the action of natural selection in accumulating... differences of structure in certain definite directions. Hence I believe a well-marked variety may be justly called an incipient species. (Darwin 1859: 51, 52; see also 1872: 41, 42)

In other places, however, Darwin appeals to notions that impede this easy march of evolution. Among points he himself provides—related to variation and inheritance—are the following: (1) Life must be exposed to new conditions during several generations before any appreciable amount of variation can be generated and ultimately maintained for some time (1859: 7, 1872: 5). This comes close to a homeostatic conception of life: at first life seems to resist change, but once change comes, it persists for many generations. (2) It is recognized that modified organisms, when compared to a parental stock, could revert in few features or many, to the ancestral condition.[17] This is seen when domesticated varieties are reintroduced in the wild (Darwin 1859: 14, 15, 25, 26, 152–154, 159, 1872: 11, 19, 120, 121, 125). (3) To Darwin, the notion of "correlation of growth" was useful when it came to thinking about change: in the embryological development, quite independent parts of the same organism are correlated in unexpected ways, since modification of one will be echoed in the modification of another (Darwin 1859: 11, 12, 85, 86, 143, 1872: 8, 9, 67, 114). Yet, as alluded to in a number of different cases, the notion of correlation of parts may also be a stabilizing factor, wherein the change of one part is impeded by its connection to other vital and unchangeable parts, thus mitigating Darwin's atomism with his own holism (biological integrity or wholeness) (Darwin 1859: 143–150, 1872: 114–118). (4) Taking "blending inheritance" for granted like most scholars of his time—the idea that some kind of intermediate feature would result from the encounter of two parents; for instance, one tall parent and one small parent would produce a middle-sized offspring—Darwin (1859: 108, 1872: 71) acknowledged that under such conditions, free intercrossing would considerably impede the preservation and spread of new variations within a single population. The issue of blending inheritance constituted a serious challenge to

[17]This issue was openly debated among Darwin's contemporaries. See Ellegard, A. (1990), *Darwin and the General Reader: The Reception of Darwin's Theory of Evolution in the British Periodical Press, 1859–1872* (Chicago: University of Chicago Press), pp. 198–215. This book was originally published in 1958.

Darwin's theory, as recognized by Darwin's contemporaries, such as William Hopkins (1860–1861) and especially Fleeming Jenkin (1867).[18]

However, more important for our thesis here are factors impeding the easy march of evolution associated with Darwin's horizontal approach, which was based on the complex interrelationships of life forms. We know that, for Darwin, proportions seen in nature between species are not (fully) pre-established ones—in contradistinction to what was argued by Linnaeus—but are rather the product of a self-regulated process involving the encounter of several variables: specific environmental conditions, species in contact, and available biological variability. At the heart of this self-regulatory process whose outcome is transmutation and adaptation one finds, of course, natural selection:

> Owing to this struggle for life, any variation, however slight and from whatever cause proceeding, if it be in any degree profitable to an individual of any species, in its infinitely complex relations to other organic beings and to external nature, will tend to the preservation of that individual, and will generally be inherited by its offspring. The offspring, also, will thus have a better chance of surviving, for, of the many individuals of any species which are periodically born, but a small number can survive. I have called this principle, by which each slight variation, if useful, is preserved, by the term of Natural Selection, in order to mark its relation to man's power of selection. We have seen that man by selection can certainly produce great results, and can adapt organic beings to his own uses.... (Darwin 1859: 61; see also 1872: 49)

This aspect of Darwin's view is too familiar for us to elaborate further upon. More interesting is the way Darwin describes the nature of the relationship of life forms between themselves and their milieus. As an heir of the tradition of natural theology, he couches the whole issue of transmutation in terms of "exquisite adaptations" and "beautiful co-adaptations" (1859: 60, 61, 1872: 48). We have already seen in the Introductory chapter how Michael Ruse (1996, 1999, 2003, 2004), under our expression "teleology-mechanism," characterized Darwin's approach to evolution: a mechanistic approach to nature with the constitutive living parts organized under the principle of teleology (adaptation), with each living form being adapted and designed to fit into a particular place in nature. Ruse positions Darwin within the British context of a natural history inherited from natural theology, permeated with such notions as design, places in the economy of nature, division of labor, and competition.

What follows constitutes a contribution to this historiographical tradition, although the emphasis is shifted toward the implication accompanying Darwin's horizontal view of evolution, a view that meshes well with Darwin's own family tradition, which was grounded in agriculture and land capitalism (Hodge 2009, 2014). It was Darwin's challenge and originality to try to generate evolutionary change out of what seemed to be a complex network of life forms mutually dependent upon each other and their environment, bound in apparently unmovable

[18]The reviews of Hopkins and Jenkin are reproduced in Hull (1973: 229–272, 303–344). On the challenge posed by blending inheritance to Darwin's theory, see also Ellegard (1990 [1958]: 243–246) and Gayon (1998: 85–102).

interlocking adaptations. Describing this "entangled bank" among unrelated life forms, Darwin (1859: 74, 1872: 58) could not but recognize the stabilizing effect this interdependence had on life, thus referring to concepts such as "checks and balances," "nature's uniformity," "war of all against all," and the "same diversity and proportion of kinds," concepts echoing those used by John Ray, Linnaeus, and William Paley (among others):

> Many cases are on record showing how complex and unexpected are the checks and relations between organic beings, which have to struggle together in the same country… [I]n several parts of the world insects determine the existence of cattle. Perhaps Paraguay offers the most curious instance of this; for here neither cattle nor horses nor dogs have ever run wild, though they swarm southward and northward in a feral state; and Azara and Rengger have shown that this is caused by the greater number in Paraguay of a certain fly, which lays its eggs in the navels of these animals when first born. The increase of these flies, numerous as they are, must be habitually checked by some means, probably by birds. Hence, if certain insectivorous birds (whose numbers are probably regulated by hawks or beasts of prey) were to increase in Paraguay, the flies would decrease—then cattle and horses would become feral, and this would certainly greatly alter (as indeed I have observed in parts of South America) the vegetation: this again would largely affect the insects; and this… the insectivorous birds, and so onwards in ever-increasing circles of complexity. We began this series by insectivorous birds, and we have ended with them… *Battle within battle* must ever be recurring with varying success; and yet *in the long-run the forces are so nicely balanced, that the face of nature remains uniform for long periods of time*, though assuredly the merest trifle would often give victory to one organic being over another… (Darwin 1859: 71–73; see also 1872: 55–57) [my emphasis]

To anyone wishing to appreciate fully the texture of Darwin's conceptual fabric, the following passage also deserves attention:

> I am tempted to give one more instance showing how plants and animals, most remote in the scale of nature, are bound together by a web of complex relations… When we look at the plants and bushes clothing an *entangled bank*, we are tempted to attribute their proportional numbers and kinds to what we call chance. But how false a view is this! Every one has heard that when an American forest is cut down, a very different vegetation springs up; but it has been observed that the trees now growing on the ancient Indian mounds, in the Southern United States, *display the same beautiful diversity and proportion of kinds* as in the surrounding virgin forests. What a struggle between the several kinds of trees must here have gone on during long centuries, each annually scattering its seeds by the thousand; what *war* between insect and insect—between insects, snails, and other animals with birds and beasts of prey—all striving to increase, and all feeding on each other or on the trees or their seeds and seedlings, or on the other plants which first clothed the ground and thus *checked* the growth of the trees! Throw up a handful of feathers, and all must fall to the ground according to definite laws; but how simple is this problem compared to the action and reaction of the innumerable plants and animals which have determined, in the course of centuries, *the proportional numbers and kinds* of trees now growing on the old Indian ruins! (Darwin 1859: 73–75; see also 1872: 57, 58). [my emphasis]

We know that Darwin thought he could manage to cut through this apparently monolithic block of life in order to generate evolutionary change. Nonetheless, the context outlined above invites the following question: was Darwin too committed to the tradition of natural theology to develop a convincing evolutionary theory? The level of interdependence recognized among various life forms seems hardly the best way, conceptually speaking, to approach evolution. Indeed, the overexploitation of "Y" upon which "Z" is dependent for its survival—for instance, parasites or forms involved in predator-prey relationships—could lead to the extinction of both forms. In this context, stability rather than evolution is more likely to prevail if life is to ensure longevity. It is certainly no accident that Darwin's sole Diagram in the *Origin* (see Fig. 2.1) shows that out of 11 distinct lineages presented, no less than 9 of them remained in a state of stagnation, with 8 of these having gone extinct after different lengths of geological time.

Darwin took a rather difficult and peculiar intellectual path to evolution. So much so that the ambiguities of his theory are revealed in the notion of coadaptation: "Slow though the process of selection may be... I can see no limit to the amount of change, to the beauty and infinite complexity of the coadaptations between all organic beings, one with another and with their physical conditions of life, which may be effected in the long course of time by nature's power of selection" (Darwin 1859: 109; see also 1872: 85). Within the same breath, it is argued that natural selection is responsible for both tying and untying the knots of the mutual interdependence of organisms. Can Darwin have it both ways?

To be well founded, Darwin's theory also had to show that life forms were organized around monophyletic groups, *monophyly* being defined as *a group of organisms descended from a single ancestor*. This condition was necessary for a theory based on the notion of related forms diverging from a common ancestor while climbing the taxonomic scale, acquiring during that process a distinctness ever more affirmed when moving from the status of variety, species, and genus. The main alternative to monophyletism could be most damaging to Darwin's view, as seen in theories founded on the general notion of hybridism, and as once contemplated, for instance, by Linnaeus (1756; see also Bowler 1989: 95–99), Paul Broca (1867, 1870; see also Blanckaert 1989: i–xliii, Delisle 2007: 52, 53), and Johannes Lotsy (1916; see also Bowler 1983: 196, 197), in the eighteenth, nineteenth, and twentieth centuries, respectively. Two theories are worth mentioning here: polyphyletism, *the derivation of a line from more than one ancestral line*, and reticulate evolution, *the origination of a lineage through the partial merging of two or more ancestral lineages.*[19]

Hybridization theories implied that life forms could not really become distinct from one another, since they are caught up in the seamless web of a reproductive network. It is in this context that Darwin (1859: 23) unequivocally embraced the monophyletic origin of pigeons: "Great as the differences are between the breeds of pigeons, I am fully convinced that the common opinion of naturalists is correct, namely, that all have descended from the rock-pigeon (Columba livia)" (see also 1872: 17). It is not clear who are the naturalists sharing Darwin's "common opinion" in this last quote. This being said, Darwin felt the need to lay out a series of arguments—difficult to evaluate unless one is a pigeon fancier—aimed at supporting

[19]Theories based on polyphyletism and reticulate evolution where quite common after Darwin, especially in the area of human evolution. See Bowler (1986) and Delisle (2007).

his monophyletic view (1859: 23–28, 1872: 17–21). Apparently, this maneuver was necessary considering that breeders and cultivators of all stripes were committed to polyphyletism:

> . . .all the breeders of the various domestic animals and the cultivators of plants, with whom I have ever conversed, or whose treatises I have read, are firmly convinced that the several breeds to which each has attended, are descended from so many aboriginally distinct species. . . I have never met a pigeon, or poultry, or duck, or rabbit fancier, who was not fully convinced that each main breed was descended from a distinct species. (Darwin 1859: 28, 29; see also 1872: 21)[20]

We know that Darwin counted on the validity of his analogy between artificial selection (domestication) and natural selection (nature) to give credence to his theory. Yet, although Darwin put up a solid fight against breeders' and cultivators' opinions regarding monophyletism, he had to concede that the whole issue was a muddled one: "In the case of most of our anciently domesticated animals and plants, I do not think it is possible to come to any definite conclusion, whether they have descended from one or several species" (1859: 17; see also 1872: 13). Still, he launched a brief review of some domesticated species, trying to decide between monophyletism and polyphyletism. At stake in this passage, contrary to what Darwin (1859: 16, 17, 1872: 12, 13) explicitly writes, is not merely the question of transmutation. Beyond this, Darwin implicitly prepares the way for a theory based on the evolutionary pattern-process of divergence: the fanning out of closely related species out of a common ancestor. The issue, therefore, is of importance for Darwin's theory. In his review, Darwin (1859: 16–20, 1872: 12–15) listed two domesticated forms as having polyphyletic origins: dogs and cattle. On the other hand, he argued that other domesticated forms were probably of monophyletic origins: horses, poultry, ducks, and rabbits. This being said, the whole issue was condemned to remain moot, by Darwin's own admission: "The origin of most of our domestic animals will probably for ever remain vague" (1872: 14). And even more clearly: "we know nothing about the origin or history of any of our domestic breeds" (1859: 40; see also 1872: 29).

It was certainly Darwin's hope that the possibility of some monophyletic cases among domesticated animals would be suggestive enough to allow him to further build the credentials of his theory. However, it is difficult for Darwin's readers to follow him when we know, on the one hand, that he thought pigeon fanciers had produced new species and genera (see the quote above) and, on the other hand, that the entire group of pigeons were bound in a single reproductive network: "the hybrids or mongrels from between all the domestic breeds of pigeons are perfectly fertile" (1859: 26; see also 1872: 19). By what means, then, can the divergence process be started at all if domestic pigeons separated at the generic level are believed to be perfectly fertile?

---

[20]The expression "all the breeders" at the beginning of this 1859 quote was only slightly modified by the sixth edition of 1872 to become "nearly all the breeders". The rest of the quote remained exactly the same in the two editions.

In fact, Darwin himself implicitly recognized that the segregation of forms that belong to the same reproductive network was difficult to achieve, as illustrated by the kind of work done by breeders:

> What English breeders have actually effected is proved by the enormous prices given for animals with a good pedigree... The improvement is by no means generally due to crossing different breeds; all the best breeders are strongly opposed to this practice, except sometimes amongst closely allied sub-breeds. And when a cross has been made, the closest selection is far more indispensable even than in ordinary cases. If selection consisted merely in separating some very distinct variety, and breeding from it, the principle would be so obvious as hardly to be worth notice; but its importance consists in the great effect produced by the accumulation in one direction, during successive generations, of differences absolutely inappreciable by an uneducated eye—differences which I for one have vainly attempted to appreciate. Not one man in a thousand has accuracy of eye and judgment sufficient to become an eminent breeder. (Darwin 1859: 31, 32; see also 1872: 23)

Must we suppose that nature works its way in such a neat and methodical fashion as to cut into the seamless fabric of reproductive networks? As seen in several places in Chapters 1–5 of the *Origin*, the reader is left with the impression that the practice of artificial selection by breeders constitutes a method more efficient and powerful than the action of natural selection in the wild. In fact, in these chapters, artificial selection is depicted as a method that allows for the creation of different species and even genera in a few hundred or thousand years of human history,[21] whereas elsewhere in the *Origin*, the mere creation of distinct varieties and species under natural selection is counted in long geological periods (1859: 116–126, 331–333, 420–422). Clearly, Darwin's argument to the effect that natural selection is much more powerful and works much more deeply than artificial selection suffers from a calibration problem throughout the *Origin*.[22]

More to the point, Darwin (1859: 96) openly recognized that crosses between *different varieties* gave more vigor and fertility to organisms than crosses between very *closely related organisms* (interbreeding): "I have collected so large a body of facts, showing... that with animals and plants a cross between different varieties, or between individuals of the same variety but of another strain, gives vigour and fertility to the offspring; and on the other hand, that *close* interbreeding diminishes vigour and fertility" [italics original] (see also 1872: 76). This self-defeating argument struck William Hopkins (1861: 80): "If we continue to breed in and in, as Mr. Darwin has somewhere remarked, the variety may in turn dwindle away and disappear; but on being crossed with another variety of the same stock, it seems immediately to regain its pristine vigour and fertility, in exact opposition to [Darwin's theory based on ongoing segregation and divergence]."

According to Darwin, nature favors new variations in the struggle for survival, but new variations found beyond the confines of close reproductive partners (1859: 105, 106, 1872: 82, 83). If so, it means that nature works her way by naturally

[21]For the case of breeded pigeons, see Darwin (1859: 22, 23, 35, 36; 1872: 17, 26).

[22]For this calibration problem, compare the following pages: Darwin (1859: 29–31, 38, 39, 61, 82–84, 109, 213, 269, 454).

bonding fairly closely related forms at very low taxonomic levels. In light of those observations, Darwin had to find a way to allow for the segregation of forms upon which the divergence process ultimately depended. In some places, Darwin (1859: 99) was inclined to suggest that the species level constituted the taxonomic level at which decisive segregation occurred: "When distinct *species* [rather than *varieties*] are crossed the case is directly the reverse, for a plant's own pollen is always prepotent over foreign pollen" [italics original] (see also 1872: 78). But we know that Darwin was not consistent in allowing crosses between species and even genera (see below). Again, what we want to point out here is the inherent epistemological limitations and constraints Darwin's theory faced as he tried to elaborate it from within the context of horizontal evolution.

Although he was not very successful at determining a decisive segregation level between entities low in the taxonomic scale (see our more detailed discussion below), one might also wonder if Darwin was more successful at proposing an evolutionary process that led to such a decisive segregation during divergence and the ascent of the taxonomic scale. As recognized long time ago by Moritz Wagner (1868) and George Romanes (1886), the *splitting* or *speciation* theory presented in Chapter 4 of the *Origin* is not very convincing (Vorzimmer 1970; Sulloway 1979; Mayr 1982: 411–417, 1991: 31–34; Bulmer 2004). Darwin was at pains to offer a clear model for species division, muddling the issue by wavering between a model based on geographic isolation (allopatric speciation) and a model founded on intense competition in the same place (sympatric speciation). He was confronted with two distinct sets of facts not easily reconcilable in light of his theory: (1) For the evolutionary dynamics to be driven and sustained, forms had to be tightly pressed against each other within limited places in the economy of nature (especially among closely related forms; see below), a situation favoring organisms and populations carrying the best variations. In this context, the sympatric model is the most suitable. (2) Considering that forms at low taxonomic levels are united in common reproductive networks, such bonds can only be broken up by isolating reproductive organisms from each other. In this other context, the allopatric model is the most suitable.

Interestingly, Darwin's indecisiveness over sympatric and allopatric speciation models is a direct consequence of his commitment toward the tradition that sees life as characterized by contiguous forms entangled in web of complex relationships. On the one hand, to elect the allopatric model (geographical isolation) was to move out of this tradition, at the cost of forsaking the notion of an evolutionary drive. On the other hand, to choose the sympatric model (evolutionary pressure created by forms in close contact) was to remain within this tradition but at the cost of seeing life being immobilized within the indivisible web of interrelationships. Apparently, Darwin entirely overlooked these explanatory tensions by treating the two models as being complementary. From the outset, Darwin (1872: 80) unambiguously underlined the problem of severing forms from their common reproductive network. Consider Darwin's own words: "In the case of methodical selection, a breeder selects for some definite object, and if the individuals be allowed freely to intercross, his work will completely fail" (see also 1859: 102). Then, building on his analogy with artificial selection, he moved to natural selection using cases in the wild. While the following passage is formally concerned with evolution within a single lineage, it is in reality implicitly preparing the way for the notion of divergence. Just as breeders manage to slowly segregate organisms from the rest of the reproductive network—the argument goes—so is natural selection in the context of sympatric speciation:

> But when many men, without intending to alter the breed, have a nearly common standard of perfection, and all try to get and breed from the best animals, much improvement and modification surely but slowly follow from this unconscious process of selection, notwithstanding a large amount of crossing with inferior animals. Thus it will be in nature; for within a confined area, with some place in its polity not so perfectly occupied as might be, natural selection will always tend to preserve all the individuals varying in the right direction, though in different degrees, so as better to fill up the unoccupied place. (Darwin 1859: 102; see also 1872: 80)

Perfectly circumscribed places in the economy of nature may not be what Darwin saw as of common occurrences in nature. Opening up space, Darwin looked for ways to prepare the application of the sympatric model in an open geographical context. The following passage shows Darwin perfectly aware that interbreeding constituted a key component of the equation, but instead of seeing this as potentially threatening for his theory, he simply assumed natural selection and interbreeding were complementary factors: natural selection was powerful enough to extract organisms from the equalizing force of a common reproductive network through the selection of the most advantageous features only, and interbreeding was the means of dispatching such advantageous features to other varieties of the same species:

> But if the area be large, its several districts will almost certainly present different conditions of life; and then if natural selection be modifying and improving a species in the several districts, there will be intercrossing with the other individuals of the same species on the confines of each. And in this case the effects of intercrossing can hardly be counterbalanced by natural selection always tending to modify all the individuals in each district; for in a continuous area, the conditions will generally graduate away insensibly from one district to another... Even in the case of slow-breeding animals, which unite for each birth, we must not overrate the effects of intercrosses in retarding natural selection; for I can bring a considerable catalogue of facts, showing that within the same area, varieties of the same animal can long remain distinct... Intercrossing plays a very important part in nature in keeping the individuals of the same species, or of the same variety, true and uniform in character... I have already attempted to show that we have reason to believe that occasional intercrosses take place with all animals and with all plants... [T]he young thus produced will gain so much in vigour and fertility over the offspring from long-continued self-fertilisation, that they will have a better chance of surviving and propagating their kind.... (Darwin 1859: 102–104; see also 1872: 80, 81)

Turning now to the allopatric model, Darwin notes how geographic isolation could play a role in severing forms from their common reproductive network, thus creating new species:

> Isolation, also, is an important element in the process of natural selection. In a confined or isolated area, if not very large, the organic and inorganic conditions of life will generally be in a great degree uniform; so that natural selection will tend to modify all the individuals of a varying species throughout the area in the same manner in relation to the same conditions. Intercrosses also, with the individual of the same species, which otherwise would have inhabited the surrounding and differently circumstanced districts, will be prevented...

> Lastly, isolation, by checking immigration and consequently competition, will give time for any new variety to be slowly improved; and this may sometimes be of importance in the production of new species. (Darwin 1859: 104, 105; see also 1872: 81, 82)

Concluding his argument on the sympatric-allopatric question, Darwin accepts the effectiveness of both, while insisting that the sympatric model was the most efficient one, since it allowed for the rise of more variations in a more competitive environment:

> Although I do not doubt that isolation is of considerable importance in the production of new species, on the whole I am inclined to believe that largeness of area is of more importance, more especially in the production of species… Throughout a great and open area, not only will there be a better chance of favourable variations arising from the large number of individuals of the same species there supported, but the conditions of life are infinitely complex from the large number of already existing species; and if some of these many species become modified and improved, others will have to be improved in corresponding degree or they will be exterminated. (Darwin 1859: 105, 106; see also 1872: 82, 83)

Darwin's argumentative strategy regarding speciation was somewhat innovative: couching life forms against a Lyellian environmental backdrop in perpetual motion (Hodge 1982: 7; Rudwick 1970: 12–21), it was argued that this would do the trick to sever populations from their common reproductive networks, thus creating new species. It will be remembered that Lyell erected a neo-Huttonian system of cyclical geology, in which aqueous factors erode the Earth's surface while igneous ones restore it. This ongoing and gradual process produces changes, with life forms being continually displaced, thus facing ecological checks and balances. Inspired by Lyell, Darwin (1859: 105–108, 1872: 82–84) imagined oscillations of sea levels (continental subsidence-elevation) continually modifying the environmental context going from isolated parts (islands) to open continental space and back. Whereas Lyell held that numerous individual species were lost forever against these ever-changing physical conditions, Darwin insisted that such conditions also allowed for replenishment by the definitive breakup of populations, as they created alternating conditions for allopatric speciation and sympatric speciation with populations being continually isolated from each other and then put back in contact again.

Darwin was fully aware that his theory might not be convincing, as can be seen in the fact that he continued to list the necessary conditions for its application. Among his listed arguments, one finds the problem of intercrossing. Interestingly, the epistemological status of this question is suddenly changed. We have already seen Darwin arguing that intercrossing was the most efficient means for dispersing selected features by natural selection to other neighboring populations part of the same species. On this previous view, intercrossing constituted an accelerator of evolutionary change. In the quote that follows, however, intercrossing implicitly becomes an impediment to evolutionary change, thus confirming what we suspected all along: intercrossing was a key challenge Darwin had to address in Chapters 1–5 of the *Origin*:

> That natural selection will always act with extreme slowness, I fully admit. Its action depends on there being places in the polity of nature, which can be better occupied by some of the inhabitants of the country undergoing modification of some kind. The existence

> of such places will often depend on physical changes, which are generally very slow, and on the immigration of better adapted forms having been checked. But the action of natural selection will probably still oftener depend on some of the inhabitants becoming slowly modified; the mutual relations of many of the other inhabitants being thus disturbed. Nothing can be effected, unless favourable variations occur, and variation itself is apparently always a very slow process. The process will often be greatly retarded by free intercrossing. Many will exclaim that these several causes are amply sufficient wholly to stop the action of natural selection. I do not believe so. On the other hand, I do believe that natural selection will always act very slowly, often only at long intervals of time, and generally on only a very few of the inhabitants of the same region at the same time. (Darwin 1859: 108; see also 1872: 84)

As seen above, Chapter 4 of the *Origin* saw Darwin proposing interpretations and scenarios much to his theoretical advantage, often remaining silent about obvious problems. Interestingly, Darwin revisited the question under the title "Hybridism"[23] in Chapter 8 of the *Origin*, this time presenting it as a special challenge to his theory. We will return to this point later in the chapter.

3.6 Natural Selection: A Force Blurring Affiliative Signs?

It is no exaggeration to say that Darwin had to fight his way through diffused and intertwined reproductive networks at low taxonomic levels to impose the pattern of divergence onto this messy situation. Two key passages of the *Origin* illustrate Darwin's struggle over this prevalent issue particularly well. In Chapter 2 of the *Origin*, he presents a few basic facts about extant plants and insects of low taxonomic levels: geographical range, number of varieties and individuals, and domination of types in specific regions. Analyzing these variables, Darwin notes that what he called *dominant* or *large* species—those with wide range and composed of numerous individuals—produced well-marked varieties or incipient species more often than others, because already successful species are in a better position than others to transmit their favorable variations. The same fact, Darwin continues, applies at a slightly higher taxonomic level with dominant or large genera being characterized by more dominant species than small and less dominant genera (1859: 53–57, 1872: 43–46; see also Browne 1980, 1983: 210–216).

Darwin (1859: 56) was quick to move from the horizontality of these biogeographical facts to some evolutionary deductions: "for wherever many species of the same genus have been formed, or where, if we may use the expression, the *manufactory of species* has been active, we ought generally to find the manufactory still in action, more especially as we have every reason to believe the process of manufacturing new species to be a slow one" [my emphasis] (see also 1872: 45). Of course, what Darwin was preparing for is a demonstration of the process of evolution based on the principle of divergence, which he refers to in these pages as the "divergence of character" (1859: 57, 1872: 46): from dominant species sprout

---

[23] In the sixth edition of 1872, the chapter on "hybridism" becomes Chapter 9.

well-marked varieties (incipient species) and from dominant genera bud new species, new forms currently being in the making and clustered around certain forms (common ancestors) with varieties being genealogically closer (or physically similar) to parent-species than species relative of parent-genera.

After having fought hard to try to impose monophyletism over polyphyletism in Chapter 1 of the *Origin*—as noted above—it is rather surprising to see Darwin in Chapter 2 suddenly contrast his view, twice, to creationism! The more explicit statement of the two is the following: "if we look at each species as a special act of creation, there is no apparent reason why more varieties should occur in a group having many species, than in one having few" (1859: 55; see also 1872: 44). Was Darwin really unable to come up with a more serious scientific alternative to his own view? Was the choice only between divergence and creationism in 1859 or, even worse, in 1872? Here, Darwin simply evaded the real issue altogether. Indeed, Darwin's most serious opponent in Chapters 1–5 is the doctrine of hybridism (polyphyletism, reticulate evolution), as facts he himself provided could easily be reinterpreted in light of models based on hybridism. For instance, a contemporary reader of Darwin might have argued that the manufactory of species does not work at all through divergences but rather through hybridization: larger genera and species generate more variability (species or varieties), because they have more opportunity of crossing. On this view, the similarity of forms living in close geographical proximity is expected under such a hybridization model, which does away with the notions of full divergence and unique common ancestors.

A short digression on Darwin's rhetorical mode of argumentation is relevant here. In a review of the *Origin of Species*, William Hopkins (1861: 86, 87) perceptively noted how Darwin introduced a logical break between the exposition of facts and his theory, bypassing altogether considerations for other theories:

> In the statement of facts, the author [Darwin] is uniformly impartial. It is difficult to conceive a fairer advocate. But when, in his judicial capacity, he comes to the discussion of facts in their theoretical bearings, we recognise a want of strict adherence to philosophical and logical modes of thought and reasoning. There is one great and plausible error of this kind which pervades nearly his whole work. He constantly speaks of his theory as explaining certain phenomena, which he represents as inexplicable on any other theory. We altogether demur to this statement... Mr. Darwin's theory can explain nothing in this sense, because it cannot possibly assign any necessary relation between phenomena and the causes to which it refers them. A great number of facts are mentioned as being only explicable on this theory, and might thus appear to an inattentive reader to constitute a large amount of inductive evidence. But all that is attempted to be done is to *assert*, not to *prove*, that the facts are consistent with the theory; and so far from being explicable, in some imperfect sense, by Mr. Darwin's theory alone, they are equally so, in general, by other theories we have mentioned [Lamarck's and Chambers's] [italics original].[24]

Returning to our main topic, more revealing of Darwin's struggle with the complex web of reproductive networks is the last section of the Chapter 5 of the *Origin*, entitled *Distinct species present analogous variations; and a variety of one*

[24]Hopkins text is reproduced in Hull (1973: 249–272) with the quote on p. 267.

species often assumes some of the characters of an allied species, or reverts to some of the characters of an early progenitor (1859: 159–167, 1872: 125–131). This section is somewhat arid and convoluted, accessible only to the most dedicated readers. It would be possible here to produce a very detailed analysis of this chapter's section. To do so, however, lies outside the scope of this chapter. It will suffice to keep our analysis to the most salient features. The section's title reveals quite clearly what Darwin's arguments are about: the fanning out of a group of closely related forms along a divergent pattern-process should be accompanied by reappearances of similar (called "analogous") features present in either allied forms or in an assumed common ancestor. Yet, again, Darwin resorts to examples that could also be consistent with theories other than his own. For instance, he writes:

> These propositions will be most readily understood by looking to our domestic races. The most distinct breeds of pigeons, in countries most widely apart, present sub-varieties with reversed feathers on the head and feathers on the feet,—characters not possessed by the original rock-pigeon; these then are analogous variations in two or more distinct races. The frequent presence of fourteen or even sixteen tail-feathers in the pouter, may be considered as a variation representing the normal structure of another race, the fantail. I presume that *no one will doubt* that all such analogous variations are due to the several races of the pigeon having inherited from a common parent the same constitution and tendency to variation, when acted on by similar unknown influences. (Darwin 1859: 159; see also 1872: 125) [my emphasis]

Why should no one have doubts? Let us review the main explanatory components presented by Darwin:

1. Allied forms living in removed countries
2. An assumed common ancestor for the whole group (rock-pigeon) not sharing the reappearances of some "analogous" traits seen in some of its descendants
3. The tendency of life forms to vary in the same direction
4. The existence of unknown influences acting on the descendants

One can forgive Darwin for his lack of knowledge of the complexity of heredity laws, a general affliction of scholars of that time. But to claim, as Darwin again does, that the alternative to divergence (common ancestry) is creationism is extremely disappointing, to say the least. In the context of domesticated vegetables, which is found in the same paragraph as the pigeon example discussed above, Darwin (1859: 159) writes: "According to the ordinary view of each species having been independently created, we should have to attribute this similarity in the enlarged stems of these three plants, not to the vera causa of community of descent, and a consequent tendency to vary in a like manner, but to three separate yet closely related acts of creation" [italics original] (see also 1872: 125). Scientific alternatives to Darwin's claims could have been found in polyphyletism and reticulate evolution. Returning to the pigeon case, Darwin continues his analysis:

> With pigeons, however, we have another case, namely, the occasional appearance in all the breeds, of slaty-blue birds with two black bars on the wings, a white rump, a bar at the end of the tail, with the outer feathers externally edged near their bases with white. As all these marks are characteristic of the parent rock-pigeon, I presume that no one will doubt that this

> is a *case of reversion, and not of a new yet analogous variation appearing in the several breeds*. (Darwin 1859: 159, 160; see also 1872: 125, 126) [my emphasis]

The case is now made more complex with (a) occasional reappearances of traits common to all the breeds, since such traits already existed in an assumed common ancestor for the whole group (rock-pigeon), and (b) the reappearances of traits in some allied breeds only, this time excluding the common ancestor (as seen above). Of course, the last thing Darwin needed was a further complexification of the case by the appearance of genuinely *new* "analogous" traits in distinct breeds. Indeed, this would have considerably weakened his argument by rendering such traits useless for tracing the affiliation of assumed related forms. New "analogous" traits, in Darwin's theory, can only be "new" in so far as they represent features unknown previously to observers but assumed to have existed in one place before spreading in the stream of an affiliated group derived from a common ancestor. In these pages, revealingly, Darwin elects not to raise a pertinent question: can natural selection repeatedly forge new analogous features independently of kinship? A positive answer to this question could potentially be devastating by confronting Darwin with the issue of "parallel evolution," that is, *the independent evolution of similar morphological traits among unrelated organisms*, an issue that caught the attention of Darwin's contemporaries (see Ellegard 1990[1958]: 256, 257). Coping with polyphyletism and reticulate evolution, albeit largely implicitly, was already more than enough for Darwin. On this issue, Darwin had to find a way to convince his readers that natural selection was powerful and directional enough to segregate low taxonomic entities living in the complex web of interrelationships, and ultimately take them down a clear divergent road.

Now, parallel evolution—real "analogies" in modern terminology, not to be confused with Darwin's "analogous" features—posed a threat of even greater magnitude to Darwin's theory: if natural selection constitutes a force whose action is also reflected in the prevalent rise of independent yet similar features, then not only would natural selection be responsible for blurring the true nature of phylogenetic connections, but, in addition, it would deprive Darwin the means of identifying segregated evolutionary entities from one another. Darwin's strategy in dealing with this problem was twofold: (1) to assume that "new" is not truly new, as explained above, and (2) to establish a distinction between two kinds of biological variations, unimportant ones seen under the form of analogous variations and important ones under the purview of natural selection. In other words, Darwin assumed that the understanding of analogous features was exclusively part of a problem concerned with the complex laws of heredity. It was a problem he believed to be confined to forms already assumed to be affiliated. Now, as far as the action of natural selection goes, Darwin restricted it to play only on variations that really count for survival, that is, variations that give forms an edge in carving out new places in the economy of nature by moving away from each other (divergence). Again, natural selection is exclusively conceived as a force severing entities from each other, and not as one blurring genealogical relations. In these pages, Darwin claims to know beforehand whether variations should be treated as important (natural selection) or as unimportant (analogous) ones. Of course, this is convincing only if one already buys into Darwin's argument. The following quote is revealing for what it simultaneously contains and evades:

> As all the species of the same genus are supposed, on my theory, to have descended from a common parent, it might be expected that they would occasionally vary in an analogous manner; so that a variety of one species would resemble in some of its characters another species... But characters thus gained would probably be of an unimportant nature, for the presence of all important characters will be governed by natural selection, in accordance with the diverse habits of the species, and will not be left to the mutual action of the conditions of life and of a similar inherited constitution. It might further be expected that the species of the same genus would occasionally exhibit reversions to lost ancestral characters. As, however, we never know the exact character of the common ancestor of a group, we could not distinguish these two cases: if, for instance, we did not know that the rock-pigeon was not feather-footed or turn-crowded, we could not have told, whether these characters in our domestic breeds were reversions or only analogous variations... Hence, though under nature it must generally be left doubtful, what cases are reversions to an anciently existing character, and what are *new* but analoguous variations, yet we ought, on my theory, sometimes to find the varying offspring of a species assuming characters (either from reversion or from analogous variation) which already occur in some other members of the same group. And this undoubtedly is the case in nature. (Darwin 1859: 161, 162; see also 1872: 126, 127) [my emphasis]

It will be seen in Chaps. 4 and 5 of this book that Darwin will not always be able to evade the issue of parallel evolution and analogies in general so easily: new traits may eventually turn out to be truly "new" trait, and entirely independent ones at that. However, in the meantime, and at this early stage of Darwin's argumentation, he could claim in a sweeping statement found in Chapter 4 of the last edition of the *Origin of Species* the following:

> Mr. H.C. Watson thinks that I have overrated the importance of divergence of character..., and that convergence... has likewise played a part. If two species, belonging to two distinct though allied genera, had both produced a large number of new and divergent forms, it is conceivable that these might approach each other so closely that they would have all to be classed under the same genus; and thus the descendants of two distinct genera would converge into one. But it would in most cases be extremely rash to attribute to convergence a close and general similarity of structure in the modified descendants of widely distinct forms... [W]ith organic beings we should bear in mind that the form of each depends on an infinitude of complex relations [intricate causes of variations, the nature of selective pressures, physical conditions, competing forms, etc.]... It is incredible that the descendants of two organisms, which had originally differed in a marked manner, should ever afterwards converge so closely as to lead to a near approach to identity throughout their whole organization. If this had occurred, we should meet with the same form, independently of genetic connection, recurring in widely separated geological formations; and the balance of evidence is opposed to any such an admission. (Darwin 1872: 100, 101)

Brave words, indeed! Let us accept Darwin's arguments for the moment. But what if we were to apply the same logic but at slightly lower taxonomic levels when two varieties of the same species or two closely related species belonging to the same genus are involved? Would convergence (analogies) not be more probable in these cases? Darwin's swift and light dismissal of this whole topic is surprising for someone whose theory is so dependent upon the investigation of low taxonomic levels.

3.7 Graduated Lower Entities or Intertwined Strains?

No matter how much Darwin invested himself in the argument that the action of natural selection would generate evolutionary divergence—as highlighted in examples such as the artificial selection of pigeons, the manufactory of species, the process of speciation (sympatric-allopatric), and the divergence of characters—the fact remains that little seemed to have been accomplished by the time the reader gets to Chapter 8 of the *Origin* entitled "Hybridism." There, the reader encounters a Darwin struggling even harder to justify how entities can be segregated from their common reproductive networks, leaving the impression that much remains to be done to ground Darwin's theory. The main message Darwin hoped to push in this chapter was the following: low entities, such as varieties, species, and genera are not completely distinct and isolated from one another; rather, they are graduated in such a way as to be organized along a continuum of gradual loss of fertility as the taxonomic scale is ascended among closely related forms. From Chapter 8, the reader gathers that Darwin would like to argue for the following results (see Darwin 1859: 248, 255, 267–272):

Families: absolute sterility
Genera: mostly sterile, not always
Species: variably fertile/variably sterile
Varieties: mostly fertile/not always

Of course, the ultimate aim of Chapter 8 consists in providing additional support for a theory founded on the two key notions of common ancestry and divergence. Unfortunately, this quest was conducted by a Darwin wrestling with background assumptions that he himself provided, and which supported alternative views of evolution based on hybridization theories (polyphyletism, reticulate evolution). Let us review these background assumptions.

To begin with, Darwin faces an interesting challenge, insofar as he needs natural selection to be powerful enough to segregate strains originally bound in common reproductive networks. By what process does one go from two strains united in reproduction to two strains separated by a sterility barrier? From the outset, Darwin recognizes that natural selection cannot *directly* act on a feature providing no positive advantage for survival such as "sterility." Natural selection can only work *indirectly* in the sense that sterility must be the consequence or a by-product of other features positively selected:

> The view generally entertained by naturalists is that species, when intercrossed, have been specially endowed with the quality of sterility, in order to prevent the confusion of all organic forms. This view certainly seems at first probable, for species within the same country could hardly have kept distinct had they been capable of crossing freely. The importance of the fact that hybrids are very generally sterile, has, I think, been much underrated by some later writers. On the theory of natural selection the case is especially important, inasmuch as the sterility of hybrids could not possibly be of any advantage to them, and therefore could not have been acquired by the continued preservation of successive profitable degrees of sterility. I hope, however, to be able to show that sterility is not a

> specially acquired or endowed quality, but is incidental on other acquired differences. (Darwin 1859: 245; see also 1872: 235)

Theoretically speaking, it is easy to understand the benefit living entities reap from being segregated from one another: they can advantageously occupy distinct places in the economy of nature, as they are differentially adapted to do so. A thought experiment here will illustrate the point. It is easy to imagine why planet Earth is inhabited by millions of distinct species organized around a relatively limited number of constituent parts (organisms) instead of only a single species composed of billions and billions of organisms united in a unique and open reproductive nexus. The single-species hypothesis would deprive life from enough biological flexibility to face most effectively the very fragmented economy of nature. Flexibility is introduced into the system when organisms are allowed to mold themselves, adaptively speaking, to the various landscapes of nature's economy. This molding process would not be achieved if organisms were profoundly united through the sharing of too many features in common with all other organisms. Darwin's theory rests on the central idea that a fragmented economy of nature is most efficiently filled by a fragmented fabric of life. It also rests on the notion that a fragmented fabric of life must be irreversibly segregated, that is, fragmentation must ultimately lead to full and complete speciation with life entities being definitively isolated from one another, reproductively speaking. Unfortunately for Darwin, Chapter 8 will be yet another place where he will struggle to superimpose his theory upon a complex and messy reality.

Two fundamental issues intersect in Chapter 8: (1) Since natural selection cannot by itself directly proceed to positively select a negative feature such as sterility, it can only achieve segregation indirectly by independently modifying the internal organization of two related forms under a distinct adaptive impetus, thus perhaps eventually instituting a full reproductive incompatibility between the two. (2) Since natural selection requires only some maneuvering room to generate some kind of fragmentation in the fabric of life—and thus respond to the most pressing adaptive demands made by a fragmented economy of nature—it may not be necessary in order to attain sufficient adaptive value to institute a full divergence process, to the point of reaching an irreversible state. To put these two issues in different light: the choice may not necessarily be between millions of fully segregated life forms or a single-unified form spread throughout the planet. One may want to argue that it would be quite advantageous, adaptively speaking, to hold a middle-road adaptive strategy between these two extreme models, with living forms being semi-isolated entities able to take advantage of their moments of relative evolutionary isolation generated under adaptive pressure, while simultaneously profiting from evolutionary innovations appearing at one extreme of a complex nexus and transmitted through reproduction. This middle-road strategy is no fantasy, as it naturally comes to mind when reading Chapter 8. Indeed, by wrestling with such topics, Darwin himself furnishes plenty of conceptual and empirical items that point in the direction of this alternative model. This has been noted by David Morrison (2015: 547, 548) who writes:

> Darwin's original book has a chapter on hybridization (Chapter 8), which he uses as part of his argument that evolutionary divergence is continuous, so that all degrees of breeding should exist, from complete interbreeding within a species, through occasional cross-species hybridization, and on to complete isolation (e.g., between phyla). However, he concurrently fails to note that his Tree of Life must, therefore, have reticulations (representing the hybridization events).

Elsewhere, Morrison (2016: 459) adds: "So, even Darwin's Tree is actually a network." This is the case because Darwin elected to ground his theory in Chapter 1–5 and 8 of the *Origin* on the horizontality of currently living entities at low taxonomic levels. Things might have been otherwise, had Darwin chosen to build a theory whose evolutionary drive is centered on middle and high taxonomic entities (families, orders, classes, phyla). At such high levels, living entities are much more clearly segregated from one another. Instead, Darwin elected to explore the jungle of low-level entities (varieties, species, and genera) for ontological-methodological reasons. At such low levels, as already alluded to, an alternative theory could propose that groups of varieties, species, and genera are, as a general rule, part of reproductive networks of fluctuating semi-isolated populations on the verge of semi-speciations under a fission-fusion process, with most of them being eventually reabsorbed into the main nexus, while rare strains emerge as fully segregated forms at higher taxonomic levels.

Differently stated, Darwin's view, which was founded on concepts such as common ancestry, monophyletism, and complete divergence, could be interpreted by a careful reader as an artifact of his theoretical expectations, that is, plain and simple theory-ladenness in the face of empirical ambiguities he himself had exposed. Apparently, his theory falls victim to an evolutionary force called "natural selection" conceived as being insufficiently powerful to directly and definitively segregate entities by cutting through the reticulate ties of horizontal reproductive networks.

In fact, Darwin will accept the general principle that *inbreeding* may be detrimental to life, whereas exchanges of variations coming from *more removed strains* may be favorable. In Chapter 8, this argument is made through the varying levels of fertility observed in different cases: "I have collected so large a body of facts, showing that close interbreeding lessens fertility, and, on the other hand, that an occasional cross with a distinct individual or variety increases fertility, that I cannot doubt the correctness of this almost universal belief amongst breeders" (1859: 248, 249; see also 1872: 237). Darwin will illustrate this principle by providing cases from both the plant and animal kingdoms: "[T]here are individual plants, as with certain species of Lobelia, and with all the species of the genus Hippeastrum, which can be far more easily fertilized by the pollen of another and distinct species, than by their own pollen" (1859: 250; see also 1872: 238). Showing the openness of reproductive networks, he writes: "The hybrids from the common and Chinese geese (A. cygnoides), species which are so different that they are generally ranked in distinct genera, have often bred in this country with either pure parent, and in one single instance they have bred *inter se*" (1859: 253; see also 1872: 240). Elsewhere, Darwin will be even more straightforward, claiming that close interbreeding has a negative effect on the chance of survival, whereas frequent intercrossing has a beneficial one (1859: 70, 71, 1872: 55). Of course, the question that immediately comes to mind is the following: how are forms supposed to break free from their common reproductive networks if one assumes not only that crosses produce more fertile and vigorous organisms but also that natural selection thrives on new variations exchanged in open geographical space, as also argued elsewhere by Darwin (1859: 103–106, 1872: 81–83)? If so, how can closely related forms ever manage to segregate themselves from each other?

How far is Darwin prepared to push his general principle, that is, up to what taxonomic level will he maintain it? Clearly, crosses between distinct families are a biological impossibility for Darwin—an "absolute zero of fertility" (1859: 255, 257, 261)—a notion clearly expressed in the following manner: "When pollen from a plant of one family is placed on the stigma of a plant of a distinct family, it exerts no more influence than so much inorganic dust" (1859: 255; see also 1872: 241). Yet, below the family level things are rather hazy:

> From this absolute zero of fertility, the pollen of different species of the same genus applied to the stigma of some one species, yields a perfect gradation in the number of seeds produced, up to nearly complete or even quite complete fertility… There are many cases, in which two pure species can be united with unusual facility, and produce numerous hybrid-offspring, yet these hybrids are remarkably sterile. On the other hand, there are species which can be crossed very rarely, or with extreme difficulty, but the hybrids, when at last produced, are very fertile. Even within the limits of the same genus, for instance in Dianthus, these two opposite cases occur. (1859: 255, 256; see also 1872: 241, 242)

Things do not get any less hazy when Darwin (1859: 22, 23, 253, 1872: 17, 240) alludes to the fact that some species belonging to different genera are able to cross, an expected and logical conclusion considering that Darwin placed the "absolute zero of fertility" at the family level. To understand Darwin's position, it must be remembered that his theory is based entirely on the idea that "microevolution" explains "macroevolution": evolution at or below the species level explains everything observed above that level (genus, family, order, and up). His theory, therefore, rests on this *extrapolation*. Now, the full and definitive segregation level, by Darwin's own admission, is only clearly achieved at the family level. This means that related entities at lower levels (varieties, species, and genera) lack clear boundaries in their separateness under the binding agent of reproduction.

This is much to the theoretical and empirical disadvantage of Darwin's theory. The complex nature of the connections between forms at lower taxonomic levels is best summarized by Darwin's (1859: 257) own formula: "But the correspondence between systematic affinity and the facility of crossing is by no means strict. A multitude of cases could be given of very closely allied species which will not unite, or only with extreme difficulty; and on the other hand of very distinct species which unite with the utmost facility" (see also 1872: 243). In other words, there is only a weak connection between the phenotype—the physical conformation upon which the taxonomic assessment is based—and the ability to reproduce (1859: 269, 1872: 256). But if the phenotype is not revelatory of the level of reproductive isolation, on what exactly is natural selection working to achieve irreversible breaks? Darwin's answer is unsatisfactory: "Now do these complex and singular rules indicate that species have been endowed with sterility simply to prevent their becoming confounded in nature? I think not. . . .[T]he sterility both of first crosses and of hybrids is simply incidental or dependent on unknown differences, chiefly in the reproductive systems, of the species which are crosses" (1859: 260; see also 1872: 245). This amounts to claiming that somehow natural selection eventually succeeds in building biological organizations distinct enough to have become incompatible, reproductively speaking.[25] Indeed, Darwin argues that "[n]ature acts uniformly and slowly during vast periods of time on the whole organisation, in any way which may be for each creature's own good; and thus she may, either directly, or more probably indirectly, through correlation [of growth], modify the reproductive system in the several descendants from any one species" (1859: 269).

The actual segregation process itself, which must ultimately be driven by the action of natural selection, thus remains largely unexplained. Again, Darwin seems unable to extricate himself from the complex web of reproductive networks, networks he explicitly describes as composed of "strains" and "sub-breeds" (1859: 267), that is, non-monolithic entities too often hidden under rigid taxonomic categories such as "varieties" and "species." Even Darwin's vocabulary seems to betray his almost unconscious commitment to seamless reproductive networks, a fabric of life made ever more real at Darwin's hand in his objections to the typological distinction established by Gärtner: varieties are always fertile; species are variably sterile. Darwin (1859: 267–276, 1872: 255–262) could not accept this rigid distinction, as he needed low-level entities to move imperceptively from full fertility to complete sterility during the divergence process, hence his defense of the following status for each taxonomic level: varieties mostly fertile, species variably fertile, genera mostly sterile, and families always sterile.

Empirically speaking, Darwin receives little support from the assumption he advances that animals and plants are characterized by identical reproductive processes, as reflected in numerous examples indiscriminately appealing to both kingdoms throughout the *Origin* (Smocovitis 2009). Indeed, plants are more open systems than animals, with complex and flexible reproductive processes such as self-fertilization, cross-fertilization, and pollination.

Many obstacles have piled up on Darwin's intellectual road. When he discussed the Diagram (see Fig. 2.1) in the *Origin* in the context of low taxonomic entities, he should have visually represented complex, semicontinuous networks of strains caught up in a reproductive nexus of various sizes and taxonomic levels (varieties, species, genera). Instead, Darwin (1859: 116–126, 1872: 90–97) chose to visually force his way by presenting neat and segregated evolutionary lines irreversibly speciating while ascending the taxonomic scale. In Chapter 8, Darwin had to impose his preferred concepts—monophyletism, common ancestry, and irreversible divergence—perhaps even more forcefully than he did in Chapters 1–5.

---

[25]On the explicit use of the notion of biological "organization," see Darwin (1859: 265, 266; 1872: 250, 251).

3.8 The Exhaustion of the Evolutionary Dynamics over Time

Let us bring our analysis of Chapters 1–5 and 8 of the *Origin* to completion. We know that Darwin used overpopulation as a precondition of evolutionary change: the mounting demographic pressure in the context of limited places and resources in the economy of nature somehow had to be dissipated; the struggle for existence was the way taken by some life forms to escape from this unbearable situation (1859: 63–71, 80–87, 1872: 50–55, 62–69). Forms unable to evolve or generate the right variations are condemned. Darwin offers us a bleak image of nature: "The face of Nature may be compared to a yielding surface, with ten thousand sharp wedges packed close together and driven inwards by incessant blows" (1859: 63). Yet, at the same time, the dissipation of the demographic pressure happens in anything but a chaotic or disordered fashion, according to Darwin, for he has in mind a very precise idea of how the evolutionary process is to be structured and organized: with the assistance of the twin principles of divergence and specialization, evolution is to proceed as an ongoing divergence pattern-process all the way up the taxonomic scale. Although the principle of divergence is largely implicit, albeit prevalent, in the first three chapters of the *Origin*, Darwin tackled the principle head-on in Chapter 4:

> A fancier is struck by a pigeon having a slightly shorter beak; another fancier is struck by a pigeon having a rather longer beak; and on the acknowledged principle that 'fanciers do not and will not admire a medium standard, but like extremes,' they go on... choosing and breeding from birds with longer and longer beaks, or with shorter and shorter beaks... As the differences slowly become greater, the inferior animals with intermediate characters... will have been neglected, and will have tended to disappear. Here, then, we see in man's productions the action of what may be called the principle of divergence, causing differences, at first barely appreciable, steadily to increase, and the breeds to diverge in character both from each other and from their common parent. But how, it may be asked, can any analogous principle apply in nature? ... [T]he more diversified the descendants from any one species become in structure, constitution, and habits, by so much will they be better enabled to seize on many and widely diversified places in the polity of nature, and so be enabled to increase in numbers. (1859: 112; see also 1872: 87)

Darwin illustrated the principle of divergence by taking the example of carnivorous quadruped diversifying in habits and conformation with some forms, for instance, feeding on new kinds of prey, with some others on inhabiting new stations (climbing trees, frequenting water), and still others on becoming less carnivorous (1859: 113, 1872: 87, 88). The principle of divergence extends, for Darwin, also to include what is known as the "principle of specialization": a greater amount of biomass can live in the same territory if life forms are diversified in such a way as to exploit better some resources over others:

> The truth of the principle, that the greatest amount of life can be supported by great diversification of structure, is seen under many natural circumstances... The advantage of diversification in the inhabitants of the same region is, in fact, the same as that of the physiological division of labour in the organs of the same individual body—a subject so well elucidated by Milne Edwards. No physiologist doubts that a stomach by being adapted to digest vegetable matter alone, or flesh alone, draws most nutriment from these substances.

> So in the general economy of any land, the more widely and perfectly the animals and plants are diversified for different habits of life, so will a greater number of individuals be capable of there supporting themselves. (1859: 114–116; see also 1872: 88–90)

Darwin then introduced in Chapter 4, and for the first time, his Diagram (1859: 116–126, 1872: 90–97). It is at the precise junction of Chapters 3 and 4 that one finds in Darwin's argumentation a powerful clue to his commitment toward a fully constituted world. To use modern terminology, this clue is revealed in the application of what would be called today the "competitive exclusion principle." The issue is best approached from a question: what kind of evolutionary dynamics can generate an ongoing divergence process all the way up the taxonomic scale? To put the question differently, why should evolution be divergent at all? After all, the demographic pressure could be dissipated in a number of different ways when the tree of life is looked at from a wider perspective: divergence, parallelism, convergence, or a mix of all these processes-patterns (a random walk). We know that Darwin considered the principle of divergence to be the keystone of his theory,[26] and nowhere is this keystone more legitimate and efficient in Darwin's eyes as when it is exemplified in the context of the competitive exclusion principle. Let us follow his clear and simple reasoning:

> But the struggle almost invariably will be most severe between the individuals of the same species, for they frequent the same districts, require the same food, and are exposed to the same dangers. In the case of varieties of the same species, the struggle will generally be almost equally severe... As species of the same genus have usually, though by no means invariably, some similarity in habits and constitution, and always in structure, the struggle will generally be more severe between species of the same genus, when they come into competition with each other, than between species of distinct genera... We can dimly see why the competition should be most severe between allied forms, which fill nearly the same place in the economy of nature. (1859: 75, 76; see also 1872: 58, 59)

This passage outlines an explicit argument to the effect that similar forms should go through evolutionary repulsion, so to speak, from each other: the best way to escape the direct and brutal competition from someone similar and exploiting similar resources is by evolving away from that other entity, hence divergence. This passage, however, also contains an inherent limitation of Darwin's theory, not recognized by Darwin himself, when he alludes to the fact that species of the same genus compete more harshly than species belonging to different genera (see also Darwin 1859: 110, 1872: 86). If so, why should related lineages continue to diverge from each other as they climb the taxonomic scale and become separated at the generic, family, order, class levels, thus exploiting ever more different resources? The logical outcome of this reasoning is that the evolutionary dynamics behind the divergence pattern-process will gradually exhaust itself as the taxonomic scale is ascended. In other words, as related forms become more and more different (divergent) from each other, the evolutionary reason for diverging should gradually be

---

[26]Again, "keystone" is the word used by Darwin in a letter to J.D. Hooker dated to June 8, 1858. See F. Burkhardt and S. Smith (eds.), *The Correspondence of Charles Darwin*, Vol. 7 (Cambridge: Cambridge University Press), p. 102.

turned off. Furthermore, when reaching a certain taxonomic level, the evolution of related forms should look more like a "random walk"—a mix of patterns-processes combining parallel evolution, divergence, convergence, stagnation—rather than exclusive divergence. Apparently, Darwin evaded the issue altogether and preferred instead to impose upon the complexity of evolution divergence all the way up. This important question will be revisited in greater detail in Chap. 5 of this book.

Our thesis in this chapter has been that Darwin imposed divergence through and through, precisely because his approach to evolution is based on a horizontal view applied at low taxonomic levels to closely related forms, where the context of competition press forms against each other. This reasoning is then applied to the other higher taxonomic levels by *extrapolation* only—by assuming that what happens at low levels should also be valid at higher ones—irrespective of the fact that the evolutionary pressure, if we follow Darwin's own logic, should be gradually reduced as the taxonomic scale is ascended. In other words: *Darwin committed the revealing mistake of thinking that the competitive exclusion principle would be effective all the way up the taxonomic scale, precisely because he took it for granted that conditions in the past or future were the same as today among related forms of low taxonomic levels, like in a perpetual succession of present times.* This is precisely what should have been expected from someone who believed in a fully constituted world. Again, a genuine historical dimension is lacking in Darwin's thought.

Furthermore, instead of tackling the problem raised here head-on, Darwin imperceptibly moved from the context of the "competitive exclusion principle" to the context of the "entangled bank," assuming that both competitive contexts would generate divergence throughout all of evolution and at all taxonomic levels, in a sort of *pan-divergent* view valid for all eternity. Indeed, Chapter 3 of the *Origin* is structured around three main parts: the first being devoted to demographic pressure (1859: 62–71, 1872: 49–55), the second to competition of unrelated forms in the entangled bank (1859: 71–75, 1872: 55–58), and the third to competition of related forms under the exclusion principle (1859: 75, 76, 1872: 58, 59). One must wait until Chapter 4 for a formal exposition of the principles of divergence and specialization. Darwin was implicit enough in the very first chapters of the *Origin* for his readers to know where he was heading. Yet, by delaying this formal presentation to Chapter 4, Darwin could easily avoid presenting a compromising discussion in Chapter 3, one specifically requiring distinguishing between various contexts of competition and their accompanying differential evolutionary outcomes: divergence, parallelism, convergence, stagnation, or a mix of these patterns-processes (a random walk). Strangely enough, there is a logical gap or discontinuity between, on the one hand, Darwin's explanations of the different contexts of competition and, on the other hand, his formal exposition of the principles of divergence and specialization.

Setting aside Darwin's inconsistencies or argumentative strategies, let us at least take the analysis to its logical conclusion. Whereas divergence all the way up the taxonomic scale was not justified under Darwin's own understanding of the competitive exclusion principle—being only valid at low levels—general divergence at any level was less justifiable still when competition occurred in the context of unrelated forms within the so-called "entangled bank." Considering that life forms are linked in complex relationships of interdependence involving a host of unrelated forms, evolution of any two forms part of the same entangled blank (if there is any evolution at all) should be closer to a random walk than widespread and sustained divergence. It is unclear whether Darwin ever realized to extent to which his theory imposed upon the complexity of evolution a single rigid pattern-process. In the following quote, we see him loosening the grip of his rigid pattern-process only very slightly, only to reinstate it under a "general rule":

> But I must remark that I do not suppose that the process ever goes on so regularly as is represented in the diagram, though in itself made somewhat irregular. I am far from thinking that the *most divergent varieties* will invariably prevail and multiply: a medium form may often long endure, and may or may not produce more than one modified descendant; for natural selection will always act according to the nature of the places which are either unoccupied or not perfectly occupied by other beings; and this will depend on infinitely complex relations. But as a *general rule*, the more diversified in structure the descendants from any one species can be rendered, the more places they will be enabled to seize on, and the more their modified progeny will be increased. (1859: 118, 119; see also 1872: 91, 92) [my emphasis]

This passage makes clear that Darwin refrained from establishing a distinction in the competition contexts—*entangled bank* versus *exclusion principle*—as the two of them are fused here. In addition, Darwin also abstained from modulating this competition context in accordance with the taxonomic scale, as seen in the application of the Diagram at various taxonomic levels throughout the *Origin*.[27] No matter the taxonomic level or the context of competition, Darwin's view of evolution is a pan-divergent one, a rigid pattern based on a dynamics of entities pushing each other away in a sort of sideways evolution, in opposition to free, open, and vertical evolution. Darwin left little room for genuine evolutionary contingency.

3.9 Conclusion

At this early stage of our investigation, it is not possible to fully grasp the nature of Chapters 1–5 and 8; this requires further consideration of the rest of the *Origin*. However, this should not prevent us from very briefly insisting on what constitutes, in our view, the main features of these chapters. Darwin tried to put in place a number of key explanatory components that implicitly follow the ideal of a science based on a vera causa. To show that natural selection was truly a real evolutionary cause—and not a hypothetical (conjectural) one—he relied on an analogy with artificial selection. To be effective, that selective or adaptive cause had to be supported by, and applied in the context of, a conceptual cluster: places in the economy of nature, demographic pressure, variations channelled along adaptive

[27]For the reading of the Diagram at various taxonomic levels, see Darwin (1859: 116–126, 331, 332, 412, 413; 1872: 90–97, 303, 304, 364).

paths, the principles of divergence and specialization, and the competitive exclusion principle.

Irrespective of what Darwin hoped to achieve, he does not present the careful reader with the necessary supporting material. Implicitly or explicitly, Darwin in these chapters unwittingly offers abundant counterfactual evidence that speaks against his theory: (a) life-resisting change for several generations; (b) reversal to the parental condition; (c) correlation of growth; (d) blending inheritance; (e) nature favoring new variations (interbreeding) over inbreeding; (f) the imposition of the monophyletic agenda over the polyphyletic view of breeders; (g) the inconsistent calibration of the effects of artificial selection and of natural selection; (h) the lack of a clear sterility barrier below the family level; (i) the gradual exhaustion of the evolutionary drive accompanying the competitive exclusion principle during taxonomic ascent; and (j) the lack of distinction in the various competition contexts and their evolutionary outcomes on the evolutionary pattern.

Darwin's neat and clear conception of life forms diverging from one another largely served the purpose of hiding the complexities of evolution the reader could gather from a close reading Chapters 1–5 and 8 the *Origin of Species*. Perhaps the most convincing theory in these chapters concerns "reticulate evolution," not "divergence."

References

Amundson R (1996) Historical development of the concept of adaptation. In: Rose M, Lander G (eds) Adaptation. Academic Press, New York, pp 11–53

Barsanti G (2005) Linné et Buffon: deux visions différentes de la nature et de l'histoire naturelle. In: Hoquet T (ed) Les fondements de la botanique: Linné et la classification des plantes. Vuibert, Paris, pp 103–127

Bartley M (1992) Darwin and domestication: studies on inheritance. J Hist Biol 25:307–333

Blanckaert C (1989) L'Anthropologie personnifiée: Paul Broca et la biologie du genre humain. In: Broca P (ed) Mémoires d'anthropologie. Jean-Michel Place, Paris, pp i–xliii

Bowler PJ (1976) Malthus, Darwin, and the concept of struggle. J Hist Ideas 37:631–650

Bowler PJ (1983) The eclipse of Darwinism: anti-Darwinian evolution theories in the decades around 1900. Johns Hopkins University Press, Baltimore

Bowler PJ (1986) Theories of human evolution: a century of debate, 1844–1944. Johns Hopkins University Press, Baltimore

Bowler PJ (1989) The Mendelian revolution. Johns Hopkins University Press, Baltimore

Bowler PJ (1990) Charles Darwin: the man and his influence. Basil Blackwell, Oxford

Bowler PJ (2003) Evolution: the history of an idea, 3rd edn. University of California Press, Berkeley

Broca P (1867) Discours de M. Broca sur l'ensemble de la question. Congrès International d'Anthropologie et d'Archéologie Préhistorique, 2e session, pp 367–402

Broca P (1870) Le transformisme. Bulletins de la Société d'Anthropologie de Paris, 2e série 5:168–239

Brooke JH (1991) Science and religion: some historical perspectives. Cambridge University Press, Cambridge

Browne J (1980) Darwin's botanical arithmetic and the 'principle of divergence,' 1854–1858. J Hist Biol 13:53–89

Browne J (1983) The secular ark: studies in the history of biogeography. Yale University Press, New Haven
Bulmer M (2004) Did Jenkin's swamping argument invalidate Darwin's theory of natural selection? Br J Hist Sci 37:281–297
Burkhardt F, Smith S (eds) (1994) The correspondence of Charles Darwin, vol 7. Cambridge University Press, Cambridge
Cannon W (1961) The bases of Darwin's achievement: a revaluation. Vic Stud 5:109–134
Carpenter WB (1860) Darwin and the Origin of Species. Natl Rev 10(January):188–214
Darwin C (1859) On the Origin of Species. John Murray, London
Darwin C (1872) The origin of species, 6th edn, with additions and corrections [1876]. John Murray, London
Delisle RG (2007) Debating humankind's place in nature, 1860–2000: the nature of paleoanthropology. Pearson-Prentice Hall, New Jersey
Egerton FN (1968) Studies of animal populations from Lamarck to Darwin. J Hist Biol 1:225–259
Egerton FN (1973) Changing concepts of the balance of nature. Q Rev Biol 48:322–350
Egerton FN (2007) A history of the ecological sciences, Part 23: Linnaeus and the economy of nature. Bull Ecol Soc Am 88:72–88
Ellegard A (1990[1958]) Darwin and the general reader: the reception of Darwin's theory of evolution in the British periodical press, 1859–1872. University of Chicago Press, Chicago
Evans L (1984) Darwin's use of the analogy between artifical and natural selection. J Hist Biol 17:113–140
Gayon J (1998) Darwinism's struggle for survival: heredity and the hypothesis of natural selection. Cambridge University Press, Cambridge
Geison G (1969) Darwin and heredity: the evolution of his hypothesis of pangenesis. J Hist Med Allied Sci 24:375–411
Greene JC (1959) The death of Adam: evolution and its impact on Western thought. Iowa State University Press, Iowa
Greene JC (1981) Science, ideology, and world view. University of California Press, Berkeley
Herbert S (1971) Darwin, Malthus and selection. J Hist Biol 4:209–217
Hodge MJS (1977) The structure and strategy of Darwin's 'long argument'. Br J Hist Sci 10:237–246
Hodge MJS (1982) Darwin and the laws of the animate part of the terrestrial system (1935–1837): on the Lyellian origins of his zoonomical explanatory program. Stud Hist Biol 6:1–106
Hodge MJS (1983) The development of Darwin's general biological theorizing. In: Bendall D (ed) Evolution from molecules to men. Cambridge University Press, Cambridge, pp 43–62
Hodge MJS (1989) Darwin's theory and Darwin's arguments. In: Ruse M (ed) What the philosophy of biology is. Kluwer, Dordrecht, pp 163–182
Hodge MJS (2009) Capitalist contexts for Darwinian theory: land, finance, industry and empire. J Hist Biol 42:399–416
Hodge MJS (2014) On Darwin's science and its contexts. Endeavour 38:169–178
Hodge MJS, Kohn D (1985) The immediate origins of natural selection. In: Kohn D (ed) The Darwinian heritage. Princeton University Press, Princeton, pp 185–206
Hooker JD (1859) [Review of the Origin of Species]. The Gardeners' Chronicle and Agricultural Gazette, December 31, pp 1051–1052
Hopkins W (1860–1861) Physical theories of the phenomena of life part I and II. Fraser's Magazine, 61(1860):739–752 62(1861):74–90
Hull DL (1973) Darwin and his critics: the reception of Darwin's theory of evolution by the scientific community. Harvard University Press, Cambridge
Jenkin F (1867) The Origin of Species. N Br Rev 46(June):277–318
Kohn D (1980) Theories to work by: rejected theories, reproduction, and Darwin's path to natural selection. Stud Hist Biol 4:67–170
Kohn D (1985) Darwin's principle of divergence as internal dialogue. In: Kohn D (ed) The Darwinian heritage. Princeton University Press, Princeton, pp 245–257

Kohn D (2009) Darwin's keystone: the principle of divergence. In: Ruse M, Richards RJ (eds) The Cambridge companion to the 'Origin of Species'. Cambridge University Press, Cambridge, pp 87–108
La Vergata A (1988) Theodicy and nature's economy. Nuncius 3:139–152
Largent M (2009) Darwin's analogy between artificial and natural selection in the Origin of Species. In: Ruse M, Richards RJ (eds) The Cambridge companion to the 'Origin of Species'. Cambridge University Press, Cambridge, pp 14–29
Limoges C (1968) Darwin, Milne-Edwards, et le principe de divergence. Actes du XIIe Congrès International d'Histoire des Sciences 8:111–115
Limoges C (1970) La sélection naturelle. Presses Universitaires de France, Paris
Limoges C (1972) Introduction. In: Linné C (ed) L'équilibre de la nature. J. Vrin, Paris, pp 7–22
Linnaeus C (1744) Oratio de Telluris habitabilis incremento. In: Linné C (ed) L'équilibre de la nature, translated from Latin to French. J. Vrin, Paris, pp 29–49
Linnaeus C (1749) Oeconomia naturae. In: Linné C (ed) L'équilibre de la nature, translated from Latin to French. J. Vrin, Paris, pp 57–101
Linnaeus C (1756) Disquisitio de Sexu Plantarum. Published in 1760 by the Imperial Academy of Science of St Petersburg
Linnaeus C (1760) Politia naturae. In: Linné C (ed) L'équilibre de la nature, translated from Latin to French. J. Vrin, Paris, pp 103–121
Lotsy J (1916) Evolution by means of hybridization. Martinus Nijhoff, The Hague
Lyell C (1832) Principles of geology, vol 2. John Murray, London
Lyell C (1833) Principles of geology, vol 3. John Murray, London
Mayr E (1982) The growth of biological thought. Belknap Press, Cambridge
Mayr E (1991) One long argument: Charles Darwin and the genesis of modern evolutionary thought. Harvard University Press, Cambridge
Morrison DA (2015) [Book review] The tree of life: evolution and classification of living organisms. Syst Biol 64:546–548
Morrison DA (2016) Genealogies: pedigrees and phylogenies are reticulating networks not just divergent trees. Evol Biol 43:456–473
Newton I (1730) Optiks: or a treatise of the reflections, refractions, inflections, and colour of light, 4th edn. William Innys, London
Olson R (1987) On the nature of God's existence, wisdom and power: the interplay between organic and mechanistic imagery in Anglican natural theology, 1640–1740. In: Burwick F (ed) Approaches to organic forms. D. Reidel, Dordrecht, pp 1–48
Ospovat D (1978) Perfect adaptation and teleological explanation: approaches to the problem of the history of life in the mid-Nineteenth century. Stud Hist Biol 2:33–56
Ospovat D (1979) Darwin after Malthus. J Hist Biol 12:211–230
Ospovat D (1980) God and natural selection: the Darwinian idea of design. J Hist Biol 13:169–194
Ospovat D (1981) The development of Darwin's theory: natural history, natural theology, and natural selection, 1838–1859. Cambridge University Press, Cambridge
Pearce T (2010) 'A great complication of circumstances'—Darwin and the economy of nature. J Hist Biol 43:493–528
Ray J (1701) The wisdom of God manifested in the works of the creation, 3rd edn. Smith & Walford, London
Ray J (1709) The wisdom of God manifested in the works of creation, 5th edn. Samuel Smith, London
Recker D (1987) Causal efficacy: the structure of Darwin's argument strategy in the Origin of Species. Philos Biol 54:147–175
Romanes GJ (1886) Physiological selection: an additional suggestion on the Origin of Species. J Linn Soc 19:337–411
Rudwick MJS (1970) The strategy of Lyell's Principles of Geology. Isis 61:5–33

Ruse M (1975) Darwin's debt to philosophy: an examination of the influence of the philosophical ideas of John F.W. Herschel and William Whewell on the development of Charles Darwin's theory of evolution. Stud Hist Phil Sci 6:159–181
Ruse M (1979) The Darwinian revolution. University of Chicago Press, Chicago
Ruse M (1996) Monad to man: the concept of progress in evolutionary biology. Harvard University Press, Cambridge
Ruse M (1999) The Darwinian revolution, 2nd edn. University of Chicago Press, Chicago
Ruse M (2000) Darwin and the philosophers: epistemological factors in the development and reception of the theory of the Origin of Species. In: Creath R, Maienschein J (eds) Biology and epistemology. Cambridge University Press, Cambridge, pp 3–26
Ruse M (2003) Darwin and design: does evolution have a purpose? Harvard University Press, Cambridge
Ruse M (2004) The Romantic conception of Robert J. Richards. J Hist Biol 37:3–23
Ruse M (2009) The history of evolutionary thought. In: Ruse M, Travis J (eds) Evolution: the first four billion years. Belknap Press, Cambridge, pp 1–48
Sargent R-M (1995) The diffident naturalist: Robert Boyle and the philosophy of experiment. University of Chicago Press, Chicago
Schweber SS (1980) Darwin and the political economists: divergence of character. J Hist Biol 13:195–289
Sloan PR (1999) The making of a philosophical naturalist. In: Hodge MJS, Radick G (eds) The Cambridge companion to Darwin, 2nd edn. Cambridge University Press, Cambridge, pp 21–43
Smocovitis VB (2009) Darwin's botany in the Origin of Species. In: Ruse M, Richards RJ (eds) The Cambridge companion to the "Origin of Species". Cambridge University Press, Cambridge, pp 216–236
Stauffer R (1960) Ecology in the long manuscript version of Darwin's Origin of Species and Linnaeus' Oeconomy of Nature. Proc Am Philos Soc 104:235–241
Sterret S (2002) Darwin's analogy between artificial and natural selection: how does it go? Stud Hist Phil Biol Biomed Sci 33:151–168
Sulloway FJ (1979) Geographic isolation in Darwin's thinking: the vicissitudes of a crucial Idea. Stud Hist Biol 3:23–65
Tammone W (1995) Competition, the division of labor, and Darwin's principle of divergence. J Hist Biol 28:109–131
Vorzimmer PJ (1969) Darwin, Malthus and the theory of natural selection. J Hist Ideas 30:527–542
Vorzimmer PJ (1970) Charles Darwin: the years of controversy. Temple University Press, Philadelphia
Wagner M (1868) Die Darwin'sche Theorie und das Migrationgesetz der Organismen. Duncker & Humblot, Leipzig
Waters CK (2009) The arguments in the Origin of Species. In: Hodge MJS, Radick G (eds) The Cambridge companion to Darwin, 2nd edn. Cambridge University Press, Cambridge, pp 120–143
Winther R (2000) Darwin on variation and heredity. J Hist Biol 33:425–455
Worster D (1994) Nature's economy: a history of ecological ideas, 2nd edn. Cambridge University Press, Cambridge
Young RM (1985) Darwin's metaphor: nature's place in Victorian culture. Cambridge University Press, Cambridge

Chapter 4
The Wild Power of Natural Selection: Vertical Evolution, Analogies, and Imaginary Scenarios (*Origin*, Chaps. 6–7)

Abstract Upon close inspection, Chapters 6 and 7 of the *Origin of Species* present a staggering spectacle. As often held, these chapters are about the postulation of intermediate evolutionary connections and the rise of complex organs from simple ones. But these explanatory components are put in the service of an entirely non-Darwinian theory of evolution, whereby life is depicted as a series of parallel and independent lines racing to acquire similar structures across evolutionary grades and taxonomic levels. Whereas Darwin's theory requires a delicate balance between the "unity of type" (descent) and the "conditions of existence" (natural selection), Chapters 6 and 7 present natural selection as an untamed force annihilating traces of affiliation and generating evolutionary analogies. We argue that at least two incompatible theories of evolution exist side by side in the *Origin of Species*.

Keywords Homologies · Analogies · Conditions of existence · Unity of type · Hypothetical scenarios · Intermediate links · Directional evolution · Evolutionary ascent · Evolutionary grades

With Chapters 6 and 7 of the *Origin*, the critical reader is presented with an entirely different worldview from that presented in prior chapters: from horizontality, one goes directly to verticality. Of course, the rhetorical Darwin was skillful enough to largely conceal this profound shift in perspective. It will be our task in this chapter to expose this shift more clearly. In the recent historiography (Ruse 1979: 192; Bowler 1990: 119–120), Chapters 6 and 7 are usually presented as addressing the problem of the existence of intermediate forms in evolution. Darwin's strategy, on that count, was twofold: first, to argue that many intermediate forms are missing if only because descendants of a lineage had to exterminate their ancestors in order to take their place in the economy of nature; and second, to seek among currently living forms structures and behaviors whose states make them susceptible to bridging evolutionary gaps between forms, as natural selection could tinker with their inherent variations.

Therefore, the essential point Darwin hoped to make in Chapters 6 and 7 of the *Origin* is the following: *natural selection as a force is powerful enough to build*

R. G. Delisle, *Charles Darwin's Incomplete Revolution*, Evolutionary Biology – New Perspectives on Its Development 1, https://doi.org/10.1007/978-3-030-17203-9_4

complex structures and behaviours that were, originally, in a simple state. Of course, the action of such a force would leave transitional evolutionary links in its wake.

Although the recent historiography has taken notice of these conceptual developments, it has largely neglected the fact that, in an attempt to reach his main explanatory goal in these two chapters, Darwin profoundly undermined the theoretical core of his theory, especially as deployed in Chapters 1–5 and 11–13. Indeed, at the heart of the theory Darwin (1859, 206; 1872, 166) tries to build, one finds a very specific relationship between the two laws he calls the "Unity of Type" and the "Conditions of Existence." On the one hand, the Unity of Type captures the reality that life comes in clusters of forms sharing similarities because of common ancestry. For Darwin, the Unity of Type is equivalent to "descent." On the other hand, the Conditions of Existence stand for the necessity of life to continue evolving in order to stay attuned to environmental changes and biotic competition. For Darwin, the Conditions of Existence are equated with the action of "natural selection." *Descent* and *natural selection* are, then, two opposite terms in Darwin's theory that had to be properly balanced. Indeed, a relationship overstressing descent at the expense of natural selection would leave evolution impotent, considering that the evolutionary dynamics would be too weak to allow for the breakup (segregation) of entities. Life forms would remain tightly packed within their reproductive networks. We have already seen Darwin wrestle with this problem in prior chapters. Conversely, a relationship overstressing natural selection at the expense of descent would dissolve, disperse, and ultimately shatter all traces of affiliation, since the urge for evolutionary change would be too great, in addition to creating the condition for the rise of similar yet independent features in unrelated groups (analogies). To avoid the problem of parallel evolution, Darwin needed descent to be strong enough to maintain the cluster of features shared within reproductive or phylogenetic networks.

This, however, is precisely Darwin's approach in Chapters 6 and 7 of the *Origin*. In the laying out of assumptions in Chapters 6 and 7, natural selection is presented in such a way as to constitute a wild and untamed force which nothing could resist. Natural selection is thus presented as an extremely powerful vertical force pushing life up across evolutionary grades, leaving in its wake traces of evolutionary analogies. The thesis defended by John C. Greene (1957, 1959: 11–13, 25–29; 1971, 1981: 128–157; 1986)—as seen in the Introduction of this book under the designation "atomism-mechanism"—captures something essential of Darwin's view in Chapters 6 and 7: the atomistic motion of biological entities channelled along a directional path under an external pressure called natural selection, a time-oriented and law-bound system of matter in motion projected forward. We would like to expand on this historiographical piece by bringing in additional considerations.

In his haste to show the power of natural selection in forging transitional links and building ever more complex structures/behaviors, Darwin destabilized the delicate balance between descent and natural selection his theory required. He may have unconsciously felt the need to overdo it with natural selection in these chapters in hopes of counterbalancing the weakness of his case in Chapters 1–5. The following passage makes the point clear:

> In North America the black bear was seen by Hearne swimming for hours with widely open mouth, thus catching, like a whale, insects in the water. Even in so extreme a case as this, if the supply of insects were constant, and if better adapted competitors did not already exist in the country, I can see no difficulty in a race of bears being rendered, by natural selection, more and more aquatic in their structure and habits, with larger and larger mouths, till a creature was produced as monstrous as a whale. (Darwin 1859: 184)

Critics responding immediately in the wake of the publication of the *Origin of Species* could not resist taking advantage of this extreme example as an object of ridicule (see Ellegard 1990 [1958]: 238–241). This forced Darwin to give up on the illustrative value of this passage. In the 1872 edition, it reduces to this: "In North America the black bear was seen by Hearne swimming for hours with widely open mouth, thus catching, almost like a whale, insects in the water" (p. 141). The rest was removed. Be that as it may, the original version seems less a slip of the tongue than the tip of the iceberg. As will be seen, life can no longer be depicted as a rigid pan-divergent pattern-process of evolution in Chapters 6 and 7 of the *Origin*.

If Darwin could not put forward a very convincing case in favor of the power of natural selection in Chapters 1–5 and 8, how could he offer an even more ambitious one in Chapters 6–7? The answer seems obvious: he could not. That is why he had to fill in the explanatory voids by resorting to speculative scenarios, strongly appealing to our imagination. We will follow Darwin through the steps that ultimately led him to move beyond the core of his theory as originally intended. Indeed, Darwin became ever more ambitious in his arguments, moving from transitional features observed within the same organisms, to transitions between closely related forms, and eventually to groups separated at high taxonomic levels (families, orders, classes, phyla). In his desire to continually widen the taxonomic scope of his analysis, Darwin entirely overlooked the fact that he went from cases appealing to common ancestry and divergence (homologies) to cases explained by parallel evolution (analogies).

We have seen in the previous chapter of this book that the argumentative structure of the *Origin of Species* was divided in three parts: (1) the first chapters concerned with the mechanism of evolution (*vera causa*); (2) the last chapters devoted to applying this mechanism to a host of disciplines (consilience of inductions); and (3) the middle chapters addressing some difficulties of the theory. For instance, Jonathan Hodge (1977, 1989) holds that Chapters 6 and 7 of the *Origin* not only address some difficulties of the theory but, in addition, aim at showing the "adequacy" or the "competence" of the case studies considered in supporting Darwin's theory. We propose an alternative view of these chapters for the following two reasons: (1) in them, Darwin is not defending his theory against difficulties but is, instead, busy conquering brand new theoretical territory; (2) in so doing, Darwin went beyond what he intended to achieve in the rest of the *Origin*. Chapters 6 and 7 are profoundly non-Darwinian, so to speak. The critical reader of Darwin's *magnum opus* is now confronted with two incommensurable evolutionary theories within the same book.

4.1 Looking for Transitions: Darwin's Explicit Method

In Chapters 6 and 7, as in all other chapters of the *Origin*, Darwin seeks to promote a conceptual cluster constituted by the following notions: monophyletism, common ancestry, divergence, gradualism, and transitional forms; all this, of course, under the impetus of selective forces (natural selection). While these notions are present implicitly throughout the *Origin*, various chapters stress some notions over others. In Chapters 6 and 7, the notions of gradualism and transitional forms take center stage. The first step Darwin will take to make the reality of transitional forms intelligible to his reader will be through the search of versatile organs and behaviors within the same organisms or species, variations assumed to be ultimately seized upon by natural selection to take life forms into new adaptive territories:

> It has been asked by the opponents of such views as I hold, how, for instance, a land carnivorous animal could have been converted into one with aquatic habits; for how could the animal in its transitional state have subsisted? It would be easy to show that within the same group carnivorous animals exist having every intermediate grade between truly aquatic and strictly terrestrial habits; and as each exists by a struggle for life, it is clear that each is well adapted in its habits to its place in nature. Look at the *Mustela vison* of North America, which has webbed feet and which resembles an otter in its fur, short legs, and form of tail; during the summer this animal dives for and preys on fish, but during the long winter it leaves the frozen waters, and preys like other polecat on mice and land animals. (Darwin 1859: 179–180; see also 1872: 138)

Among other examples of alleged intermediate features useful to natural selection, Darwin (1859: 190, 210; 1872: 147, 207) lists the following:

1. The same organ performing distinct functions: in the Hydra, the animal may be turned inside out, so that the exterior surface will perform the task of digesting and the stomach of breathing.
2. Two distinct organs in the same organism performing the same function: fish able to breathe both in free air and in water.
3. The same species of animal having different repertoires of instincts varying in accordance to the year's season or the periods of life.

Darwin's method in Chapters 6 and 7 is made quite explicit:

> In looking for the gradations by which an organ in any species has been perfected, we ought to look exclusively to its lineal ancestors; but this is scarcely ever possible, and we are forced in each case to look to species of the same group, that is to collateral descendants from the same original parent-form, in order to see what gradations are possible, and for the chance of some gradations having been transmitted from the earlier stages of descent, in an unaltered or little altered condition. (Darwin 1859: 187; see also 1872: 144)

On Darwin's own terms, the soundness of his method is based on *providing analyses of forms that are both truly closely related and at fairly low taxonomic levels*. Indeed, the affiliative signs indicative of such transitional states or gradations would get blurred with both taxonomic distance and separation time, hence undermining the utility of a theory based on a delicate balance between descent and natural selection.

4.2 Case Studies: Squirrels, Ants, and Flying Creatures

The reader is certainly prepared to extend Darwin a measure of freedom in his hunt for intermediate evolutionary stages if he remains within the methodological confines imposed by his theory. Cases presented within this purview are, at least, not beyond plausibility. The example of squirrels is a good illustration:

> Look at the family of squirrels; here we have the finest gradation from animals with their tails only slightly flattened, and from others... with the posterior part of their bodies rather wide and with the skin on their flanks rather full, to the so-called flying squirrels; and flying squirrels have their limbs and even the base of the tail united by a broad expanse of skin, which serves as a parachute and allows them to glide... (Darwin 1859: 180; see also 1872: 139).

Another case presented by Darwin (1859: 219–224; 1872: 216–220) involves closely related species of ants of the genus *Formica* and for which a gradation in their slave-making instincts is postulated. Similarly, the case of transition in instincts for the economy of wax used by bees in three genera of the family Apidae, going from *Bombus* (humble bees) to *Apis* (honey bees), through *Melipona* (stingless bees) is not without argumentative force (1859: 224–235; 1872: 220–227). Soon enough, however, the reader is presented with more ambitious and ambiguous cases moving beyond the strict resolution power of Darwin's method, as least as originally formulated:

> If about a dozen genera of birds had become extinct or were unknown, who would have ventured to have surmised that birds might have existed which used their wings solely as flappers, like the logger-headed duck (Micropterus of Eyton); as fins in the water and front legs on the land, like the penguin; as sails, like the ostrich; and functionally for no purpose, like the Apteryx... *It must not be inferred from these remarks that any of the grades of wing-structure here alluded to... indicate the natural steps by which birds have acquired their perfect power of flight; but they serve, at least, to show what diversified means of transition are possible* (Darwin 1859: 181–182; see also 1872: 140) [my emphasis].

"What diversified means of transition are possible"? Is Darwin suggesting here that the various transitions in wing function of birds adapted to live in the water, on the ground, or in the air could somehow be supported, or explained, by some kind of phylogenetic connections between them? In the next example, Darwin is much less convincing, as he hints at transitions between mammalian forms belonging to three distinct orders: bats (Chiroptera), flying lemurs (Dermoptera), and lemurs (Primates):

> Now look at the Galeopithecus or flying lemur, which formerly was falsely ranked amongst bats. It has an extremely wide flank-membrane, stretching from the corners of the jaw to the tail, and including the limbs and the elongated fingers... Although no graduated links of structure, fitted for gliding through the air, now connect the Galeopithecus with the other Lemuridae, yet *I can see no difficulty in supposing that such links formerly existed*, and that each had been formed by the same steps as in the case of the less perfectly gliding squirrels... Nor can I see any insuperable difficulty in further believing it possible that the membrane-connected fingers and fore-arm of the Galeopithecus might be greatly lengthened by natural selection; and this, as far as the organs of flight are concerned, would convert it into a bat. In bats which have the wing-membrane extended from the top of the shoulder to

> the tail, including the hind-legs, we perhaps see traces of an apparatus originally constructed for gliding through the air rather than for flight (Darwin 1859: 181; see also 1872: 139–140) [my emphasis].

As much as Darwin might have wanted to suggest the existence of transitional links uniting three distinct orders of mammals (descent; homologies) on their way to evolving from terrestrial abilities to flying ones, a very serious case could have been made for the independent rise of similar features in unrelated forms (natural selection; analogies). Was Darwin aware of the analogy problem? Let us briefly postpone answering that question while keeping in mind that Darwin may well have voluntarily entertained an ambiguity he thought was to his advantage. It is at this precise explanatory junction that Darwin modified the nature of the relationship he had established between "descent" and "natural selection." Unstoppable in his reflection on transitional links, at that point Darwin stresses natural selection over descent:

> [S]eeing that we have flying birds and mammals, flying insects of the most diversified types, and formerly had flying reptiles, it is conceivable that flying-fish, which now glide far through the air, slightly rising and turning by the aid of their fluttering fins, might have been modified into perfectly winged animals. If this had been effected, who would have ever imagined that in an early transitional state they had been inhabitants of the open ocean, and had used their incipient organs of flight exclusively... to escape being devoured by other fish? When we see any structure highly perfected for any particular habit, as the wings of a bird for flight, we should bear in mind that animals displaying early transitional grades of the structure will seldom continue to exist to the present day, for they will have been supplanted by the very process of perfection through natural selection... Thus, to return to our imaginary illustration of the flying-fish, it does not seem probable that fishes capable of true flight would have been developed under many subordinate forms... until their organs of flight had come to a high stage of perfection, so as to have given them a decided advantage over other animals in the battle for life. Hence the chance of discovering species with transitional grades of structure in a fossil condition will always be less, from their having existed in lesser numbers, than in the case of species with fully developed structures. (Darwin 1859: 182–183; see also 1872: 140–141)

What is Darwin trying to achieve here? He is certainly not defending the idea that flying creatures seen among birds, mammals, insects, reptiles, and hypothetical fish are all related through a unique series of intermediate links. After all, the few possible homologies (descent) related to the flying habits binding some of these forms are largely overshadowed by their superficial similarities (analogies). But by showing-off the power of natural selection in its capacity to forge organs for similar functions across a wide range of taxonomic forms, Darwin overlooks the fact that he undermines the theory he has worked to establish in the rest of the *Origin of Species*. Under the new theory seen in Chapters 6 and 7, the notions of common ancestry and divergence are jeopardized by the ease with which natural selection can forge analogical features, thus blurring affiliative bonds and offering the picture of a series of lines independently acquiring similar organs and functions. Furthermore, the neat pattern-process of divergence that Darwin tried so painstakingly to impose over the blurred reality of complex reproductive networks is replaced by a pattern-process assuming linear or vertical evolution, a directional drive climbing the levels of complexity.

4.3 Case Study: The Rise of Complex Eyes

Our thesis can be substantiated through a closer examination of the case Darwin presents concerning the rise of complex eyes. Darwin's argument is deployed in two parts. The first part is devoted to finding among extant forms some possible transitional steps on the evolutionary road from a simple to a complex eye. To begin with, Darwin (1859, 187; 1872, 144) is not interested in speculating about the appearance of the simplest form of an eye: "How a nerve comes to be sensitive to light, hardly concerns us more than how life itself first originated." As we know, Darwin had always been concerned with studying evolution on the march, that is, among taxonomic groups that are already well constituted. Yet, in Chapters 6 and 7 of the *Origin*, Darwin could hardly conceal his temptation to use *analogies* (parallel evolution) as a mean to probe deeply and broadly, geologically and taxonomically speaking, in search of *homologies* (descent). This, however, could only be achieved at the high epistemological price of implicitly proposing an inconsistent theory. In the last edition of the *Origin*, Darwin restates the basis of his investigative method with even greater precision than before. It is worth quoting this important passage at length, highlighting the changes and additions to its content:

> In searching for the gradations through which an organ in any species has been perfected, we ought to look exclusively to its lineal progenitors; but this is scarcely ever possible, and we are forced to look to other *species and genera* of the same group, that is to the collateral descendants from the same parent-form, in order to see what gradations are possible, and for the chance of some gradations having being transmitted in an unaltered or little altered condition. *But the state of the same organ in distinct classes may incidentally throw light on the steps by which it has been perfected.* (Darwin 1872: 144) [my emphasis]

Darwin is quite precise about the taxonomic levels at which one should work to investigate transitional forms legitimately: related species or genera. Not only does this make perfect sense, it is actually entirely expected. After all, related forms separated at high taxonomic levels have been independently subjected to selective forces long enough for their affiliative signs to have been blurred or confused. A meaningful quest for transitional links must be conducted in the context of phylogenetic clarity (descent), as seen in significant homologies. A balanced relationship between descent and natural selection is thus required. Strangely enough, however, Darwin concludes his methodological *exposé* by appealing to the utility of looking for possible transitional steps in forms belonging to different classes:

> The simplest organ which can be called an eye consists of an optic nerve, surrounded by pigment-cells and covered by translucent skin, but without any lens or other refractive body… In certain starfishes, small depressions in the layer of pigment which surrounds the nerve are filled… with transparent gelatinous matter, projecting with a convex surface, like the cornea in the higher animals… In the great class of the Articulata, we may start from an optic nerve simply coated with pigment, the latter sometimes forming a sort of pupil, but destitute of a lens or other optical contrivance. With insects it is now known that the numerous facets on the cornea of their compound eyes form true lenses, and that the cones include curiously modified nervous filaments. But these organs in the Articulata are so much diversified that Müller formerly made three main classes with seven subdivisions, besides a fourth main class of aggregated simple eyes… Within the highest division of the

> animal kingdom, namely, the Vertebrata, we can start from an eye so simple, that it consists, as in the lancelet, of a little sack of transparent skin, furnished with a nerve and lined with pigment, but destitute of any other apparatus. In fishes and reptiles, as Owen remarked, 'the range of gradations of dioptric structures is very great.' It is a significant fact that even in man... the beautiful crystalline lens is formed in the embryo by an accumulation of epidermic cells, lying in a sack-like fold of the skin; and the vitreous body is formed from embryonic sub-cutaneous tissue. (Darwin 1872: 144–146)

Of course, in his time, Darwin could not have suspected the extent to which parallel evolution is believed to have impacted the rise of eyes in the history of life, with numerous independent occurrences. This being said, Darwin could not have been so naive as to overlook the significant impact of parallel evolution in this particular case study in which he refers to no less than three distinct phyla: starfish, insects, and vertebrates.[1] Darwin's contemporary, St. George Jackson Mivart, in his *On the Genesis of Species* (1871: 76–78) stressed this very point. Apparently, Darwin was only too happy to entertain the ambiguity between analogies and homologies, hoping, perhaps, it would bring support to the view that life is ultimately traceable to a few or perhaps a single common ancestor. *For Darwin,*

[1]To put this discussion in context, it should be noted that the issue over the rise of eyes has become quite sophisticated in recent decades, as far as the monophyletic and polyphyletic perspectives are concerned. However, it seems fair to summarize the whole matter in the following way: just as all life on Earth is based on the same conservative and monotonous DNA molecular code (presumably pointing toward some kind of monophyletism), the diversity and disparity of life at the phenotypic level are significantly influenced by selective factors, thus inevitably introducing various amounts of analogies (polyphyletism) as similar phenotypic solutions were found to similar adaptive problems across the taxonomic spectrum. The fact that packs of genes or regulatory genes (and their accompanying developmental networks and pathways) associated with eyes such as *Pax 6* are much more widespread, taxonomically speaking, than expected, does not fundamentally change the issue. At one point or another, the conservative and widespread molecular and genotypic elements such as the opsin proteins must encounter selective pressure at the phenotypic level. It is at the junction of these two levels that eyes with many shapes, sizes, and designs have evolved. The terminology used to describe such eyes varies in the literature. A basic distinction is made between the various types of simple and complex eyes. Among complex eyes, further basic distinctions are made between the complex eyes in Arthropoda, the inverse eyes in Vertebrata, and the reverse eyes in Cephalopoda; or between the compound eyes of insects, the camera eyes of vertebrates, and the pigment-cup eyes of many invertebrates; or still between the ciliary type of photoreceptor cell and the rhabdomeric type. Many other distinctions are proposed for specific evolutionary groups such as pinholes (*Nautilus*); spherical lenses (many aquatic animals); multiple lenses (copepod crustaceans, *Pontella*, *Copilia*); corneal refraction (spiders, penguins, seals, diving birds); concave reflectors (rotifers, Platyhelminthes; scallops, *Pecten*; copepod crustaceans, ostracod crustacean, *Gigantocypris*); apposition eyes (found in all three arthropod subphyla); neural superposition eyes (dipteran flies); afocal apposition (butterflies); refracting superposition eyes (many nocturnal insects, crustaceans); reflecting superposition eyes (decapod crustaceans); parabolic superposition eyes (swimming crab: *Portunus*), and so on. It is also at the junction of the genotypic and phenotypic levels that analogies, parallel evolution, and the independent rise of similar structures become possible. In short, monophyletism is more prevalent at the lower levels of matter (molecules, genes, and proteins.), whereas polyphyletism is expressed more often at the high levels (phenotypes). Consult, for instance, Salvini-Plawen and Mayr (1977), Land and Fernald (1992), Mayr (2001: 205–207), Arendt (2003), Oakley (2003), Gehring (2004), Salvini-Plawen (2008), Land and Nilsson (2012: 7–10), and Glaeser and Paulus (2015: 146–149).

however, the line is porous between actual phylogenetic propositions, possible phylogenetic suggestions, and indirect light being thrown on steps taken by evolution.

It should be noted that the eye case is not the only instance in which Darwin selects examples from across the taxonomic board. In Chapter 7, he also investigated the case of parasitic behavior of species laying eggs in the nests of other organisms. His example appeals to three different orders of birds (Cuculiformes, Struthioniformes, and Galliformes) and two distinct phyla (birds and bees). This example is also a mixed bag of analogies and homologies (Darwin 1859: 216–218; 1872: 215–216).

This takes us to the second part of Darwin's argument regarding the rise of complex eyes. This time, he will appeal to an analogy between the telescope and the eye, comparing the work of natural selection to the work of an engineer, thus blurring the distinction between nature and human artifice in favor of a single and unified mechanistic outlook (Cornell 1984; Inkpen 2014). By itself, the reference to engineering is interesting: engineers work with an end result in mind (teleology), but even more importantly for our thesis here is the fact that engineers establish a series of linear steps to get to the final result. Moving away from the real world of biology and from the various kinds of eyes he described, Darwin proceeds, in a thought experiment, to an abstraction of the various evolutionary pathways in order to reach a unique, ideal, or linear evolutionary sequence leading from a simple to a complex eye. Suddenly, all the complexity of potentially different evolutionary lineages is condensed and simplified, perhaps suggesting that apparent cases of parallel evolution could well be, ultimately, resolvable by conceiving a single and unified evolutionary stem (descent) driven under the considerable power of natural selection:

> It is scarcely possible to avoid comparing the eye to a telescope. We know that this instrument has been perfected by the long-continued efforts of the highest human intellects; and we naturally infer that the eye has been formed by a somewhat analogous process... If we must compare the eye to an optical instrument, we ought in imagination to take a thick layer of transparent tissue, with a nerve sensitive to light beneath, and then suppose every part of this layer to be continually changing slowly in density, so as to separate into layers of different densities and thicknesses, and with the surfaces of each layer slowly changing in form. Further we must suppose that there is a power always intently watching each slight accidental alteration in the transparent layers; and carefully selecting each alteration which... tend to produce a distincter image... [N]atural selection will pick out with unerring skill each improvement. (Darwin 1859: 188–189; see also 1872: 146)

On the basis of this example, one may want to imagine a directional road opening up under an all-powerful agent called natural selection, slowly climbing all the levels of complexity across evolutionary grades. In order to help us conceive the nearly infinite number of missing transitions forged by natural selection on the road from simplicity to complexity—in an organ as perfect as an eagle's eye, for instance—Darwin had this say to his reader: "His reason ought to conquer his imagination" (1859: 188; see also 1872: 146). The eye-telescope case struck Darwin's contemporary, William Whewell (1864: 15), as merely allowing us to *imagine* evolutionary transitions and not *actual* transitions (see also Hull 2009: 188–189).

On the one hand, Darwin may have wanted to argue, as he did no less than four times in Chapters 6 and 7, that *Natura non facit saltum*, that is, we see everywhere in nature continuities or gradations in structures and behaviors (Darwin 1859: 194, 206, 210, 243; 1872: 156, 166, 234). On the other hand, he could well have tried to display the immense power of natural selection in seizing such gradations, as he did, in cases such as the building of efficient and economical honey cells by bees or the rise of apparently non-adaptive behaviors like sterility, worker castes, and altruism in social insects (1859: 224–242; 1872: 220–233). In fact, in a passage explicitly addressing the "power" of natural selection (Darwin actually used that word), he writes: "But I am bound to confess, that, with all my faith in this principle, I should never have anticipated that natural selection could have been efficient in so high a degree, had not the case of these neuter insects convinced me of the fact" (1859: 242; see also 1872: 233). Armed with the brute material (gradations) as well as the all-powerful force (natural selection) to mold it, Darwin's "imagination" (rather than "reason") was boundless in his search for possible evolutionary roads. As we have seen, Darwin was perhaps only too happy to entertain some ambiguities as to the real phylogenetic significance (descent) of his presented cases. Indeed, the least that can be said is that Chapters 6 and 7 of the *Origin* are not lacking in confusions for the reader. Here's why:

1. Cases appealing to quite different taxonomic levels are mixed up within the same page or few pages.
2. Many cases are presented without any clear precision as to the exact taxonomic levels implicated.
3. The nature of the relationship between "descent" and "natural selection" is modulated from case to case, oscillating between a delicately balanced one and one overstressing natural selection over descent.
4. Analogies and homologies are confusingly or indiscriminately exploited throughout the chapters.
5. The reader is presented with various cases illustrating either likely transitions among very closely related forms, possible transitions among related forms, unclear transitions among unrelated forms, or merely "diversified means of transition" among entirely unrelated forms (to use Darwin's own cryptic words).

4.4 The Homology-Analogy Spectrum

Was this ambiguity, or confusion, engendered by design? A good case could be made that it was Darwin's strategy, among other objectives, to mix the explanatory variables presented above in the hope of transferring, in the mind of the reader, the plausibility of cases devoted to closely related forms at low taxonomic levels to the improbability (if not impossibility) of cases concerned with unrelated forms separated at high taxonomic levels. Against this confused background, the way was open for Darwin to make suggestions as to the transitional steps taken in evolution. These suggestions came under the shape of simplified illustrations: the evolutionary complexity associated with analogies and the lack of relatedness in general was neglected in favor of directional pathways whose explanatory goals could only have been to indicate possible phylogenetic roads taken by natural selection. Several of Darwin's suggestions developed in Chapters 6 and 7, albeit sometimes implicitly only, are presented below. Let it be noted that the sequence of Darwin's presentation is respected in these selected cases, beginning with Chapter 6:

1. From ground/tree-living squirrels to gliding squirrels (also called flying squirrels) (Darwin 1859: 180–181; 1872: 139).
2. From terrestrial Lemuridae, to gliding Galeopithecus (flying lemurs), to flying bats (Darwin 1859: 181; 1872: 139–140).
3. Transitions in birds from wings with no purpose (Apteryx) to wings used as sails (ostriches). At that junction, wings could eventually be used either as flying devices for most birds, or as devices adapted to aquatic milieus in a sequence going from flappers (logger-headed ducks) to fins (penguins) (Darwin 1859: 182; 1872: 140).
4. From entirely aerial petrels to petrels adapted to the aquatic milieu (diving and swimming), or the reverse (Darwin 1859: 184–185; 1872: 142).
5. The case of the rise of complex eyes from an original simple state. The argumentative strategy comes in two parts: a selection of various levels of complexity among widely different groups of living animals (1859: 186–188; 1872: 144–146), followed by an entirely hypothetical reconstruction of a directional evolution of the eye's architecture based on the eye-telescope analogy (1859: 188–189; 1872: 146).

Turning now to some cases from Chapter 7 of the *Origin*, the following examples are presented:

1. The transition from most birds laying their eggs in their own nests to those that have developed the parasitic behavior of exploiting nests of other organisms, either imperfectly (as in ostriches) or more perfectly (as in cuckoos) (1859: 216–219; 1872: 212–216)
2. The transition in the genus *Formica* going from ants taking no slaves to ants receiving some help from their slaves to ants doing no work at all since it is entirely performed by their slaves (1859: 219–224; 1872: 216–220)
3. The transition from humble-bees that merely make cocoons to hive-bees that make perfect and optimal comb cells, with an intermediate stage seen in the species *Melipona domestica* (1859: 224–235; 1872: 220–227)[2]

The phylogenetic value of these examples varies greatly, ranging from cases significantly based on homologies to cases substantially appealing to analogies, with

---

[2]Interestingly, Darwin's strategy in his bee case is similar with the one used in the eye-telescope analogy. First, he looked for some possible transitions in nature and then turned to an intellectual abstraction which consisted in proposing a linear sequence based on geometrical considerations and experimentation.

some other cases being mere imaginary scenarios. Interestingly, Darwin did acknowledge the reality of parallel evolution (analogies) in Chapter 6 of the *Origin*. In a passage so typical of Darwin's convoluted way of presenting arguments, he writes:

> The electric organs offer another and even more serious difficulty [when it comes to finding transitions]; for they occur in only about a dozen fishes, of which several are widely remote in their affinities. Generally when the same organ appears in several members of the same class, especially if in members having very different habits of life, we may attribute its presence to inheritance from a common ancestor; and its absence in some of the members to its loss through disuse or natural selection. But if the electric organs had been inherited from one ancient progenitor thus provided, we might have expected that all electric fishes would have been specially related to each other. Nor does geology at all lead to the belief that formerly most fishes had electric organs, which most of their modified descendants have lost. The presence of luminous organs in a few insects, belonging to different families and orders, offers a parallel case of difficulty. Other cases could be given; for instance in plants, the very curious contrivance of a mass of pollen-grains, borne on a foot-stalk with a sticky gland at the end, is the same in Orchis and Asclepias—genera almost as remote as possible amongst flowering plants. In all these cases of two very distinct species furnished with apparently the same anomalous organ, it should be observed that, although the general appearance and function of the organ may be the same, yet some fundamental difference can generally be detected. I am inclined to believe that in nearly the same way as two men have sometimes independently hit on the very same invention, so natural selection, working for the good of each being and taking advantage of analogous variations, has sometimes modified in very nearly the same manner two parts in two organic beings, which owe but little of their structure in common to inheritance from the same ancestor. (Darwin 1859: 193–194; see also 1872: 150–151)

This rich passage can be broken down into distinct arguments following the logical sequence below (with the addition of comments from the present author):

1. The independent rise of similar organs is a biological reality. This, Darwin recognized.
2. Such occurrences can affect forms as closely related as those belonging to orders or even families. Whereas Darwin could probably live with analogies in forms separated by a very wide taxonomic gap, he would not want affiliative signs to be blurred at such low taxonomic levels.

Analogies could thus be a real problem for Darwin's theory. That is why he offered mitigating arguments:

3. Generally, so-called "similar" organs forged by natural selection are not as similar as it might seem, considering that they are separated by some detectable "fundamental differences." Implicitly, Darwin here argues that natural selection cannot forge absolutely identical organs in unrelated forms. That admission would have opened the door to claims such as: it is impossible to trace descent since homologies may just be analogies disguised by natural selection. Darwin thus attempts to reinstitute the importance of the notion of "descent" for his theory.
4. Analogies exist, but they affect unusual or "anomalous" organs. Darwin's implicit argument here is most convenient, since analogies suddenly become more easily detectable by their unusual and marginal state. By extension

"normal" organs—that is, the large majority of them according to Darwin—can only be homologies, thus allowing us to follow genealogical connections.
5. Generally, the same organs that appear among members of the same class are explainable by common ancestry. By definition, continues Darwin, class members are largely bound in homologies, thus confining analogies to marginality.

4.5 The Unity of Type (Descent) Versus the Conditions of Existence (Natural Selection)

What clearly emerges from the last passage cited above is Darwin's desire to limit the impact of parallel evolution on his theory of evolution, hence his attempt at reinstituting the explanatory power of the notion of "descent" over "natural selection" within each class. If this is indeed the case, Darwin's so-called suggestions about some possible transitional steps taken by evolution become much more than mere suggestions in his mind. Rather, they are seen as powerfully illustrative of phylogenetic trends. Clearly, we are far from the stated method that proofs for transitional forms must be sought among closely related forms, especially among related species and genera (see Darwin 1859: 187, 210; 1872: 144, 207). The problem for Darwin's theory, however, comes from the fact that this argumentation is made in the two chapters devoted to documenting numerous transitional links and directional evolutionary sequences (as listed above), which showcase the power of natural selection across a wide taxonomic range. By so doing, Darwin had to rely more often than he cared to admit on collections of analogies, thus allowing "natural selection" to overcome "descent." Darwin concludes Chapter 6 of the *Origin* by stating that:

> It is generally acknowledged that all organic beings have been formed on two great laws—Unity of Type, and the Conditions of Existence. By unity of type is meant that fundamental agreement in structure, which we see in organic beings of the same class, and which is quite independent of their habits of life. On my theory, unity of type is explained by unity of descent. The expression of conditions of existence... is fully embraced by the principle of natural selection. For natural selection acts by either now adapting the varying parts of each being to its organic and inorganic conditions of life; or by having adapted them during long-past periods of time... Hence, in fact, the law of the Conditions of Existence is the higher law; as it includes, through the inheritance of former adaptations, that of Unity of Type. (Darwin 1859: 206; see also 1872: 166–167)

I share Timothy Shanahan's (2011) puzzlement at this remarkable statement. In this chapter, we have tried to outline its logical and epistemological implications for Darwin's theory. In several places of Chapters 6 and 7, Darwin certainly seems to think that natural selection is the higher law of the two. But it is mistaken on his part to assume, as suggested in his last quote, that his theory in the *Origin* is founded on such a hierarchy of laws. On the contrary, most of his *magnum opus*, as will be seen, is devoted to showing that natural selection constitutes a fairly weak force, one that presumably allows the observer to trace genealogical connections within each class. Darwin wanted to have it both ways in Chapters 6 and 7: directional-transitional series *and* traces of affiliation. This, however, he achieved at the price of constantly meddling with the nature of the relationship between the Unity of Type (descent) and the Conditions of Existence (natural selection).

4.6 Conclusion

It is worth remembering Darwin's horizontal approach to evolution: to travel in geographical space is to travel in geological time. In Chapters 6 and 7 of the *Origin*, therefore, Darwin achieves another of his rhetorical tour de force: the reader leaves them with the impression that evolution is filled with vertical evolutionary lines ascending the scale of complexity through a series of transitional steps. In reality, Darwin had no recourse to the fossil record. Instead, he selected among currently living forms organs or species which, when aligned next to each other, might be indicative of a temporal evolutionary sequence. But this temporal illusion was fragile. We can certainly grant Darwin that transitional states observed among very closely related forms—as seen in cases involving the squirrels, the petrels, the ants, and the bees—are suggestive of phylogenetic sequences. In a sense, this is unsurprising, considering that these forms are united in reproductive networks at low taxonomic levels (varieties, species, and genera), as documented in Chapters 1–5 and 8 of the *Origin*. The difficulty for Darwin was in explaining how entities bound in reproduction could irreversibly break free from one another. To be successful in creating a temporal or geological dimension, Darwin's theory had to be applied beyond the related entities at such low taxonomic levels. At the same time, the more Darwin tried to cast his taxonomic net wide by climbing the taxonomic scale of related forms (such as families, orders, classes, and phyla), the more he moved beyond the obvious homologies shared within reproductive networks, thus entering the territory of analogies. Darwin faced a double peril: (1) the more he extended the horizontality of his view (taxonomic breadth), the less its verticality became credible; (2) the more he stressed vertical cases by insisting on an all-powerful natural selection, the more he undermined his theory, which was in critical need of a delicate balance between "descent" and "natural selection."

Although Darwin wanted to convince himself of the relative insignificance of analogies and parallel evolution, the fact remains that the issue of the independent rise of similar features could not be evaded. A careful analysis of Chapters 6 and 7 of the *Origin* offers a picture of evolution quite different from Darwin's pan-divergent view. Indeed, the overall picture looks more like a series of parallel and independent evolutionary lines rushing to acquire similar features across the taxonomic spectrum: different land animals acquiring flying abilities (flying lemurs, bats, and lemurs); various developments of parasitic behaviors in birds and bees; the rise of variously complex eyes in various lines (starfish, insects, reptiles); the appearance of electric organs in a dozen of remotely related fish; and the presence of luminous organs in few insects also remotely related. This time, the most convincing theory in these chapters concerns "parallel and directional evolution," not "divergence."

Chapters 1–7 of the *Origin of Species* offer to the careful reader two competing and incompatible theories. It is no surprise, then, that the recent historiography has captured important pieces of both theories, as seen in the Introduction of this book in "teleological/mechanism" (Ruse) and "atomism-mechanism" (Greene). Although these two historiographical interpretations are not necessarily incompatible with one another, the way Darwin exploited his two theories in the *Origin* make them so. In the end, unable to extract himself from the horizontality of his view, Darwin resorted to the subterfuge of creating a vertical one. The Gordian Knot of the seamless fabric of reproductive networks was thus cut with the knife of a vertical illusion. This was the inevitable epistemological constraint encountered on an intellectual path taking us from John Ray and Carl von Linnaeus to Charles Darwin, one that could only lead to an incomplete transition to evolutionism as will be seen in the following chapters of this book.

References

Arendt D (2003) Evolution of eyes and photoreceptor cell types. Int J Dev Biol 47:563–571

Bowler PJ (1990) Charles Darwin: the man and his influence. Basil Blackwell, Oxford

Cornell J (1984) Analogy and technology in Darwin's vision of nature. J Hist Biol 17:303–344

Darwin C (1859) On the origin of species. John Murray, London

Darwin C (1872) The origin of species, 6th edn, with additions and corrections [1876]. John Murray, London

Ellegard A (1990 [1958]) Darwin and the general reader: the reception of Darwin's theory of evolution in the British periodical press, 1859-1872. University of Chicago Press, Chicago

Gehring W (2004) Historical perspective on the development and evolution of eyes and photoreceptors. Int J Dev Biol 48:707–717

Glaeser G, Paulus HF (2015) The evolution of the eye. Springer, Dordrecht

Greene JC (1957) Objectives and methods in intellectual history. Miss Val Hist Rev 44:58–74

Greene JC (1959) The death of Adam: evolution and its impact on Western thought. Iowa State University Press, Iowa

Greene JC (1971) The Kuhnian paradigm and the Darwinian revolution in natural history. In: Roller D (ed) Perspectives in the history of science and technology. University of Oklahoma Press, Norman, pp 3–25

Greene JC (1981) Science, ideology, and world view. University of California Press, Berkeley

Greene JC (1986) The history of ideas revisited. Rev Synth 3:201–227

Hodge MJS (1977) The structure and strategy of Darwin's 'long argument'. Br J Hist Sci 10:237–246

Hodge MJS (1989) Darwin's theory and Darwin's arguments. In: Ruse M (ed) What the philosophy of biology is: essays dedicated to David Hull. Kluwer, Dordrecht, pp 163–182

Hull DL (2009) Darwin's science and Victorian philosophy of science. In: Hodge MJS, Radick G (eds) The Cambridge companion to Darwin, 2nd edn. Cambridge University Press, Cambridge, pp 173–196

Inkpen SA (2014) 'The art itself is nature': Darwin, domestic varieties and the scientific revolution. Endeavour 38:246–256

Land MF, Fernald RD (1992) The evolution of eyes. Annu Rev Neurosci 15:1–29

Land MF, Nilsson DE (2012) Animal eyes, 2nd edn. Oxford University Press, Oxford

Mayr E (2001) What evolution is. Basic Books, New York

Mivart SG (1871) On the genesis of species. Macmillan, London

Oakley TH (2003) On homology of arthropod compound eyes. Integr Comp Biol 43:522–530

Ruse M (1979) The Darwinian revolution. University of Chicago Press, Chicago
Salvini-Plawen LV (2008) Photoreception and the polyphyletic evolution of photoreceptors (with special reference to Mollusca). Am Malacol Bull 26:83–100
Salvini-Plawen LV, Mayr E (1977) On the evolution of photoreceptors and eyes. Evol Biol 10:207–263
Shanahan T (2011) Phylogenetic inertia and Darwin's higher law. Stud Hist Phil Biol Biomed Sci 42:60–68
Whewell W (1864) Astronomy and general physics considered with reference to natural theology, 7th edn. Pickering, London

Chapter 5
An Attempt at Taming Natural Selection with Convergence Backward in Time, Part I (*Origin*, Chaps. 11 and 12)

Abstract In Chapters 11 and 12 of the *Origin of Species* (as well as in Chapters 9 and 10), Darwin is busy refocusing his theory on the principle of divergence, after his theoretical drift in Chapters 6 and 7. This is done through investigations of closely related forms at low taxonomic levels assumed to have been derived from unique centers of dispersal. Yet, a close inspection of Darwin's approach reveals his pan-divergent view of evolution was too rigid to accommodate the full complexity of evolution, forcing him to give up on the strong contingency thesis in favor of a much weaker version. This is seen in Darwin's attempt to fit the pattern-process of evolution in an ideal model in which explanatory variables are symmetrically organized, whereby a certain quantity of geological time is proportional to a certain level of taxonomic separation; a certain distance in geographical space is proportional to a certain amount of selective pressure; a certain level of taxonomic separation is proportional to a certain distance in geographical space, and so on. What Darwin was pursuing was the preservation of the core of his theory: the twin principles of divergence-gradation. Owing to evolution's complexity, however, it was not always possible for Darwin to subscribe to his ideal model, forcing him to adjust explanatory variables. An analysis of his utilization of this ideal model allows us to establish that Darwin's theory contains two different components: (1) unmovable or irrefutable explanatory variables that could not be modified, since they formed part of the theory's core (divergence and gradation), and (2) flexible variables that could be adjusted to protect the theory's core through the postulated taxonomic level, the assumed geological depth, the geographical distance evaluated, and the selective pressure postulated. Darwin failed to make sense of the full complexity of evolution because he tried to save the core of his theory at all costs. For the same reason, we argue that "natural selection" is not part of his theory's core, being merely an auxiliary explanatory component that should be jettisoned if necessary to protect that core.

Keywords Biogeography · Principle of divergence · Principle of gradation · Core of theory · Auxiliary explanations · Imre Lakatos · Ideal explanatory model · Proportional explanatory model · Symmetrical explanatory model · Strong contingency thesis · Weak contingency thesis

R. G. Delisle, *Charles Darwin's Incomplete Revolution*, Evolutionary Biology – New Perspectives on Its Development 1, https://doi.org/10.1007/978-3-030-17203-9_5

By the time the reader enters the last part of the *Origin*, Darwin's theory of evolution centered on vertical ascent vanishes entirely. From now on he will focus exclusively on proposing a pan-divergent view of evolution. Skipping over Chapters 9 and 10 of the *Origin* (1859), which have already been considered in Chap. 2 of this book and revisited in Chap. 7, let us go straightaway to Chapters 11 and 12. In the recent historiography (Ruse 1979: 195–196; Bowler 1990: 122), these are often presented as intended to support Darwin's theory with biogeography. Nowhere in the *Origin* is the motto "to travel in geographical space, is to travel in geological time" more obviously applied than in these two chapters (see also Browne 1983; Moore 2005; Bowler 2009). Darwin is busy travelling the world compiling arguments in favor of few fundamental notions. First, a clinal distribution is observed when one travels across continents with closely related variety or species replacing each other in geographical space. Second, the geographical distribution of closely related varieties or species can often be explained by postulating the existence of a unique center of dispersal from which they all radiated. Third, geographical obstacles of all sorts (lakes, mountains, oceans, islands, landscapes, etc.) impact the distribution of varieties or species and the nature of the evolutionary pressure they face.

The essential point Darwin tries to make in Chapters 11 and 12 is the following: *the geographical distribution of closely related entities at low taxonomic levels observed in the world today can be explained by postulating the existence of a common ancestor confined in space and from which they sprung.*

This historiography, however, should be expanded also to include other important and largely implicit explanatory components. For instance, the implied polarity of the time dimension is reversed in these two chapters. In Chapters 1–5 of the *Origin*, Darwin worked to convey the impression of an evolutionary process going forward in time with the divergence of closely related entities ascending the taxonomic scale (varieties, species, genera), as illustrated, for instance, in the work of pigeon fanciers. There, Darwin seems to tell the reader that evolution is on the march right before our eyes. This sense of evolution going forward in time is powerfully reinforced in Chapters 6 and 7 through cases of transitional traits and forms transcending the evolutionary grades or escalating the levels of complexity, irrespective of the fact that it undermined Darwin's pan-divergent view of evolution. By the time the reader gets to the two paleontological Chapters (9 and 10), the perspective on time is reversed. Darwin now projects himself backward in time, assuming that everything important about the past would be registered in today's taxonomy, with extinct forms either being aligned with or falling in-between extant forms. With Chapters 11 and 12 of the *Origin*, the reversal of the time polarity becomes even more obvious, as Darwin travels in geographical space in search of unique centers of dispersal among closely related varieties and species. The same approach is applied in Chapter 13 with some specificities of its own, as will be seen in the next chapter of this book.

Chapters 11 and 12 are therefore quite a contrast when compared to Chapters 6 and 7. Whereas in the latter chapters Darwin presented "natural selection" as a brute force which nothing could resist, the former two chapters re-establish the delicate balance between "natural selection" and "descent," with the power of natural selection being severely tamed under a reaffirmed explanatory component called descent. Although Darwin implicitly wavered between competing and incompatible theories in Chapters 1–7 of the *Origin*, the last part of his book (Chapters 9–13) slightly clarified the issue by discarding the theory that appealed to a vertical approach by reinforcing a horizontal one. In fact, the notions of evolutionary grades and directionality are so weakened in Chapters 11 and 12 that Darwin offers a remarkably shallow view of the evolutionary past. As will be seen, Darwin reviewed cases of closely related varieties and species in search of their unique center of dispersal. As he got ever more ambitious in reuniting varieties or species spread over ever-larger geographical space, Darwin's evolutionary time depth seemed to become shallower and shallower. To illustrate this point, let us turn to hypothetical scenarios.

Let us begin with a question: how can one explain the fact that two very closely related forms separated, taxonomically speaking, only at the variety level can share so many similarities, irrespective of the fact that they live in quite separate regions of the world and in quite distinct environments? Were these two varieties not subjected to the differential action of natural selection ever since they independently radiated from their common cradle, in which case they should have been modified more profoundly than currently observed? In this hypothetical scenario, the more one reunites very closely related varieties and species spread over large territories, the more one is constrained to weaken the explanatory power of natural selection, if the coherence of the theory is to be preserved. Another way to adjust the same scenario to fit the facts would be to claim that the two varieties in question had shared a common ancestor so recently in time that natural selection did not have the time to modify them to any significant degree. What is gained in favor of space in this other version of the scenario is compensated by a reduction in geological depth.

An entirely different scenario could be imagined involving, this time, two closely related species (and not merely two varieties) living in close geographical proximity (rather than far apart). To be consistent with Darwin's theory as presented in Chapters 11 and 12 of the *Origin*, this second scenario could be adjusted either by claiming that the power of natural selection was slightly stronger than postulated in the first scenario—thus explaining that the two forms are separated at the specific level—or by holding that they share a common ancestor that is slightly more ancient, a factor that provides more time for the accumulation of dissimilarities.

These two hypothetical scenarios illustrate clearly how Darwin proceeded in Chapters 11 and 12: *by adjusting a set of explanatory variables to meet his theoretical expectations*. Considering that Darwin mainly worked in these two chapters with closely related forms separated at very low taxonomic levels, it is no surprise that the time depth he postulated in his various examples could only have been a fairly shallow one. Indeed, under Darwin's understanding, differences separating related families and orders require much more geological time to arise than those separating related varieties and species.

Darwin therefore worked by adjusting explanatory variables to meet the expectations of his theory. Before making those expectations more explicit and listing the various explanatory variables used, let us state the thesis that will be defended in this chapter: *Darwin's theory does not genuinely propose an open evolutionary process-*

pattern based on evolutionary contingencies, but rather one elaborated on a fairly rigid framework organized around the principles of divergence and gradualism-gradation, exclusively understood from the present time, and to which all the other explanatory variables are merely subsidiary. Once again, it will be necessary to go beyond Darwin's rhetoric (see Table 2.1) so often repeated in the *Origin of Species* under the implicit or explicit heading of "there is no law of development,"[1] an expression designed to capture the idea that biological evolution is not predetermined, internally organized, driven independently of external conditions, or equally applied to all evolutionary entities:

> These several facts accord well with my theory. I believe in *no fixed law of development*, causing all the inhabitants of a country to change abruptly, or simultaneously, or to an equal degree. The process of modification must be extremely slow. The variability of each species is quite independent of that of all others. Whether such variability be taken advantage of by natural selection, and whether the variations be accumulated to a greater or lesser amount, thus causing a greater or lesser amount of modification in the varying species, depends on many *complex contingencies*—on the variability being of a beneficial nature, on the power of intercrossing, on the rate of breeding, on the slowly changing physical conditions of the country, and more especially on the nature of the other inhabitants with which the varying species comes into competition. Hence it is by no means surprising that one species should retain the same identical form much longer than others; or, if changing, that it should change less. (Darwin 1859: 314; see also 1872: 291) [my emphasis]

This passage constitutes a perfect example of the kind of interpretative distortion that can be introduced into the mind of the modern reader. Schooled in twentieth-century evolutionary biology, the modern reader tends to read into the text what is not necessarily there. Contrary to what we usually assume today, Darwin's view is not about evolutionary contingency in the fullest sense of that expression. If it were, how could the principle of divergence be the irrefutable core of Darwin's theory? The answer to that question will become obvious as our analysis proceeds. For the moment, let us insist on the fact that in the *Origin*, Darwin provides many strong arguments in favor of the notion of "common ancestry." However, considering that he links this notion with another assumed to be strongly complementary, if not synonymous—the principle of divergence—Darwin ends up proposing a rigid view of evolution. Indeed, a group of species may well have sprung from a common ancestor, but why assume that they all had to diverge from each other? Some of the time, species may have only slightly or very significantly moved apart from each other (divergence), at other times, they may have remained at a fairly constant phenotypic distance from each other (parallelism), and at still other times they may have moved toward each other (convergence) to the point of intersecting. Undoubtedly, that real complexity or contingency would considerably blur the phylogenetic relationships of the entities involved, especially after very long geological periods. This, Darwin will not allow. To admit it would be to recognize that

---

[1]See Darwin (1859: 313–315, 318, 331–332, 343, 351, 408–409; 1872: 290–292, 294, 303–304, 313–314, 319, 360–361).

the present time is significantly blind to the evolutionary past.[2] Instead, Darwin allows for *some* evolutionary contingency but only provided that the separateness of the considered entities is largely registered in today's taxonomy under the organizing principle of divergence.

Another principle considered irrefutable in Darwin's theory is the notion of gradualism-gradation. Darwin repeatedly claimed that his theory would fail if innumerable transitional states did not pervade nature everywhere: in connecting various related groups, in crossing evolutionary grades, or in the rise of complex organs.[3] He even used the formula "principle of gradation" to express that reality. Ultimately, the epistemological function of this principle is to ensure that the present time is connected in a contiguous manner, without any interruption, to the entire evolutionary past (at least within classes). Armed with the principle of gradualism-gradation, Darwin pursues his investigations under the assumption that there is no gap in the fabric of life: transitional forms must exist today or must have existed in the past. On that principle, Darwin cannot accept the idea that two forms sharing similarities are not, genealogically speaking, somehow connected through a series of past evolutionary transitions, therefore dismissing the significance of evolutionary analogies. Similarly, Darwin could not accept that a phenotypic gap observed between two closely related extant forms might be a real one. That gap had to be filled by as yet unknown extant forms or, ultimately, by unknown forms now extinct. One way or another, Darwin (1859: 409–410; 1872: 361) refuses to contemplate the possibility that the evolutionary pattern-process can generate real gaps in nature: as he often claim, *Natura non facit saltum*. His subscription to gradualism-gradation is yet another reason why Darwin wrestled with the problem of closely related entities bound in common reproductive networks in Chapters 1–5 and 8 of the *Origin*.

5.1 The Structure of Darwin's Pan-Divergent Theory: Levels of Explanation

The notions of common ancestry, unique centers of dispersal, the principle of divergence, and the principle of gradualism-gradation all fuse together at the core of Darwin's theory, a core treated by him as irrefutable by methodological decision—something like the "negative heuristic" of Imre Lakatos (1970: 132–134), an epistemology which forbids any attack on a theory's hard core—and to which all the other explanatory variables must be adjusted in one way or another (see below). As alluded to above, that core ultimately aims at keeping the evolutionary past under the control of the present.

---

[2]This question has already been discussed in Chap. 2 of this book from a different viewpoint.

[3]Darwin (1859: 179, 189–190, 225, 280, 282, 296, 301; 1872: 138, 146–147, 220, 264–266, 278, 282).

Table 5.1 The *Origin of Species*: the structure of Darwin's pan-divergent theory

The unmovable and irrefutable core: the principle of divergence and the principle of gradation. To modify such explanatory variables is to attack the core of the theory.
Variables offering some explanatory flexibility: Every time it was possible, Darwin worked within a *shallow geological timeframe* and at *low taxonomic levels.*
Variables offering substantial explanatory flexibility: It never was a problem for Darwin to postulate migrations over *short or large geographical distances* and to modulate *selective pressure along a weak-strong* spectrum.
An explanatory "bag of tricks" offering endless explanatory flexibility: Among a long list of *ad hoc* explanations and bold speculations, one encounters in the *Origin*: (a) Migrational routes always open or close because of sunken or rising land, strong wind and water currents, and floating icebergs; (b) Selective pressure modulated at will to the point of being totally absent; (c) Appeals to authority.

The nature of Darwin's pan-divergent theory constitutes a collection of explanatory variables playing different roles

When Darwin claims, therefore, that "there is no law of development," this expression must be understood differently than we would today. Given the core of Darwin's theory, as just defined, the notion of evolutionary "contingency" must be seriously qualified. Our suggestion here is to distinguish between:

1. A *strong contingency thesis* characterized by a genuinely open evolutionary pattern-process, as understood today.
2. A *weak contingency thesis* that consists in adjusting a number of explanatory variables in order to fit the facts to the theoretical expectations of Darwin's theory, thus preserving the theory's core.

As already alluded to, Darwin's theory did provide some maneuvering room to allow for some kind of evolutionary contingency, but one incorporated into a fairly rigid interpretative framework comprising four main explanatory components, each with a different level of explanatory flexibility (see Table 5.1).

The Core of Darwin's Theory The core is composed of the principle of divergence, the notions of common ancestry (monophyletism) and a unique center of dispersal, and the principle of gradualism-gradation. These variables offer no flexibility in interpretation. To modify them is to attack the core of Darwin's theory. As such, they are fixed variables solidly anchoring the evolutionary past to the present.

The Geological Timeframe and the Taxonomic Level Precisely because the core of Darwin's theory is so tightly riveted to the present, its natural comfort zone applies to closely related forms at low taxonomic levels (varieties, species, and genera) that are not very ancient, geologically speaking. A large portion of the *Origin of Species* is devoted to studying that reality. The higher Darwin climbs the taxonomic scale (families, orders, classes, phyla), the less precise he can be about phylogenetic connections (descent). Let us remember that the notion of "common ancestry" need not necessarily be tightly connected to "divergence," a connection that Darwin wants to maintain at all costs. As the taxonomic scale is ascended, traces of affiliations (homologies) become diluted with analogies, as seen in Chapters 6 and 7 of the *Origin*. Since ascending the taxonomic scale is like travelling in geological time, the more Darwin ventures into the geological past, the less the core of his theory can convincingly apply. The innumerable gradations that are assumed to bind closely related forms are lost in the mist of deep time. Indeed, the more two related forms are separated by geological time, the more time natural selection would have had to modify them, thus blurring to some extent their affiliative connections. For these reasons, Darwin applies the two explanatory variables of "time" and "taxonomic level" in preferred states whenever possible: (1) a shallow timeframe is privileged to a deep time and (2) low taxonomic levels are preferred to high ones.

The Geographical Distance and the Amount of Selective Pressure With a set of fixed variables at its core (divergence-gradation) and marked preferences for the parameters of both time (shallow) and taxonomy (low), Darwin's theory lacks explanatory flexibility when it encounters any significant degree of evolutionary contingency. That is why Darwin had no other choice but to shift the weight of explanatory flexibility on two other variables concerned with "geographical distance" and the amount of "selective pressure," respectively. Indeed, Darwin will freely play with the polarity of these two variables to fit the theoretical predictions of his theory. For instance, if two forms look alike, regardless of whether or not they are separated by short or long geographical distances, Darwin will assume that they arose by divergence from a common ancestor originally situated in a localized cradle. As will be seen, the question of how allied forms could have travelled over very long distances is never a serious issue for Darwin. Similarly, Darwin will modulate with ease the selective pressure on related forms in order to offer a credible explanation. When necessary, a strong pressure from natural selection is postulated to explain the differences between two related forms. On the other hand, when two related forms are separated by but very few differences, then a much weaker evolutionary pressure is assumed.

Ad hoc Explanations and Bold Speculations When forced in an awkward position in making the "facts" fit the theory, Darwin will avail himself of even more explanatory flexibility by turning to bolder explanations. For instance, it will be claimed that migratory routes were opened or closed through various means[4]: (a) oscillations in water levels (sunken and rising land bridges and islands[5]; rising rivers flowing into each other); (b) strong water currents or wind currents carrying seeds, insects, and birds or preventing them from reaching some regions; (c) floods, rivers, and sea currents carrying seeds; (d) seeds contained in floating tree trunks; (e) eggs or larvae of shells attached to seaweed or floating timber; (f) seeds or ova carried by birds and fish in their

---

[4]Darwin (1859: 355–365, 370, 381, 384–387, 391, 393–394, 396–397, 399, 401; 1872: 322–330, 333, 341, 344–346, 349, 351, 353–354, 356).

[5]Darwin's view is not as extreme as, for instance, that of Edward Forbes, who argued that all the islands in the Atlantic Ocean were recently connected to Europe, Africa, and America or the view of other scholars who maintained that all continents were recently united in one way or another. On the basis of his gradualism or uniformitarianism, Darwin could not accept such profound geological changes in so little time. See Darwin (1859: 357–358, 396; 1872: 323–324, 352).

digestive tract or in their earth-encrusted feet and beaks; (g) birds eating fish carrying seeds in their digestive tract; (h) floating icebergs carrying seeds or foxes; and (i) whirlwinds blowing fish over some distances. In addition to these various means of dispersal, Darwin (1859: 351, 402; 1872: 319, 356) will also push the variable concerned with selective pressure to its explanatory limit by holding, for instance, that phenotypic stability between two species could be maintained over huge amounts of geological time under the assumption that they are in a perpetual state of competitive equilibrium with each other. If necessary, Darwin (1859: 363, 381; 1872: 328–329, 341) will even fill voids left by his theory by appealing to authorities such as Charles Lyell. Whether or not such speculative explanations are credible is of no concern to us here. Rather, the point is epistemological: Darwin resorted to these explanations to compensate for the theory's lack of explanatory flexibility.

We are now in a better position to evaluate our proposed rapprochement with Lakatos's epistemology. We hope the reader will not be offended if we quote from our own work:

> [I]t could be argued that in light of Chaps. 11 and 12 of the *Origin*, the real explanatory core of Darwin's theory is composed of the principles of divergence and gradation, since both are treated as inflexible, with other variables (such as selective pressure) being used in a very loose way only, in an attempt to make the theory fit the factual reality. To borrow Imre Lakatos's... terminology, I would hold that the dual principle of divergence + gradation represents the irrefutable hard core of Darwin's theory... by methodological decision, with natural selection merely constituting an auxiliary hypothesis, a part of the protective belt whose function is to protect the hard core through a series of adjustments. These adjustments range from a complete lack of selective pressure to a very strong selective pressure. Darwin's competing interpretations of the tree of life are so central to his understanding of evolution that he had to continually adjust the mechanisms accordingly. (Delisle 2017: 154)

Darwin's theory was, therefore, ill-designed to fully accommodate a truly open evolutionary pattern-process. The genuinely flexible variables of his theory (ad hoc explanations, bold speculations, geographical distance, and selective pressure) are constrained in their explanatory function by other variables that are often used in a preferred state of polarity—shallow geological depth over a profound one and low taxonomic levels over higher ones—variables that are themselves located within an entirely inflexible theoretical core composed of the principle of divergence, the notion of common ancestry, the idea of a unique center of dispersal, and the principle of gradualism-gradation. A theory thus structured can accommodate evolutionary contingency only in an incomplete manner. When a theory assumes that the present constitutes the ultimate time horizon for the understanding of the past, it cannot be otherwise.

5.2 A Research Program on Biogeography

In order to see how this interpretative framework was articulated, let us now turn to Darwin's case studies. Darwin's biogeographical arguments are deployed in the following way. First, he attacks the idea that similar environmental conditions house similar life forms, thus breaking the one-to-one relationship sometimes assumed to exist between the conformation of living entities and the milieu in which they live (Darwin 1859: 346–347; 1872: 316–317). As similar as some land conditions are in Australia, South Africa, and western South America between the latitudes 25° and 35°, insists Darwin, it is difficult to find three faunas and floras more dissimilar from each other. Conversely, Darwin continues, it is common to encounter life forms sharing many similarities irrespective of the fact they live under quite distinct physical conditions. By contesting that kind of environmental determinism, Darwin positioned himself against the theory that holds similar life forms are regrouped in geographical provinces (see Bowler 1996: 379–389; Delisle 2007: 135–136).

Second, having broken the life-milieu link, Darwin sought to re-establish a new order by arguing that life forms that share many similarities often live in close geographical proximity, suggesting that such similarities are attributable to their sharing of genealogical connections and common center of dispersal. In ideal cases, this reality is expressed under a clinal distribution, wherein very closely related forms replace each other as one travels across a continent:

> [T]he naturalist in travelling, for instance, from north to south never fails to be struck by the manner in which successive groups of beings, specifically distinct, yet clearly related, replace each other. He hears from closely allied, yet distinct kinds of birds, notes nearly similar, and sees their nests similarly constructed, but not quite alike, with eggs coloured in nearly the same manner. The plains near the Straits of Magellan are inhabited by one species of Rhea (American ostrich), and northward the plains of La Plata by another species of the same genus; and not by a true ostrich or emeu, like those found in Africa and Australia under the same latitude. On these same plains of La Plata, we see the agouti and bizcacha, animals having nearly the same habits as our hares and rabbits and belonging to the same order of Rodents but they plainly display an American type of structure. We ascend the lofty peaks of the Cordillera and we find an alpine species of bizcacha; we look to the waters, and we do not find the beaver or musk-rat, but the coypu and capybara, rodents of the American type... If we look to the islands off the American shore, however much they may differ in geological structure, the inhabitants, though they may be all peculiar species, are essentially American. We may look back to past ages... and we find American types then prevalent on the American continent and in the American seas. We see in these facts some deep organic bond, prevailing throughout space and time, over the same areas of land and water, and independently of their physical conditions... This bond, on my theory, is simply inheritance, that cause which alone, as far as we positively know, produces organisms quite alike, or, as we see in the case of varieties nearly like each other. (Darwin 1859: 349–350; see also 1872: 318–319)

Third, and in addition to inheritance, another factor must be added to the biogeographical equation in order to explain the nature of the distribution of variation: the modulation of selective pressure in accordance with the presence or absence of geographical barriers (Darwin 1859: 347–348, 350; 1872: 317–319). In regions where migration is not obstructed, as in the nearly connected lands of the northern parts of Earth, related forms show many similarities. In regions separated by obstacles—as seen on land by continuous mountain ranges, large rivers, great deserts, and isolated islands; in the sea by the general configuration of the lands—related forms show more dissimilarities. Topography, widely construed, creates the preconditions for modifying life by cutting off, in part or entirely, the expansion and migratory routes of related forms and by allowing natural selection to act on them under these modulated conditions. This being said, Darwin makes it clear that he believes that selective pressure is less affected by the physical milieu itself than by the competition of forms living under such topographical conditions. Darwin was quite explicit about the objectives he pursued in Chapters 11 and 12 of the *Origin*. His research program was thus described:

> ...I will discuss a few of the most striking classes of facts; namely, the existence of the same species on the summits of distant mountain ranges, and at distant points in the arctic and antarctic regions; and secondly... the wide distribution of freshwater productions; and thirdly, the occurrence of the same terrestrial species on islands and on the mainland, though separated by hundreds of miles of open sea. (Darwin 1859: 354; see also 1872: 321)

On the basis of such case studies, Darwin aims at investigating the following:

> [W]hether the several distinct species of a genus, which on my theory have all descended from a common progenitor, can have migrated (undergoing modification during some part of their migration) from the area inhabited by their progenitor. If it can be shown to be almost invariably the case, that a region, of which most of its inhabitants are closely related to, or belong to the same genera with the species of a second region, has probably received at some former period immigrants from this other region, my theory will be strengthened; for we can clearly understand, on the principle of modification, why the inhabitants of a region should be related to those of another region, whence it has been stocked. (Darwin 1859: 354–355; see also 1872: 322)

It is worth keeping in mind that throughout Chapters 1–5 and 8 of the *Origin*, Darwin wrestled with the problem of closely related entities bound in common reproductive networks, an issue he evaded more than he confronted. Should we be surprised, then, that closely related forms often live in contiguity or geographical proximity? Darwin certainly has a point about unique centers of dispersal. But this fact alone is not sufficient to confirm his pan-divergent view. After all, an alternative theory based on reticulate evolution—the perpetual fission-fusion process of closely related entities at low taxonomic levels, with occasional occurrences of irreversible reproductive breakups at higher levels—is not entirely incompatible with the notion of centers of dispersal.

More important for our thesis here is the fact that Darwin explicitly presents his research agenda as being concerned with very closely related forms at low taxonomic levels. It may well be argued that it was Darwin's intent, as it surely was, to use that narrow taxonomic basis as a mere illustration of what could eventually be achieved in more ambitious cases devoted to higher taxonomic levels. When speaking of Chapters 11 and 12 of the *Origin*, Peter Bowler (2009: 154) is certainly right when he says that "[t]here was to be no attempt to reconstruct the origins and migrations of all the major groups, no comprehensive survey of the geography of life. What Darwin did was to choose a small number of examples of distribution about which enough evidence had become available, and to show that the patterns made perfectly good sense in terms of his theory." But as we have been attempting to show here, it is precisely in the extrapolation from low taxonomic cases to high ones that distortions are introduced into Darwin's theory, in two complementary ways: (1) by overlooking the fact that to travel in geographical space across taxonomic levels is to move along the homology-analogy axis and (2) that to travel in geological time is to move away from today's variations. An evolutionary theory centered entirely on closely related extant forms separated at low taxonomic levels could only act as a distorting lens.

5.3 Case Study: Alpine Plants in the Northern Hemisphere

The explanatory strain put on Darwin's theory in Chapters 11 and 12 is revealed when one analyses the various case studies he presents. While some are more advantageous in the light of his theory, others are more challenging, requiring the calibration of the explanatory variables listed above. Let us begin by considering a favorable case. In it, Darwin tries to explain why so many identical species of alpine plants live both in Europe and North America: "It is indeed a remarkable fact to see so many of the same plants living on the snowy regions of the Alps or Pyrenees, and in the extreme northern parts of Europe; but it is far more remarkable, that the plants on the White Mountains, in the United States of America, are all the same with those of Labrador, and nearly all the same... with those on the loftiest mountains of Europe" (Darwin 1859: 365; see also 1872: 330). These facts, continues Darwin, can easily be explained by a recent geological event he calls the Glacial period: "We have evidence of almost every conceivable kind, organic and inorganic, that within a very recent geological period, central Europe and North America suffered under an Arctic climate" (Darwin 1859: 366; see also 1872: 330). As the climate was modified under that glaciation, opportunities were created for this arctic flora to migrate southward in the valleys and in the mountains. The end of the glaciation forced this flora to retreat back northward, leaving its remnants as prisoners on the tops of mountains, as can be seen today. The identity of this flora across Europe and North America is explained in these terms:

> Thus we can understand the identity of many plants at points so immensely remote as on the mountains of the United States and of Europe. We can thus also understand the fact that the Alpine plants of each mountain-range are more especially related to the arctic forms living due north or nearly due north of them: for the migration as the cold came on, and the re-migration on the returning warmth, will generally have been due south to north. The Alpine plants, for example, of Scotland... and those of the Pyrenees... are more especially allied to the plants of northern Scandinavia; those of the United States to Labrador; those of the mountains of Siberia to the arctic regions of that country... In illustrating what, as I believe, actually took place during the Glacial period, I assumed that at its commencement the arctic productions were as uniform round the polar regions as they are at the present day. (Darwin 1859: 367–369; see also 1872: 331–332)

This case study constitutes a "textbook" example of Darwin's theoretical structure: all the key explanatory variables are adjusted as perfectly as he might have hoped for:

1. The core of the theory is deployed in the fullness of its rigidity: (a) the principle of divergence and the notions of common ancestry and unique center of dispersal explain why an identical arctic flora exists at the top of mountains in Europe and North America; (b) the principle of gradualism-gradation makes sense of the clinal and nearly continuous distribution of neighboring forms, as seen in the greater geographical/biological proximity in three main regions: Scotland-Pyrenees, United States-Labrador, and Siberia (North and South).
2. The two key variables concerned with the geological timeframe and the taxonomic level are presented in their preferred state of polarity: (a) a shallow time horizon restricted to the recent Glacial period and (b) a taxonomic level optimally applied to very closely related forms at very low taxonomic levels. Although Darwin was not very explicit about such levels, one can gather from his usage of the word "identity" that he was talking about related organisms belonging to the same variety or separated at the variety or specific levels (as confirmed in the passage cited below).
3. The other two key variables used by Darwin—geographical distance and amount of selective pressure—were applied, as one would expect, in a loose or flexible manner: (a) whether talking about short distances (flora within Siberia) or considerable distances (flora ranging from the White Mountains to the Pyrenees), the explanation is maintained throughout; (b) having at his theoretical disposition the possibility of applying a weak or a strong selective pressure in order to adjust to the kind of evolutionary change observed, Darwin resorted to postulating both:

> The arctic forms, during their long southern migration and re-migration northward, will have been exposed to nearly the same climate, and, as is especially to be noticed, they will have kept in a body together; consequently their mutual relations will not have been much disturbed, and, in accordance with the principles inculcated in this volume, they will not have been liable to much modification. But with our Alpine productions, left isolated from the moment of the returning warmth, first at the bases and ultimately on the summits of the mountains, the case will have been somewhat different; for it is not likely that all the same arctic species will have been left on mountain ranges distant from each other, and have survived there ever since; they will, also, in all probability have become mingled with ancient Alpine species, which must have existed on the mountains before the commencement of the Glacial epoch, and which during its coldest period will have been temporarily driven down to the plains; they will, also, have been exposed to somewhat different climatal influences. Their mutual relations will thus have been in some degree disturbed; consequently they will have been liable to modification; and this we find has been the case; for if we compare the present Alpine plants and animals of the several great European mountain-ranges, though very many of the species are identically the same, some present varieties, some are ranked as doubtful forms, and some few are distinct yet closely allied or representative species. (Darwin 1859: 368–369; see also 1872: 332)

In this first case study, it should be noted that Darwin made no recourse to bold speculations or ad hoc explanations concerning either migratory routes (means of dispersal) or unusual selective pressure (such as a complete lack of it). This first case is straightforward: nearly contiguous lands, a relatively limited geographical space when compared to the rest of the world, and a clinal distribution in taxonomy correlated with a modulation of selective pressure as forms moved away from their postulated center of dispersal.

5.4 Case Study: The Galapagos Islands

The second case study reviewed here—the now classic example of Darwinian studies about the Galapagos Islands (Darwin 1859: 397–403; 1872: 353–357; see also Lack 1947)—will put some strain on Darwin's theory, requiring him to adjust several explanatory variables. In this other case, Darwin tried to explain the biological facts surrounding the migration of floras and faunas to the Galapagos, an archipelago of ten islands in the Pacific Ocean 500 or 600 miles off the shores of South America (Ecuador). Darwin's explanation can be broken down in the following manner:

1. The inflexible core of Darwin's theory is applied by means of its two main components: (a) the principle of divergence and the notions of common ancestry (monophyletism) and unique center of dispersal are exemplified in the claim that the Galapagos Islands were populated by forms originating in the American continent; (b) the principle of gradualism-gradation is respected, as seen, for instance, in the fact that very closely related species of birds live on both sides of this oceanic divide, with some species remaining identical: "There are twenty-six land birds, and twenty-five of these are ranked by M. Gould as distinct species, supposed to have been created here [the Galapagos]; yet the close affinity of most of these birds to American species in every character, in their habits, gestures, and tones of voice, was manifest" (Darwin 1859: 398). In the last edition of the *Origin* (1872: 354), the level of gradation was increased slightly by recognizing that perhaps as many as five species out of twenty-six are identical across the oceanic divide. That kind of clinal distribution was further reinforced by the recognition that the distinct forms inhabiting the separate Galapagos Islands were more closely related to each other than to other forms living on the continent or elsewhere (Darwin 1859: 400; see also 1872: 355).
2. Darwin provides no geological timeframe for this case study in the *Origin*. Nonetheless, he apparently knew from geologists that the islands had emerged quite recently from the ocean. As stated by Thomas Glick and David Kohn (1996: 21), "Darwin understood that the Galapagos Islands—those isolated volcanoes only recently emerged from the Pacific—were a living laboratory in the early stages of the origin of new species." Indeed, speaking of the formation of the Galapagos Islands and of their relation to the continent, Darwin himself wrote in the second edition of his *Journal of Researches* (1845): "Seeing every height crowned with its crater, and the boundaries of most of the lava-streams still

distinct, we are led to believe that within a period geologically recent the unbroken ocean was here spread out" (Glick and Kohn 1996: 27). This geological shallowness complies with what Darwin prefers for such an explanatory variable. Likewise, the taxonomy appeals to living forms consigned to lower levels (identical or closely related species) rather than at higher ones.

3. We know that Darwin felt at liberty to adjust the explanatory variables concerned with geographical distance (short or long) and the amount of selective pressure (weak or strong) as needed to fit this case study to his theory. He did so by establishing a proportional rapport between geographical distribution and taxonomic proximity, wherein the greater distance between the continent and the Galapagos Islands is reflected in taxonomy when compared to the much shorter distances separating the various islands from each other:

   > Thus the several islands of the Galapagos Archipelago are tenanted... in a quite marvellous manner, by very closely related species; so that the inhabitants of each separate island, though mostly distinct, are related in an incomparably closer degree to each other than to the inhabitants of any other part of the world. And this is just what might have been expected on my view, for the islands are situated so near each other that they would almost certainly receive immigrants from the same original source, or from each other. (Darwin 1859: 400; see also 1872: 355)

   Since islands are separated by water, Darwin had to address the means by which the migratory process was sustained, keeping in mind that they were populated by birds, reptiles, mammals, and plants, all assumed to ultimately have arrived from the nearest continental land: "...it is obvious that the Galapagos Islands would be likely to receive colonists, whether by occasional means of transport or by formerly continuous land, from America" (Darwin 1859: 398–399). Unfortunately, Darwin was anything but explicit in this case study about the means of dispersal of this flora and fauna (a list of possible means was already provided above). To close his case, Darwin needed to tackle the issue of the evolutionary drive behind this entire colonization process. For that, he felt obliged to acknowledge a peculiarity accompanying this case study: although species living on the Galapagos Islands are very closely related to each other (excluding those that are identical), their dissimilarities prove difficult to explain considering that several islands are situated within sight of each other and are similar in their physical characteristics (in geology, climate, and altitude). Should not these species all be identical too? Here, Darwin avails himself of the explanatory flexibility that comes with the notion of "amount" of selective pressure. Biotic conditions are always more important than abiotic ones for Darwin, and so he argues that each island was randomly colonized by different kinds of organisms, thus creating as many different selective milieus as they are islands (Darwin 1859: 400–401; 1872: 355–356). The selective pressure on the various islands was, the argument goes, stronger than could have been deduced by looking exclusively at their geographical proximity, which explains why forms are separated at the species level rather than being somewhat identical.

4. The Galapagos Islands did put some strain on Darwin's theory, however. As some biological explanations were more obvious than others, Darwin had to

calibrate explanatory variables in such a way that he was forced toward more elaborate speculations and ad hoc explanations. In this case study, issues concerning geographical distance and selective pressure are stretched a bit. In the first edition of the *Origin*, Darwin (1859: 398–399) claimed that the Galapagos Islands had received colonists from the American continent by the following means: "…it is obvious that the Galapagos Islands would be likely to receive colonists, whether by occasional means of transport or by formerly continuous land, from America." The prior existence of some kind of land bridge would constitute a more robust and convincing explanation than one appealing to occasional means of dispersal. But where are the facts for this "better" explanation? Are we aware of sunken islands in intermediate position between the continent and the Galapagos Islands that could have been exposed during fluctuations of sea levels? Unfortunately for Darwin, the most convincing explanation was also the most speculative. This may explain why Darwin (1872: 354) clearly retreated from this explanation, reproducing the same passage in the last edition of the *Origin* but with the following addition: "…it is obvious that the Galapagos Islands would be likely to receive colonists from America, whether by occasional means of transport or (though I do not believe in this doctrine) by formerly continuous land." It is unclear whether Darwin ever seriously contemplated the idea of a land bridge in the case of the Galapagos Islands. The ambiguity is even found in the second edition of his *Journal of Researches* (1845), where Darwin wrote: "Seeing every height crowned with its crater, and the boundaries of most of the lava-streams still distinct, we are led to believe that within a period geologically recent the unbroken ocean was here spread out" (Glick and Kohn 1996: 27). Spread out up to what point? Beyond the Galapagos Islands and toward the mainland? One thing is certain: Darwin eventually dropped any allusion to the idea of a land bridge, at once discarding what would have been a most convincing yet speculative explanation. In so doing, he chose to place all the explanatory stress on the sole explanation of occasional means of dispersal.

There is more to the issue of elaborate speculations and ad hoc explanations in this case study. Darwin will state the problem in these terms: "The really surprising fact in this case of the Galapagos Archipelago… is that the new species formed in the separate islands have not quickly spread to the other islands" (1859: 401; see also 1872: 356). We have already seen how Darwin explained the dissimilarities between species living on different islands: by asserting different biotic selective contexts on each island. Yet, throughout Chapters 11 and 12, Darwin had been highly imaginative, inquisitive, and ingenuous in his suggestions regarding means of dispersal. For him, clearly, the geographical spread of life had never been a problem. Now, Darwin had to offer an explanation as to why a number of species did not freely cross from one island to another, despite living in such close proximity. Several arguments were deployed. First, the physical conditions separating the different islands:

> But the islands, though in sight of each other, are separated by deep arms of the sea, in most cases wider than the British Channel, and there is no reason to suppose that they have at any former period been continuously united. The currents of the sea are rapid and sweep across

> the archipelago, and gales of wind [that could carry organisms?] are extraordinarily rare; so that the islands are far more effectually separated from each other than they appear to be on the map. (Darwin 1859: 401; see also 1872: 356)

It is rather strange that the same conditions that so easily allowed species to colonize the Galapagos Islands from the American continent now prevented them from doing so within the Archipelago itself! Moving away from geographical speculation, Darwin will propose a new set of explanations appealing, this time, to selective issues and adaptation. His key point is the following: species already well adapted to their home ground cannot easily be dislodged:

> But we often take, I think, an erroneous view of the probability of closely allied species invading each other's territory, when put into free intercommunication. Undoubtedly if one species has any advantage whatever over another, it will in a very brief time wholly or in part supplant it; but if both are equally well fitted for their own places in nature, both probably will hold their own places and keep separate for almost any length of time... In the Galapagos Archipelago, many even the birds, though so well adapted for flying from island to island, are distinct on each; thus there are three closely-allied species of mocking-thrush, each confined to its own island... I think we need not greatly marvel at the endemic and representative species, which inhabit the several islands of the Galapagos Archipelago, not having universally spread from island to island. In many other instances, as in the several districts of the same continent, pre-occupation has probably played an important part in checking the commingling of species under the same conditions of life. (Darwin 1859: 402–403; see also 1872: 356–357)

Let us assume for a moment that the reader is willing to accept Darwin's argument. What, then, should he or she do with the numerous other claims found throughout the *Origin of Species* stating that:

(a) No matter how places in the economy of nature are supposedly filled, there is always room for improvement.
(b) Eventually biological variation will arise that will tip the balance in favor of one species over another.
(c) Because closely related species depend on similar resources, fierce competition between them will create the selective conditions for the extinction of some of them, unless they find a way to reduce that competition by continuing to move apart from each other in climbing the taxonomic scale (the competitive exclusion principle)?

Of course, it is within the nature of evolutionary biology, epistemologically speaking, to be flexible enough to accommodate many explanations. This comes with the inherent narrative structure of the field (see discussion below). But the relevant point here is that, convincing or not, Darwin opted to put the explanatory stress of the Galapagos case study on the variables of geographical distance and selective pressure. This left the core of his theory (divergence-gradation) untouched, while the time dimension and the taxonomic level were adjusted in their optimal polarity state.

5.5 Case Study: A Worldwide Dispersal from Around the North Pole

The third case study analyzed here is an extension of Darwin's first case study, which originally concerned the migration of alpine plants from a center of dispersal situated around the Arctic Circle. By this time, Darwin (1859: 369–382; 1872: 332–342) had become much more ambitious, expanding the case to nearly the entire world. In doing so, he took advantage of two basic biogeographical facts. First, the main lands on Earth are nearly connected around the North Pole (1859: 370–371; 1872: 333), with South America, South Africa, and Indonesia-Australia constituting three distinct and isolated continental dead ends. Second, most of the lands and biomass on Earth are housed in the northern hemisphere, the land and biomass south of the Equator being much more marginal (1859: 379; 1872: 340). The stage was set for an attempt to explain a wide range of biogeographical facts observed today by appealing to a dispersal from a common and unique cradle positioned northward in the northern hemisphere.

Before proceeding with our analysis, it is worth remembering the ultimate goal of Chapters 11 and 12 of the *Origin*: to demonstrate that closely related forms seen today have evolved from, and can be traced back to, common ancestors originally localized in single places. In other words, Darwin wished to support the two core components of his theory: (1) the principle of divergence (which includes the notions of common ancestry, monophyletism, and single center of dispersal) and (2) the principle of gradualism-gradation. Epistemologically speaking, this second component is especially important for Darwin's ontology, since gradualism-gradation is the only notion at Darwin's disposal capable of ensuring that the evolutionary past is directly connected to the present (at least within classes). With this notion in place, one can search among extant entities (closely related forms at low taxonomic levels) for traces of *contiguous* connections—should they be called gradations or intermediate links—so that from the present time it becomes possible to trace back an evolutionary past assumed to be tightly knit, that is, a continuous fabric of life deprived of evolutionary gaps, disruptions, or interruptions. Of course, the entire explanatory apparatus is part of Darwin's epistemological agenda, which consists in keeping the evolutionary past under the control of the present, in a world assumed today to be fully constituted. At this stage of our analysis, another question is in order: how deep in geological time can Darwin probe while confining his research to closely related extant forms of low taxonomic levels? The answer is obvious: not very deeply.

This is exactly the impression the reader of Chapters 11 and 12 of the *Origin* receives. The more Darwin attempts to connect extant forms worldwide to unique centers of dispersal situated in the Earth's northern hemisphere, the more his stretched biogeography looks shallow or flat, geologically speaking. What Darwin gained in space, he lost in time. Once again, Darwin's theory seems unable to generate a truly historical or vertical dimension. Life forms do not seem to emerge from a deep past at various localities on the planet; instead, one gathers that they arose from a fairly recent and unique localized common cradle before spreading horizontally across the planet. The main explanatory features of this global case study are the following:

1. With the exception of a single and vague allusion to the pre-Pliocene period (1859: 372; 1872: 334), Darwin's case study is confined to the following geological time horizons: Recent, Glacial period, and Pliocene.
   A distinction is sometimes established between the Older and Newer Pliocene. This means that Darwin's analysis left out most of the geological periods recognized around the mid-nineteenth century (Bowler 2003: 117): Miocene, Eocene, Cretaceous, Wealden, Oolitic, Lias, New Red Sandstone, Carboniferous, Old Red Sandstone, Silurian, Cambrian, and Primary rocks.
2. Darwin tied together as many extant forms as he possibly could to a unique center of dispersal situated northward in the northern hemisphere, with life forms migrating in parallel in the Old and New Worlds in direction of the south.
3. Darwin argued that the main evolutionary drive behind this migratory process was a major shift in worldwide climate that went from hot (Pliocene) to cold (Glacial) to warm (Recent).
4. Darwin organized worldwide floras and faunas along natural zones of habitation: arctic, subarctic, temperate, and tropical.

We have seen that in order to bolster his theory, Darwin had to adjust or calibrate a set of explanatory variables. Let us take one more step toward our thesis in this chapter. With a rigid and irrefutable dual theoretical core founded on divergence and gradation, *the ideal adjustment of all the other variables constitutes a perfectly symmetrical or proportional arrangement between geographical space, geological time, taxonomic level, and selective pressure* (see Fig. 5.1). For instance, as two closely related forms diverge from their common ancestor and unique center of dispersal, the separation created by the geographical and temporal distance should ideally be symmetrical and reflected in a perfect "V" pattern wherein the taxonomic scale is slowly ascended—from a common ancestor to two closely related varieties and to two closely related species—under moderate and continuous selective pressure. Or to put things differently, as two closely related forms diverge from their common ancestor and unique center of dispersal, a moderate and continuous selective pressure should gradually push away the two related lineages, thus instituting a taxonomic ascent that is directly proportional to both the geographical space covered and the geological time elapsed, as seen in a perfect "V" process.

In this *ideal model of symmetrical variables*, pattern and process are so completely fused together that all explanatory variables are directly proportional to one another: a certain quantity of geological time is proportional to a certain level of taxonomic separation; a certain distance in geographical space is proportional to a certain amount of selective pressure; a certain level of taxonomic separation is proportional to a certain distance in geographical space; and a certain amount of selective pressure is proportional to a certain quantity of geological time.

Of course, we know that Darwin was aware that evolution did not comply with that kind of ideal model. That is why his theory allows for some kind of evolutionary

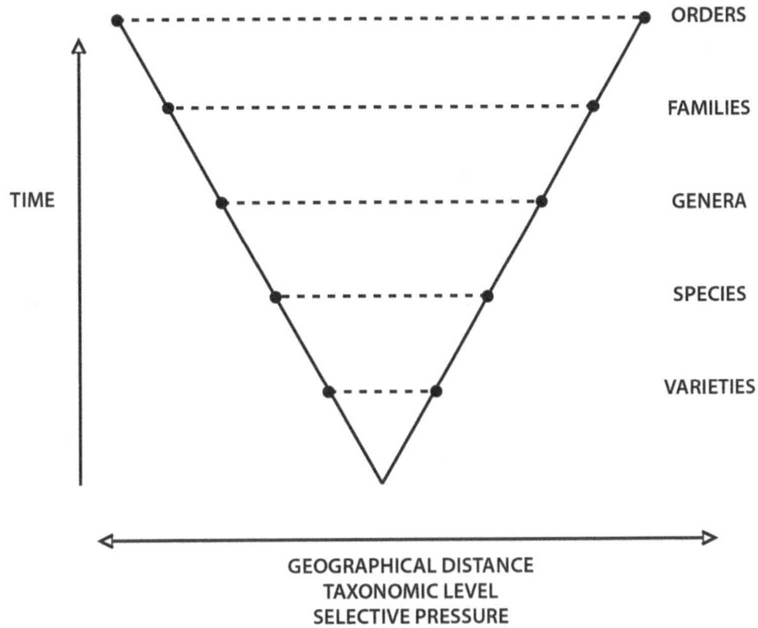

Fig. 5.1 *Charles Darwin's Ideal Model of Symmetrical Variables for Supporting Divergence and Gradation.* In this model designed to support the theory's core (divergence + gradation), pattern and process are so perfectly fused together that all explanatory variables are directly proportional to one another: geological time, geographical distance, taxonomic separation, and amount of selective pressure

contingencies under the *weak thesis* presented above and which consists in adjusting a number of explanatory variables to meet the expectations of the theory. However, this will not prevent Darwin from trying to present matters in terms of an ideal model as often as he can, if only because he is convinced that it is under such conditions that his theory is the strongest or the most persuasive. Knowing this will allow us to take one more step in our understanding of the nature of Darwin's theory:

> The default position of Darwin's theory consists in applying it in the best light of an ideal model with explanatory variables being symmetrical or proportional to each other. When this is not possible because the theory is not complying with the facts, only then will Darwin resort to adjusting some of the variables (knowing that some are fixed and others are flexible but to varying degrees).

It is certainly interesting to note that we ultimately find at the foundation of Darwin's theory an explanatory structure strongly reminiscent of the static world-view and of natural theology as held in the seventeenth and eighteenth centuries—as seen in the notions of order, equilibrium, and symmetry-proportion—and as promoted under different guises by Newton, Robert Boyle, John Ray, and Linnaeus, among others.[6]

It is precisely this kind of ideal model of symmetrical variables that Darwin will use as a template when expanding his case study to a worldwide migration originating from a northern cradle. First is the core of the theory. The idea of a unique center of dispersal incarnates de facto the principle of divergence. Then the notion of gradation is preserved in the idea of a clinal distribution radiating from that center in parallel in both the Old and New Worlds, in direction of the south: identical forms being found around the common Arctic region, different yet closely related varieties being found in the subarctic regions of both continents; different yet closely related species being found in the temperate regions of both continents; and finally a host of forms belonging to allied forms randomly distributed in the rest of the world, such as in South America, Cape of Good Hope, Ceylon, Java, and Australia. That is why Darwin could write: "Hence we see that throughout the world, the plants growing on the most lofty mountains, and on the temperate lowlands of the northern and southern hemispheres, are sometimes identically the same; but they are much oftener specifically distinct, though related to each other in a most remarkable manner" (Darwin 1859: 375–376; see also 1872: 332–342).

In this ideal model, the other explanatory variables are kept symmetrical or proportional to each other. For instance, the more forms migrate away from the northern cradle under the impetus (selective pressure) of a shifting climate going from hot to cold, the further away related forms living in the Old and New Worlds are from each other, spatially speaking, but also taxonomically speaking (from identify, to varieties, to species). Considering that Darwin's case study is not entirely deprived of a geological dimension—albeit a fairly shallow one restricted to the Pliocene and after—proportions prevailing on the strict plane of today's biogeographical distribution are also applied in the time dimension: related extant forms in the Old and New Worlds that are already separated at the species level are held to be so separated on the principle that they diverged from each other in more ancient times than extant forms only separated at the variety level. From such an ideal model based on symmetry-proportion, Darwin will either adjust variables to say attuned to it or will resort to more speculative and ad hoc explanations. Let us investigate this case study more closely by following Darwin on the intellectual road that leads from more reasonable explanations to more speculative ones.

1. When possible, Darwin will take advantage of taxonomic debates and uncertainties surrounding groups of floras and faunas:

   > It should be observed that the northern species and forms found in the southern parts of the southern hemisphere, and on the mountain-ranges of the intertropical regions, are not arctic, but belong to the northern temperate zones... Many of the forms living on the mountains of the warmer regions of the earth and in the southern hemisphere are of doubtful value, being ranked by some naturalists as specifically distinct, but others as

[6]This question is also discussed in Chaps. 3, 6, and 7 of this book.

> varieties; but some are certainly identical, and many, though closely related to northern forms, must be ranked as distinct species. (Darwin 1859: 376)

In order to stay as close as possible to his ideal model, it is understandable why Darwin would prefer to interpret the northern forms living either on the mountains of the intertropical regions or in the southern hemisphere as related distinct species rather than merely distinct varieties. After all, such extant forms are quite far from the postulated original center of dispersal situated in the northern regions of the northern hemisphere. To avoid giving the impression that such space travel had been deprived of selective pressure—being reflected in taxonomic stagnation—Darwin was only too happy to take sides in the taxonomic debate in favor of a specific distinction.

2. For Darwin, calibrating the variables in compliance with an ideal model also meant to incorporate the time dimension, time and space being intertwined. Related extant forms that share more similarities should ideally be more recent than extant forms sharing less similarities, hence a difference in taxonomic levels:

> In illustrating what, as I believe, actually took place during the Glacial period, I assumed that at its commencement the arctic productions [plants] were as uniform round the polar regions as they are at the present day... As on the land, so in the waters of the sea, a slow southern migration of a marine fauna, which during the Pliocene or even a somewhat earlier period, was nearly uniform along the continuous shores of the Polar Circle, will account, on the theory of modification, for many closely allied forms now living in areas completely sundered. Thus, I think, we can understand the presence of many existing and tertiary representative forms on the eastern and western shores of temperate North America; and the still more striking case of many closely allied crustaceans..., of some fish and other marine animals, in the Mediterranean and in the seas of Japan—areas now separated by a continent and by nearly a hemisphere of equatorial ocean. (Darwin 1859: 369, 372; see also 1872: 332, 334)

Here, Darwin established an equation between, on the one hand, *identical* extant plants that sprung from uniform ancestors of the *early Glacial period* and, on the other hand, *closely allied* extant forms exhibiting differences derived from uniform ancestors of the *Pliocene or before*. In both cases, Darwin postulated the hypothetical existence of uniform ancestors formerly living around a unique and common center of dispersal near the Arctic Circle.

3. Among the conceptual tools available to Darwin for trying to meet his ideal model was the calibration of the selective pressure. For instance, extant arctic plants are, for Darwin, quite uniform in conformation, as explained by the presumed existence of a recent common ancestor (early Glacial period) and by their proximity to the original center of dispersal (Polar Circle). At the same time, Darwin had to explain how the uniformity of arctic plants was maintained throughout the long and parallel migration in the Old and New Worlds from north to south (and back) during the climatic shift from hot (Pliocene), to cold (Glacial period), to warm (Recent). Darwin resorted to the concept of an entire flora migrating as a single body, thus maintaining the equilibrium of the selective forces as the process went on:

> The arctic forms, during their long southern migration and re-migration northward, will have been exposed to nearly the same climate, and, as is especially to be noticed, they will have kept in a body together; consequently their mutual relations will not have been much disturbed, and, in accordance with the principles inculcated in this volume, they will not have been liable to much modification. (Darwin 1859: 368; see also 1872: 332)

Precisely because selective pressure is a very flexible variable in Darwin's theory, he is at liberty also to claim, when speaking of temperate floras and faunas, that the differences observed among them are to be explained by the disturbance of their mutual relations. In this example, Darwin imagined a simultaneous migration from the North and South Poles toward the equator (and beyond) under a cooling climate, mixing up floras and faunas and creating new selective conditions:

> Thus, as I believe, a considerable number of plants, a few terrestrial animals, and some marine productions, migrated during the Glacial period from the northern and southern temperate zones into the intertropical regions, and some even crossed the equator... [B]ut the forms, chiefly northern, which had crossed the equator, would travel still further from their homes into the more temperate latitudes of the opposite hemisphere... These being surrounded by strangers will have had to compete with many new forms of life; and it is probable that selected modifications in their structure, habits, and constitutions will have profited them. Thus many of these wanderers, though still plainly related by inheritance to their brethren of the northern or southern hemispheres, now exist in their new homes as well-marked varieties or as distinct species. (Darwin 1859: 378–379; see also 1872: 339–340)

Thus far, Darwin managed to calibrate the explanatory variables in such a way as to stay close to his ideal model of symmetrical variables. But this model will show signs of strain in the face of some "anomalies." As we are about to see, Darwin confronted such anomalies with either silence or ad hoc explanations.

4. In Darwin's northern cradle scenario, the original uniformity of the temperate faunas and floras of the Old and New Worlds that used to live around the Arctic Circle in the Pliocene was broken, as the southward migration began on both sides of the Atlantic Ocean under a cooling climate. Indeed, insists Darwin, not only were temperate forms geographically cut off from each other, but they also had to face new selective conditions when they mingled with other types of flora and fauna already inhabiting the rest of the Old and New Worlds. Under this scenario, temperate forms lost their uniformity by becoming distinct varieties and species. Nevertheless, Darwin was confronted with a greater level of uniformity than expected, an anomaly to which he did not respond: "...when we compare the now living productions of the temperate regions of the New and Old Worlds, we find very few identical species (*though Asa Gray has lately shown that more plants are identical than was formerly supposed*)" (Darwin 1859: 371–372; see also 1872: 333–334) [my emphasis]. How can this greater number of identical plants be explained?
5. Darwin included as many extant forms worldwide as he could in his northern cradle scenario. We have seen that two key biogeographical facts played in favor of this scenario: nearly united lands at the Arctic Circle and isolated dead ends in South America, South Africa, and Indonesia-Australia (even under fluctuating

sea levels). Unfortunately for Darwin, some biogeographical facts in the south hemisphere could not be incorporated into this scenario, as some extant forms in the southern hemisphere could not be linked to northern ones. Darwin (1859: 377–382; 1872: 339–342) therefore had to resort to postulating a more modest and more localized migration from the vicinity of the South Pole in a northerly direction.[7] We say "resort to" here because of the considerable geographical isolation of lands in the region: Antarctica, Australia, South Africa, and South America. For Darwin, then, similarities of floras and faunas had to be explained by more speculative and bolder explanations. For instance, discarding the possibility of shifting continents—"our continents have long remained in nearly the same relative position" (1859: 370; see also 1872: 333)—Darwin turned to drifting icebergs to explain widely isolated plants: "[A]s far as regards the occurrence of identical species at points as enormously remote as Kerguelen Land, New Zealand, and Fuegia, I believe that towards the close of the Glacial period, icebergs, as suggested by Lyell, have been largely concerned in their dispersal" (1859: 381; see also 1872: 341). Darwin's appeal to authority here should be noted in his reference to Lyell. Presumably, Darwin hoped to cushion the impact of criticisms toward such ad hoc explanations devised to support a theory—sharing the risk with Lyell—a theory requiring the adjustment of its explanatory variables for it to meet an ideal scenario based on symmetry.

In other words, to explain how identical forms could have spread to isolated lands separated by large distances within a short amount of time and/or under a lack of selective pressure, Darwin turned to an out-of-the ordinary means of dispersal. Darwin's ad hoc explanation appealing to icebergs had the precise explanatory function of maintaining the application of his theory within an ideal model, as did his postulation of some Antarctic lands and some unspecified southern lands in intermediate geographical position assumed to have formerly housed forms responsible for colonizing lands as far apart as New Zealand and South America:

> New Zealand in its endemic plants is much more closely related to Australia, the nearest mainland, than to any other region: and this is what might have been expected; but it is also plainly related to South America, which, although the next nearest continent, is so enormously remote, that the fact becomes an anomaly. But this difficulty almost disappears on the view that both New Zealand, South America, and other southern lands were long ago partially stocked from a nearly intermediate though distant point, namely from the antarctic islands, when they were clothed with vegetation, before the commencement of the Glacial period. (Darwin 1859: 399; see also 1872: 254–355)

The affinity between some flora found in the Cape of Good Hope and in the south-western part of Australia remains, according to Darwin (1859: 399; 1872: 355), a remarkable fact seemingly defying explanation.

6. Even more revealing of Darwin's epistemological strategy and ontological choices in Chapters 11 and 12 of the *Origin* are the difficulties he encountered

[7]See also Darwin (1859: 399; 1872: 354–355).

in trying to explain a group of quite distinct species of plants currently restricted to the southern hemisphere. Before quoting this important passage in full, it is worth outlining some preliminary issues it raised:

(a) The explicit recognition of a profound problem for Darwin's theory.
(b) The implicit recognition that explanatory variables such as the time dimension and the selective pressure could not be adjusted to support anything close to an ideal southern cradle scenario, hence the obligation for Darwin to turn to a third, vague, and entirely new scenario (the northern cradle scenario being the first one and the southern cradle scenario the second one).
(c) The addition of ad hoc explanations concerning migratory means appealing not only to icebergs this time but also to unknown sunken islands.

After having insisted upon the remoteness of places like Kerguelen Land, New Zealand, and Fuegia, in the southern hemisphere, Darwin writes:

> But the existence of several quite distinct species, belonging to genera exclusively confined to the south, at these and other distant points of the southern hemisphere, is, on my theory of descent with modification, a far more remarkable case of difficulty. For some of these species are so distinct, that we cannot suppose that there has been time since the commencement of the Glacial period for their migration, and for their subsequent modification to the necessary degree. The facts seem to me to indicate that peculiar and very distinct species have migrated in radiating lines from some common centre; and I am inclined to look in the southern, as in the northern hemisphere, to a former and warmer period, before the commencement of the Glacial period, when the antarctic lands, now covered with ice, supported a highly peculiar and isolated flora. I suspect that before this flora was exterminated by the Glacial epoch, a few forms were widely dispersed to various points of the southern hemisphere by occasional means of transport, and by the aid, as halting-places, of existing and now sunken islands, and perhaps at the commencement of the Glacial period, by icebergs. By these means, as I believe, the southern shores of America, Australia, New Zealand have become slightly tinted by the same peculiar forms of vegetable life. (Darwin 1859: 381–382; see also 1872: 241–342)

What this passage reveals above all is Darwin's inability to countenance extant forms without other modern day equivalents. It should be remembered that Darwin did his best to organize extant related forms along a clinal distribution in both the Old and New Worlds, relating them to common ancestors originally found around the Arctic Circle. When extant forms could not be explained under this northern cradle scenario, Darwin turned to a more modest and limited scenario based on a radiation from a southern cradle. In both cases, Darwin could apply the full horizontality of his approach: extant forms could be related to each other through an uninterrupted gradation of contiguous forms, as one travels away from a presumed center of dispersal: from uniformity, to distinct varieties, to distinct species. In these two scenarios, Darwin appealed to a geological time dimension merely as a means to explain how related forms could have travelled in geographical space: the further away from the cradle, the greater amount of time required, and the greater the taxonomic difference.

In his case studies, Darwin made sure that the horizontality of the present time would be the organizing principle of the time dimension. This is precisely what seems to have gone wrong with this unexpected and scattered group of quite distinct species of plants confined to the southern hemisphere. They seemed to have made a random and sudden irruption from a deeper and unclear past. Indeed, this group of quite distinct species of plants introduced a disruption and discontinuity in the gradation and contiguity of life's fabric, contesting Darwin's core principle of gradualism-gradation. That principle, it will be remembered, serves the key epistemological role of keeping the fabric of life tightly connected to the present time. If one can detect among extant forms isolated species having no genealogical ties to other such forms, then the present becomes powerless to investigate the evolutionary past of those species. In opposition to Darwin's *horizontal approach* as exemplified in his first two scenarios, one may have to turn to a *vertical approach* to give sense to the evolutionary past of such isolated and unusual species. In fact, one need not go very far to get a good description of Darwin's horizontal approach in Chapters 11 and 12 of the *Origin*. At the very end of Chapter 11, he writes:

> I believe that the world has recently felt one of his great cycles of change; and that on this view, combined with modification through natural selection, a multitude of facts in the present distribution both of the same and of allied forms of life can be explained. The living waters may be said to have flowed during one short period from the north and from the south, and to have crossed at the equator; but to have flowed with greater force from the north so as to have freely inundated the south. As the tide leaves its drift in horizontal lines, though rising higher on the shores where the tide rises highest, so have the living waters left their living drift on our mountain-summits, in a line gently rising from the arctic lowlands to a great height under the equator. (Darwin 1859: 382; see also 1872: 342)

Darwin's living waters truly flow horizontally on a flat Earth wherein the geological past is a shallow one. It should be mentioned that Chapter 12 of the *Origin* contains other case studies devoted to the migration of freshwater fish and saltwater fish, shells, and plants. These will not be analyzed here, if only because Darwin presented them in general terms only, lacking in necessary precision with regard to the taxonomic level, the time dimension, the geographical range, and the center of dispersal (Darwin 1859: 383–388; 1872: 343–347). Apparently, Darwin's only goal in presenting these other case studies consisted in showing the relative ease with which currently existing life forms could migrate to removed areas.

5.6 The Weak Contingency Thesis Versus the Strong Contingency Thesis

The main feature of Darwin's horizontal approach in Chapters of 11 and 12 of the *Origin* consists in a search for very recent centers of dispersal—the Pliocene or after—to the exclusion of earlier geological periods. This can be clearly seen whenever Darwin provides a series of examples in which fauna and flora are separated by profound differences for which no common centers of dispersal can be identified.[8] Are we to suppose, then, that such forms did not exist prior, as if they arose out of nowhere? What constitutes a proposed unique center of dispersal at the beginning of the Glacial period for allied forms is, in reality, merely a point in place and time among many other geological moments and geographical spaces in the evolutionary history of such forms. The same allied forms (or most of them) probably lived elsewhere before the Pliocene. In fact, Darwin faces an interesting problem in the two biogeographical chapters of the *Origin*. He proposes so many means of dispersal to support large-scale migrations that the reader becomes convinced that populating the entire Earth is rarely an impassable problem for life, even within a relatively short time period. Should we not expect, then, that many extant forms had always been fairly widely distributed at all geological times? One cannot exclude the possibility that the combined notions of unique center of dispersal, common ancestry, and divergence so central in Chapters 11 and 12 are artifacts of a theory founded on the assumption that the present reveals much about the evolutionary past. Once again, Darwin faces the limitations of a theory ill-designed for contemplating the historical profundity or verticality of evolution, no matter how much he may have wanted to broaden his reasoning to higher taxonomic levels (subgenera, genera, and families) (Darwin 1859: 407–408; 1872: 359).

Let us conclude this section by revisiting the question of evolutionary contingency. At the very end of Chapter 12 of the *Origin*, Darwin (1859: 410) concludes: "On my theory these several relations throughout time and space are intelligible... the more nearly any two forms are related in blood, the nearer they will generally stand to each other in time and space." This quotation well summarizes the kind of ideal model—based on proportional-symmetrical explanatory variables—that Darwin pursued throughout Chapter 11. As we have seen, Darwin went to great pains to avoid straying from this ideal model in his case studies to the extent of even being prepared to take liberties with elaborate speculations and ad hoc explanations to stay within its boundaries. Why should Darwin have cared so much about its ideal conditions of application? The answer takes us back to the distinction introduced earlier between the *weak contingency thesis* and the *strong contingency thesis*. Darwin repeatedly declared in the *Origin*: "there is no law of development."[9] As a subscriber to the weak thesis, Darwin was more than willing to recognize that the

[8]These differences between faunas and floras point to a deep geological past beyond which Darwin's theory is powerless to penetrate. Among the most important examples are the following: (a) the terrestrial productions separating the Old and New Worlds in general; (b) the inhabitants of Australia, Africa, and South America under the same latitude; (c) mammals in Europe, Australia, and South America; (d) marine fauna living on each side of South and Central America; (e) marine fauna separated by wide space of open sea such as the westward shores of America and the eastern islands of the Pacific, all of which are pointing to three distinct marine faunas in Eastern America, Western America, and Eastern Pacific islands; (f) the great difference in the marine faunas on the opposite sides of almost every continent; and (g) finally two widely distinct mammalian faunas separated by the ocean in the Malay Archipelago. See Darwin (1859: 347–348, 352, 358, 395).

[9]This was explicitly or implicitly claimed. See Darwin (1859: 313–315, 318, 331–332, 343, 351, 408–409; 1872: 290–292, 294, 303–304, 313–314, 319, 360–361).

pace of evolution varied, that forms diverged from each other to a smaller or greater extent, and that selective pressures greatly differed along a weak-strong spectrum. This offered Darwin epistemological flexibility for fitting facts and theory together. We understand better today that it is in the nature of the field of evolutionary biology to have explanations couched in historical narratives, requiring that we accept a level of flexibility or uncertainty about the pattern-process of evolution. After all, biological evolution is the product of the interaction of too many unpredictable variables to deny its openness.

Darwin did recognize this when calibrating the explanatory variables in his case studies. This being said, it is important to realize that Darwin was not prepared to go all the way in his recognition of a fully contingent biological evolution. To do so would have jeopardized his theory as he conceived it. As a proponent of a fully constituted world—a strong ontological commitment—Darwin wanted to make sure that the present time would always remain the ultimate epistemological horizon for building knowledge in evolutionary biology. Darwin's time machine was geared toward the horizontality of the present time, guided by the principle of divergence: projecting himself in the future, he assumed that forms with similarities observed today are in the process of moving apart from each other (divergence); projecting himself into the past, he thought that forms with similarities seen today shared a common ancestor (convergence). In order to solidly anchor all of evolution in the present, Darwin assured his reader that gradation is seen everywhere among currently living forms—with no gaps, no disruptions, and no discontinuities—the principle of gradation being the warranty that the fabric of life is knit in such a tight way as to reveal most of the evolutionary past by merely pulling on it from the present, so to speak.

Of course, what we have just described refers to the core of Darwin's theory: the principle of divergence and the principle the gradualism-gradation. *Darwin could not embrace evolutionary contingency to its full extent, precisely because his theory's core prevented him from questioning the epistemological validity of the dual notion of divergence-gradation.* To do so would have constituted a frontal attack on his theory, by opening the door to the recognition that the present time may only be revealing a small and insignificant portion of that evolutionary past. Among problematic issues, Darwin's theory was ill-equipped to face the following: (1) the difficulty of extracting closely related forms of low taxonomic levels from common reproductive networks, (2) the parallel evolution of two lineages independently developing similar structures (the homology-analogy spectrum), (3) the difficulty of tracking forms separated by real evolutionary gaps, and (4) the impossibility of using today's variability as a representative scale of past's variability. Those evolutionary manifestations significantly blur what a contemporary observer can gather from the past.

Darwin's theory was constructed in such a way as to deny the importance of those manifestations. The fixed and irrefutable core of Darwin's theory—divergence and gradation—has the precise explanatory function of downplaying such evolutionary possibilities. Yet, the genuine contingency of evolution would have required from Darwin's theoretical core much more flexibility in order to make sense of the full complexity and openness of evolution. Instead, Darwin's theory was condemned to face that complexity with the calibration of a limited number of variables, leaving others fixed and unmovable. No surprise, then, that Darwin had to resort to some elaborate speculations and ad hoc explanations to fit the facts to the theory. To say it simply, Darwin's theory lacked in epistemological flexibility. Only a theory founded on a *strong contingency thesis* could have handled that complexity adequately.

5.7 Conclusion

In the traditional historiography, natural selection is seen as central to the *Origin of Species*. It is true that the reader encounters this notion on nearly every page. Because this is the case, the mind of the uncritical reader (thanks in part to Darwin's rhetorical skill) moves swiftly to the modern Neo-Darwinian view of evolution giving contingency full recognition, thus classifying the field of evolutionary biology among historical disciplines (Simpson 1964: 176–189; Mayr 1982: 21–82). In that modern view, biological evolution is the product of the encounter of two main independent series of events: (1) the unpredictable changes in the biotic and abiotic environment and (2) the unpredictable rise of biological variation (genetical and somatical) among populations. The arbiter of this encounter is natural selection: populations with the proper variations in the good environmental context at the right geological time will be positively selected, others will be eliminated. Using philosophical language, the outcome of evolution is seen as the product of the encounter between two independent causal chains. The result of that encounter can only be true and open evolutionary contingency. In that modern view, natural selection is undoubtedly a central explanatory piece regulating the evolutionary faith of separate populations (Gayon 1993); the overall picture of evolution being reducible to a myriad of independent and unpredictable pathways of numerous modalities. Unfortunately, this is not the view revealed by a critical reading of the *Origin* and for several reasons.

1. Natural selection is no arbiter of the evolutionary outcome because of its weakness. In Chapters 1–5 and 8, its force is presented as being so weak as having difficulty in irreversibly segregating forms united in reproductive networks. In Chapters 9 and 10, the same weak force only manages to deploy our extant classes since the beginning of life, classes retaining strange similarities over their entire evolutionary history.
2. Natural selection is no arbiter of the evolutionary outcome, because of its wild power. In Chapters 6 and 7, its blistering force shatters the milieu-population relationship which is bypassed in favor of abstract vertical ascents across evolutionary grades and taxonomic levels, a wild drive *à la* Lamarck, Chambers, and Spencer.
3. Natural selection has no part in the theoretical core, since it is restricted to an auxiliary epistemological function. In Chapters 11 and 12, its force is modulated

at will by playing the role of an explanatory buffer devised to protect the theory's core centered on the dual notion of divergence-gradation. Indeed, we have seen Darwin applying an ideal model that consisted in adjusting explanatory variables in a symmetrical or proportional way to comply with a pre-established pattern-process of evolution, so that the taxonomic level, the geological time, the geographical distribution, and the level of selective pressure would all conspire to impose a nearly perfect "V" (divergence) shape to evolution. When the ideal model could not be arranged to fit the facts, natural selection (among other variables) was sacrificed in order to re-establish the desired evolutionary explanation.

Although Darwin is sometimes celebrated for having introduced into evolutionary biology the idea that evolution's main characteristic is evolutionary contingency (Gould 1986, 1989: 277–291), this view cannot be substantiated. It seems that he was, at best, the promoter of a weak contingency thesis, leaving the development of a strong contingency thesis to twentieth-century evolutionary biologists.

References

Bowler PJ (1990) Charles Darwin: the man and his influence. Basil Blackwell, Oxford

Bowler PJ (1996) Life's splendid drama: evolutionary biology and the reconstruction of life's ancestry. University of Chicago Press, Chicago, pp 1860–1940

Bowler PJ (2003) Evolution: the history of an idea, 3rd edn. University of California Press, Berkeley

Bowler PJ (2009) Geographical distribution in the Origin of Species. In: Ruse M, Richards RJ (eds) The Cambridge companion to the "Origin of Species". Cambridge University Press, Cambridge, pp 153–172

Browne J (1983) The secular ark: studies in the history of biogeography. Yale University Press, New Haven

Darwin C (1859) On the Origin of Species. John Murray, London

Darwin C (1872) The Origin of Species, 6th edn, with additions and corrections [1876]. John Murray, London

Delisle RG (2007) Debating humankind's place in nature, 1860–2000: the nature of paleoanthropology. Pearson-Prentice Hall, New Jersey

Delisle RG (2017) From Charles Darwin to the evolutionary synthesis: weak and diffused connections only. In: Delisle RG (ed) The Darwinian tradition in context: research programs in evolutionary biology. Springer, Cham, pp 133–167

Gayon J (1993) La biologie entre loi et histoire. Philosophie 38:30–57

Glick TF, Kohn D (eds) (1996) Darwin on evolution: the development of the theory of natural selection. Hackett, Indianapolis

Gould SJ (1986) Evolution and the triumph of homology, or why history matters. Am Sci 74:60–69

Gould SJ (1989) Wonderful fife: the Burgess Shale and the nature of history. W.W. Norton, New York

Lack D (1947) Darwin's finches. Cambridge University Press, Cambridge

Lakatos I (1970) Falsification and the methodology of scientific research programmes. In: Lakatos I, Musgrave A (eds) Criticism and the growth of knowledge. Cambridge University Press, Cambridge, pp 91–196

Mayr E (1982) The growth of biological thought. Belknap Press, Cambridge

Moore J (2005) Revolution of the space invaders: Darwin and Wallace on the geography of life. In: Livingston D, Withers C (eds) Geography and revolution. University of Chicago Press, Chicago, pp 106–132
Ruse M (1979) The Darwinian revolution. University of Chicago Press, Chicago
Simpson GG (1964) This view of life. Harcourt, Brace & World, New York

Chapter 6
An Attempt at Taming Natural Selection with Convergence Backward in Time, Part II (*Origin*, Chap. 13)

Abstract More clearly than elsewhere in the *Origin of Species*, Chapter 13 attempts to propose a "consilience of inductions," that is, an effort to gather support from a host of disciplines assumed to be pointing at the same explanatory core. Far from being a success on close inspection, Darwin's attempt will assemble a long list of contradictory claims he himself provides by imposing upon these disciplines a number of simplifying assumptions about the pattern-process of evolution: (1) "divergence" and "common ancestry" are tightly associated; (2) generally speaking, the most divergent forms are said to be selected over less divergent ones; (3) it is assumed that divergence today is at its maximum relative to the evolutionary past of each class/phylum; and (4) the taxonomic arrangement observed among extant forms is believed to have persisted throughout the entire evolutionary history. Darwin's assumptions overlooked the full complexity and openness of evolution by putting it in a conceptual straitjacket. Chapter 13 is the place where Darwin will deliver his ultimate fight against an evolutionary engine that erases its genealogical traces as it rolls on, thus generating all sorts of complexities, including the rise of similar features (analogies) and the complete deletion of traces in apparently ancestorless forms. We will conclude our analysis by providing an overview of Part II of our book (Chaps. 3–6), arguing that a close reading of the *Origin* reveals a series of different pictures of evolution, some of which are squarely inconsistent. The *Origin of Species* is far from offering a unified and all-encompassing theory of evolution.

Keywords Consilience of inductions · Systematics · Morphology · Embryology · Comparative anatomy · Analogies · William Whewell · Strong contingency thesis · Weak contingency thesis · Pan-divergent view

R. G. Delisle, *Charles Darwin's Incomplete Revolution*, Evolutionary Biology – New Perspectives on Its Development 1, https://doi.org/10.1007/978-3-030-17203-9_6

The last opportunity at Darwin's disposal to seize the evolutionary past rests on Chapter 13 of the *Origin of Species* (1859).[1] Just as Chapters 11 and 12 exploited biogeography to investigate recently acquired differences at low taxonomic levels, Darwin will attempt to probe the geological past more deeply by climbing the taxonomic scale among currently existing forms, as already attempted in Chapters 6 and 7. He will do so by turning to fields like systematics, morphology, embryology, and comparative anatomy. Chapter 2 of this book already briefly analyzed this part of the *Origin*, but mainly from the perspective of the limitations Darwin faced when thinking about ancient progenitors. This time, we will address more directly the limitations of Darwin's theory in assessing the actual pattern and process of evolution.

It will be seen that although Darwin hoped to show that one of the two pillars at the core of his theory—the principle of divergence—renders evolution completely intelligible, the inherent complexity of evolution itself will not be easily confined to this rigid and pre-established outlook. "Through a glass, darkly" may well describe the experience of looking through the distorting lens of Darwin's theory, unless one prefers the expression "through the past, darkly." A fascinating by-product of Darwin's theory—largely neglected by Darwin himself—could be summarized this way: *evolution carries its own evolutionary time, having little consideration for the past or future*. Indeed, "descent" (divergence) and "natural selection" are interestingly tied together in Darwin's theory: the pattern-process of divergence is generated by the selective pressure under, for instance, the competitive exclusion principle, which favors related forms as far removed from each other as possible. The more this goes on in geological time, the more adaptive features independently acquired in the related lineages blur the affiliative signs originally uniting them. Differently stated, the *evolutionary engine erases its traces as it runs*.

Darwin's theory can only offer limited precision when dealing with phylogenetic hypotheses, an imprecision that increases with temporal extension. Let us keep in mind that Darwin's theoretical core is both ambitious and restrictive, asking evolution to comply with a closely packed and indivisible conceptual bundle clustered around the principle of divergence (excluding from our analysis here gradualism-gradation): divergence itself, common ancestry, unique center of dispersal, monophyletism, and homologies. Are we to assume that this conceptual straitjacket is tailor-made to accommodate the full complexity and contingency of evolution? Why is Darwin asking so much of his theory, not remaining content with notions like common ancestry, unique center of dispersal, and monophyletism; notions not necessarily linked with a strict and exclusive application of the principle of divergence?

The answer is to be found in the built-in horizontality of Darwin's theory, as seen in his commitment to a fully constituted world: the present time cannot fail to be the faithful witness of the evolutionary past, unless it is argued that it registers most of the past. Of course, that assumption would be unfounded if polyphyletism, reticulate

[1]The last chapter of the first edition of the *Origin of Species* is Chapter 14. It is devoted to presenting a recapitulation of the book and thus contains a limited amount of truly new material.

evolution, parallelism, and evolutionary convergence were to play an important part in the pattern-process of evolution. In Chapter 13 of the *Origin*, as in Chapters 9–12, Darwin will try to discipline an overflowing, exuberant, and complex evolution under the principle of divergence. Chapter 13 is undoubtedly about divergence, common ancestry, monophyletism, and the selective forces pushing living entities apart, as presented in the recent historiography (Ruse 1979: 196, 197; Bowler 1990: 122–124). It is also, in addition, the place where Darwin delivers his ultimate fight against evolutionary analogies and deleted affiliation.

6.1 Blurred Phylogenetic Connections: Facing Analogies and Deleted Affiliation

Fully aware of the potential problems of evolutionary analogies for his theory, Darwin tackles the issue head-on by repeatedly making the theoretical distinction between "homologies" and "analogies." Let us follow him in the order of his presentation in Chapter 13:

> It might have been thought. . .that those parts of the structure which determined the habits of life, and the general place of each being in the economy of nature, would be of very high importance in classification. Nothing can be more false. No one regards the external similarity of a mouse to a shrew, of a dugong to a whale, of a whale to a fish, as of any importance. These resemblances, though so intimately connected with the whole life of the being, are ranked as merely 'adaptive or analogical characters'. . . We must not, therefore, in classifying, trust to resemblances in parts of the organisation, however important they may be for the welfare of the being in relation to the outer world. (Darwin 1859: 414, 415; see also 1872: 365)

A few pages down we again read: "We have no written pedigrees; we have to make out community of descent by resemblances of any kind. Therefore, we choose those characters which, as far as we can judge, are the least likely to have been modified in relation to the conditions of life to which each species has been recently exposed" (1859: 425; see also 1872: 372). Only two pages later the reader is greeted with the same remark:

> On my view of characters being of real importance for classification, only in so far as they reveal descent, we can clearly understand why analogical or adaptive character, although of the utmost importance to the welfare of the being, are almost valueless to the systematists. For animals, belonging to two most distinct lines of descent, may readily become adapted to similar conditions, and thus assume a close external resemblance; but such resemblances will not reveal—will rather tend to conceal their blood-relationship to their proper lines of descent. (Darwin 1859: 427; see also 1872: 374)

And as if the issue was not already obvious enough, Darwin assails us again on the very next page:

> As members of distinct classes have often been adapted by successive slight modifications to live under nearly similar circumstances,—to inhabit for instance the three elements of land,

> air, and water,—we can perhaps understand how it is that a numerical parallelism has sometimes been observed between the sub-groups in distinct classes. A naturalist, struck by a parallelism of this nature in any one class, by arbitrary raising or sinking the value of the groups in other classes... could easily extend the parallelism over a wide range.... (Darwin 1859: 428; see also 1872: 375)

We must extend our apologies for imposing upon the reader this succession of quotations essentially repeating the same thing. Our purpose is admittedly, didactic: it is intended to show that Darwin felt the need to seriously confront what he knew was a serious threat to his theory. That potential threat, as already alluded to, presented itself under a dual aspect:

1. Through the problem of analogy wherein selective forces forge similar structures and functions in organisms that are not necessarily closely related
2. Through the problem of selective forces modifying organisms in such a profound and novel way as to not only blur all affiliative traces but also make these organisms seem without modern-day equivalents

This latter case was made explicit by Darwin when providing explanations for his Diagram (see Fig. 2.1): "If, however, we choose to suppose that any of the descendants of A or of I have been so much modified as to have more or less completely lost traces of their parentage, in this case, their places in a natural classification will have been more or less completely lost,—as sometimes seems to have occurred with existing organisms" (1859: 421). In the last edition of the *Origin*, Darwin (1872: 370) added two words to this last sentence in the hope, presumably, of diminishing its potential impact on his theory: "as seems to have occurred with *some few* existing organisms" [my emphasis].

The dual nature of this threat merely constitutes the extreme faces of the action of natural selection: perfect imitation or complete novelty. Of course, Darwin's theory is all about promoting an ideal middle road that sees natural selection as a force powerful enough to institute evolutionary change but weak enough to not blur the traces of affiliation.

Having presented the dual threat of natural selection only in theoretical terms, Darwin now had to face the real world. In order to impose the principle of divergence over the genuine complexity and contingency of evolution, Darwin fought his way at all taxonomic levels. Going beyond Chapter 13 for a moment, let us begin with the lower levels. We know the importance of the analogy between artificial selection and natural selection for Darwin's theory: it establishes the credibility of an explanation in one area (domestication), which is then transferred to another area (nature). Because nothing is straightforward in the biological world, Darwin (1859: 15) had to recognize that even domesticated forms revert to the progenitor's state: "[O]ur [domestic] varieties certainly do occasionally revert in some of their characters to ancestral forms. . . ."[2] In the hands of fanciers, Darwin continues, races of pigeons could be manipulated to nearly replace extinguished forms:

> We can clearly understand why a species when once lost should never reappear, even if the very same conditions of life, organic and inorganic, should recur. For though the offspring of one species might be adapted. . . to fill the exact place of another species in the economy of nature, and thus supplant it; yet the two forms—the old and the new—would not be identically the same; for both would almost certainly inherit different characters from their distinct progenitors. For instance, it is just possible, if our fantail-pigeons were all destroyed, that fanciers, by striving during long ages for the same object, might make *a new breed hardly distinguishable* from our present fantail; but if the parent rock-pigeon were also destroyed. . . it is quite incredible that a fantail, identical with the existing breed, could be raised from any other species of pigeon, or even from the other well-established races of the domestic pigeon, for the newly-formed fantail, would be almost sure to inherit from its progenitor *some slight characteristic differences*. (Darwin 1859: 315, 316; see also 1872: 292) [my emphasis]

Apparently, fanciers' techniques are powerful enough to institute parallelism and evolutionary convergence. Of course, it could be argued that this is possible only because fanciers work on such closely related forms that they are barely detached from their common ancestor. There we find Darwin's main problem when dealing with evolutionary events at low taxonomic levels. Transferring the application of the case of very closely related forms from domestication to nature, Darwin refines the analysis by distinguishing between reversal to a previous state and the rise of similar traits:

> As all the species of the same genus are supposed, on my theory, to have descended from a common parent, it might be expected that they would occasionally vary in an analogous manner; so that a variety of one species would resemble in some of its characters another species; this other species being on my theory only a well-marked and permanent variety. But characters thus gained would probably be of an unimportant nature, for the presence of all important characters will be governed by natural selection, in accordance with the diverse habits of the species, and will not be left to the mutual action of the conditions of life and of a similar inherited constitution. It might further be expected that the species of the same genus would occasionally exhibit reversions to lost ancestral characters. . . Hence, though under nature it must generally be left doubtful, what cases are reversions to an anciently existing character, and what are new but analogous variations, yet we ought, on my theory, sometimes to find the varying offspring of a species assuming characters (either from reversion or from analogous variation) which already occur in some other members of the same group. (Darwin 1859: 161, 162; see also 1872: 126, 127)

As expected, Darwin's position consists in limiting the impact of evolutionary reversals and what he calls "analogous variations"[3] on his theory by qualifying such manifestations as "occasional" or "unimportant." From Darwin's viewpoint, affiliation (descent) must remain clearly traceable. While Darwin cannot be blamed for interpreting facts to his advantage, the reader is not obliged to take his words at face value, especially when there are two ways to understand them. In the quotation that follows, Darwin focuses on traces remaining under common ancestry:

---

[2]Only the first part of this passage is reproduced in the last edition of the *Origin*. See Darwin (1872: 11).

[3]On analogous variations among closely related forms, see also Darwin (1872: 375).

> An indefinite repetition of the same part or organ is the common characteristic. . .of all low or little-modified forms; therefore we may readily believe that the unknown progenitor of the vertebrata possessed many vertebrae; the unknown progenitor of the articulata, many segments. . . We have formerly seen that parts many times repeated are eminently liable to vary in number and structure; consequently it is quite probable that natural selection, during a long-continued course of modification, should have seized on a certain number of the primordially similar elements, many times repeated, and have adapted them to the most diverse purposes. And as the whole amount of modification will have been effected by slight successive steps, we need not wonder at discovering in such parts or organs, a certain degree of fundamental resemblance, retained by the strong principle of inheritance. (Darwin 1859: 437, 438; see also 1872: 385)

Another way of reading this passage would be to focus on the implications flowing from evaluating "low or little-modified" forms, also referred to as "little specialized" forms in the last edition of the *Origin.* Since such forms are typically characterized by a repetition of undifferentiated parts, it becomes difficult to distinguish successfully between traits attributable either to common ancestry or to parallel evolution when they diverge on separate evolutionary roads from a common progenitor. Strangely enough, in the passage just quoted above and in the previous, these forms are presented as being so similar to each other that Darwin explicitly relies on natural selection to institute an evolutionary separation (differences) between them! We know that Darwin repeatedly claimed that traces of affiliation (descent) could not be established by following adaptive features forged under natural selection, as these reflect lifestyle (adaptation) more than descent. Pushing our thesis to its logical conclusion, it seems Darwin will have greater difficulty distinguishing between homologies and analogies when dealing with unspecialized evolutionary groups than with specialized ones.

To impose the principle of divergence over the genuine complexity and contingency of evolution, Darwin also had to contend with cases at higher taxonomic levels. At such levels, he needed to explain both why forms would continue to diverge from each other rather than converge toward each other (parallel evolution and convergent evolution) and why the independent action of the selective forces responsible for generating adaptations would not succeed in significantly blurring the phylogenetic connections. Let us keep in mind that Darwin was not content with merely defending the notions of common ancestry, unique center of dispersal, and monophyletism. For him, the pattern-process of evolution had to comply with the principle of divergence all the way up to the highest taxonomic levels. In so arguing, however, Darwin could not deny that evolution's way was not always as straightforward as he would have hoped. Let us quote again a part of this important passage:

> The electric organs offer another and even more serious difficulty; for they occur in only about a dozen fishes, of which several are widely remote in their affinities. . . The presence of luminous organs in a few insects, belonging to different families and orders, offers a parallel case of difficulty. . . I am inclined to believe that in nearly the same way as two men have sometimes independently hit on the very same invention, so natural selection, working for the good of each being and taking advantage of analogous variations, has sometimes

> modified in very nearly the same manner two parts in two organic beings, which owe but little of their structure in common to inheritance from the same ancestor. (Darwin 1859: 193, 194)

In the last edition of the *Origin*, Darwin (1872: 150–152, 375) did his best to reduce the significance of such analogical cases by stressing, more than in the first edition, the lack of perfect identity between such cases. Because of the dual nature of the problem faced by Darwin, parallelism-convergence constitutes only one threat to his theory. The other is seen in the apparent emergence of ancestorless features:

> No doubt many instincts of very difficult explanation could be opposed to the theory of natural selection,—cases, in which we cannot see how an instinct could possibly have originated; cases, in which no intermediate gradations are known to exist; cases of instinct of apparently such trifling importance, that they could hardly have been acted on by natural selection; cases of instincts almost identically the same in animals so remote in the scale of nature, that we cannot account for their similarity by inheritance from a common parent, and must therefore believe that they have been acquired by independent acts of natural selection. I will not here enter on these several cases.... (Darwin 1859: 235, 236; see also 1872: 228)

Here, one could argue that Darwin's challenge is confined to the rise of particular traits or groups of traits and not to entire organisms or species, thus reducing the impact of such cases on Darwin's theory. This point is well taken. But in continually pretending that problems associated with both deleted affiliation and analogies are no major impediments to the scientific robustness of his theory, Darwin's self-assurance is deceptive. In the following example and quotation, we see Darwin once again confident enough to proceed to the distinction between homologous and analogous features in species belonging to the phylogenetic groups of marsupials and rodents, the former a marsupial mammal and the latter a placental mammal. Unfortunately for Darwin, who assumed a close phylogenetic connection between them (see Archibald 2012), we know today that the two groups are not particularly close, phylogenetically speaking, when other placentals are added to the analysis (Kemp 2005: 190–290; Feldhamer et al. 2015: 50–68). Not only does this render Darwin's distinction between homologous and analogous features among marsupials and rodents uncertain, it also shows that many similarities between these two groups are largely due to parallelism and convergent evolution:

> Mr. Waterhouse has remarked that, when a member belonging to one group of animals exhibit an affinity to a quite distinct group, this affinity in most cases is general and not special: thus, according to Mr. Waterhouse, of all Rodents, the bizcacha [viscacha] is most nearly related to Marsupials; but in the points in which it approaches this order, its relations are general, and not to any one marsupial species more than to another. As the points of affinity of the bizcacha to Marsupials are believed to be real and not merely adaptive, they are due on my theory to inheritance in common. Therefore we must suppose either that all Rodents, including the bizcacha, branched off from some very ancient Marsupial, which will have had a character in some degree intermediate with respect to all existing Marsupials; or that both Rodents and Marsupials branched off from a common progenitor, and that both groups have since undergone much modification in divergent directions. On either view we may suppose that the bizcacha has retained, by inheritance, more of the character of its ancient progenitor than have other Rodents; and therefore it will not be specially related to any one existing Marsupials, but indirectly to all or nearly all Marsupials, from having partially retained the character of their common progenitor, or of an early member of the group. On the other hand, of all Marsupials, as Mr. Waterhouse has remarked, the

> phascolomys [wombat] resembles most nearly, not any one species, but the general order of Rodents. In this case, however, it may be strongly suspected that the resemblance is only analogical, owing to the phascolomys having become adapted to habits like those of a Rodent. (Darwin 1859: 429, 430; see also 1872: 379)

Darwin's level of confidence and precision in this phylogenetic case is impressive. Among Darwin's contemporaries, St. George Mivart explicitly pointed out at the analogy problem between placental and marsupial mammals in his *On the Genesis of Species* (1871: 67–70).[4] Having presented the issue in general terms, let us now follow Darwin as he gets down to specifics.

6.2 The Limitations of Systematics

Darwin's unfounded confidence in his handling of deleted affiliation and analogies is easily explained by his commitment to a very specific evolutionary process: "I attempted also to show that there is a constant tendency in the forms which are increasing in number and diverging in character, to supplant and exterminate the less divergent, the less improved, and preceding forms" (1859: 412; see also 1872: 364). This crucial passage of Chapter 13 states that not only is evolution a divergent process but, more importantly, that it is a process with a "constant tendency" to select the most divergent forms, driving less divergent ones to extinction. The epistemological implication of this statement is absolutely fundamental to the understanding of Darwin's theory. The rationale behind this statement has already been laid out in Chapter 4 of the *Origin* (1859: 111–116, 1872: 86–90). In Darwin's view, the natural outcome of evolution is the positive selection of the most divergent forms, as it logically follows from the application of two complementary principles: (1) the competitive exclusion principle, which states that the more forms are identical, the more they directly compete with each other, and (2) the principle of specialization, which states that a greater number of forms can be sustained within the same territory when each occupies distinct and specialized places in the economy of nature. From Darwin's viewpoint, this rationale constitutes a double warranty against the dual threat his theory faces to natural selection. Here's why:

1. Analogies (parallelism and convergent evolution) could not prevail in evolution, since forms occupying the same places in the economy of nature as a result of their similarities would lead to the extinction of some of those forms.
2. Traces of affiliation could rarely be entirely erased in a view of evolution that assumes that past forms are either directly aligned with, or fall in-between, currently existing forms (Darwin 1859: 329, 1872: 301). Indeed, Darwin assumed that divergence (separation) since the beginning of life is at its maximum today.

[4]While more examples of the problems of deleted affiliation and analogies in the *Origin of Species* could be cited, in the interests of avoiding repetition the reader is redirected to Chap. 4 of this book for additional examples.

The diversity of the past can thus only fall under the purview of today's variation, as reflected in groups nested in ever more inclusive taxonomic groups.

It is worth keeping in mind that taming the power of natural selection under the principle of divergence is what Darwin ultimately attempted in Chapters 9 to 13 of the *Origin*. That quest took on a more precise meaning in Chapter 13 with Darwin combatting the dual threat of natural selection (analogies and deleted affiliation) while simultaneously promoting the theoretical generalization that more divergent forms are often better off than less divergent ones. This is illustrated in Darwin's Diagram (see Fig. 2.1). About two-thirds of the speciation events involving lineages A and I are successfully accomplished by the most divergent forms during each time period. This being the case, it should be noted how this goes against the *strong contingency thesis*, for how could an evolutionary tendency favoring the most divergent forms be compatible with the full complexity and contingency of evolution? This is yet another indicator of Darwin's approach, which imposes an idealized picture onto evolution consistent with the irrefutable core of his theory (divergence-gradation). To put it differently, it is the default position of Darwin's theory to assume that the most divergent forms will prevail; only when facts are strongly against this assumption will he retreat in small measure from it, resorting to various levels of divergence, varying paces of evolution, some stagnation, and a *small degree* of parallelism and convergent evolution.

This compromise between Darwin's default position and the complexity of evolution is nicely captured in the Diagram. In it, Darwin avoids presenting descendants of lineages A and I as converging toward each other during their separate evolutionary history to the point of crisscrossing. To do so would have put Darwin's theory at risk in two different ways: (1) when the Diagram is read at low taxonomic levels (varieties, species, genera), the pattern-process would have resulted in promoting reticulate evolution and polyphyletism, and (2) when the Diagram is exploited at higher taxonomic levels (families, orders, classes), the pattern-process would have been the outcome of full parallelism and convergent evolution.

Instead, Darwin applies the principle of divergence in the context of lineages that are neatly kept separate from each other at all times, thus ensuring that taxonomy would today register the most important divergences in the past. As seen in his Diagram, the three main lineages in existence in the Silurian period under three allied genera (A, F, and I) have persisted until today, with descendants of A and I now being separated at the order level under evolutionary divergence (Darwin 1859: 420–422, 1872: 369, 370). Yet, Darwin continued, these three distinct lineages remained separate from each other and kept their position relative to one another ever since the Silurian: "their genealogical *arrangement* remains strictly true, not only at the present time, but at each successive period of descent" (1859: 421; see also 1872: 370) [italics original]. So Darwin's theory claims. To demonstrate the limitations of Darwin's view, it is useful to follow a series of steps that gradually take us away from Darwin's idealized picture:

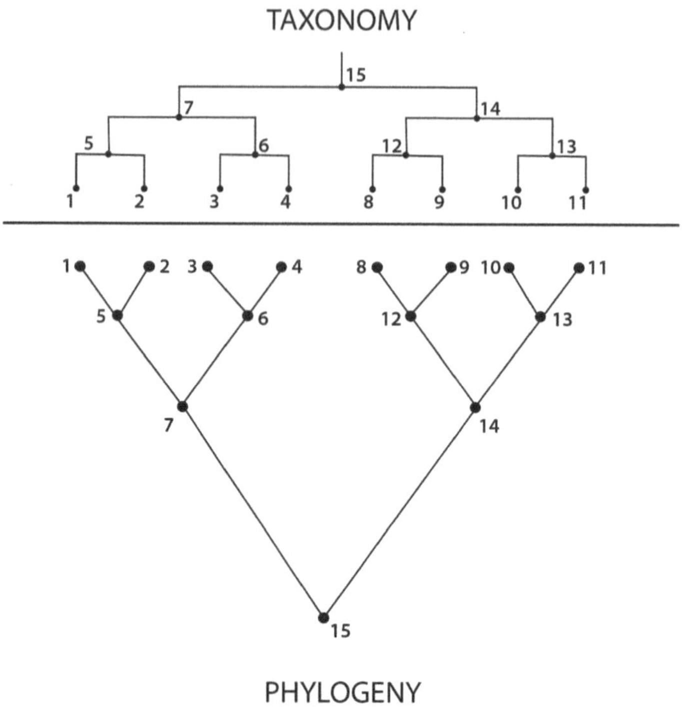

Fig. 6.1 An idealized view of Darwin's theory in which the evolutionary past is entirely reflected in the taxonomy of extant forms

1. Figure 6.1 shows an idealized view of Darwin's theory in which the evolutionary past is entirely reflected in taxonomy today, in a sort of mirror image, wherein each phylogenetic divergence is perfectly registered under a distinct taxonomic category (species, genus, family, or order), with the level of proximity seen in clusters of categories organized around more inclusive clusters.
2. Figure 6.2 presents a slightly more complex situation, with the addition of extinct forms unknown to scientists. In this case, taxonomy today cannot register all of the speciation events, in contradistinction to the previous case. Darwin had already anticipated this kind of situation, as he himself claimed that a number of evolutionary events would leave no traces in our taxonomy.[5] But Darwin would be perfectly content with the case presented here, as the essentials of evolutionary history are captured by taxonomy. Indeed, the extant and surviving forms constitute the genealogical armature around which all extinct and diverging forms are organized.
3. Figure 6.3 takes us one step closer to a realistic picture, by relinquishing the perfect symmetry of speciation events, as presented in the first two cases. This

[5]See Darwin (1859: 413, 414, 421, 429–431, 434, 1872: 365, 378–380).

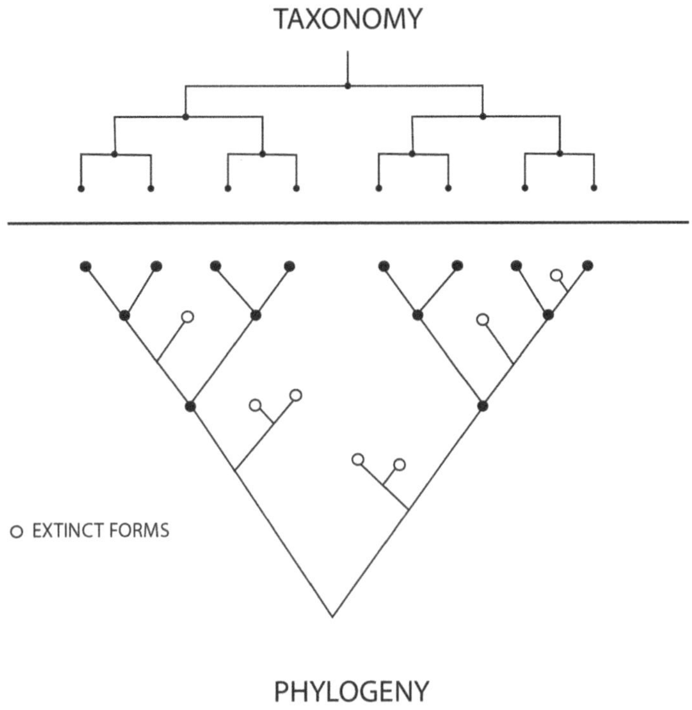

Fig. 6.2 A slightly more complex view with the addition of extinct forms unknown to scientists. Nonetheless, the essential of evolutionary history is captured by taxonomy, since extant forms constitute the genealogical armature around which all extinct and diverging forms are organized

allows us to move away from Darwin's default position, which assumes that the most divergent forms will persist, driving less divergent ones to extinction. This new level of complexity introduces various degrees of divergence, some evolutionary stagnation, and a small amount of parallel or convergent evolution. Of course, Darwin's theory can rely on the *weak contingency thesis* to make sense of this complexity. Even though taxonomy today manages to register something essential about the evolutionary history of that group, more information is still being lost as compared to the previous case. The application of the *weak contingency thesis* forces us to recognize that taxonomy reflects less and less perfectly phylogeny.

4. Figure 6.4 shows Darwin's theory to be in a difficult position with respect to the *strong contingency thesis* that allows for the full complexity and contingency of evolution, introducing the new patterns-processes of reticulate evolution, polyphyletism, and full parallelism or convergent evolution. All the major threats to Darwin's theory are encountered here: the blurring of affiliation under common reproductive networks; selective forces forging similarities among not necessarily closely related forms; and profound evolutionary changes that created apparently isolated or ancestorless groups. Furthermore, two key complementary

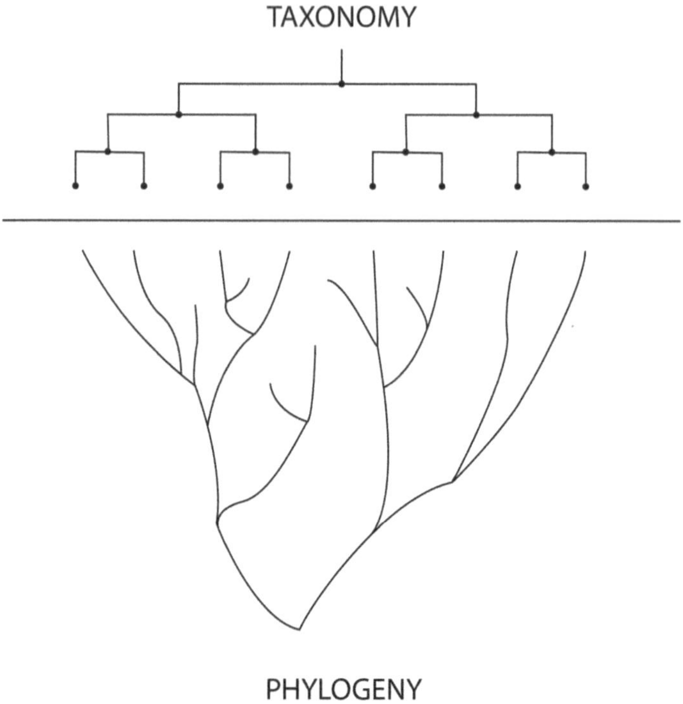

Fig. 6.3 A slightly more realistic picture that abandons the perfect symmetry of the speciation events as seen in Figs. 6.1 and 6.2. This level of complexity introduces various degrees of divergence, some evolutionary stagnation, and a small amount of parallel or convergent evolution (analogies). Although Darwin's commitment to the *weak contingency thesis* allows him to make sense of this level of complexity, this example illustrates how taxonomy reflects phylogeny less and less clearly

assumptions accompanying Darwin's theory cannot be respected: (1) the taxonomic *arrangement* currently seen among living forms is not preserved throughout the entire evolutionary history of the group; on the contrary, taxonomy has completely lost any trace of this original arrangement, and (2) extant forms are not at the maximum divergence level of the entire evolutionary history of the group; it is therefore not true that all extinct forms are either aligned with or fall in-between currently existing forms. On the contrary, a phylogenetic portion of this evolutionary group falls outside and is not registered in taxonomy of extant forms.

5. Figure 6.5 takes one last step toward a more realistic picture of evolution. We have repeatedly claimed that Darwin's horizontal approach to evolution generated a shallow view lacking in geological depth, preventing him from penetrating what we will here call the "thickness of the past." In contradistinction to Darwin,

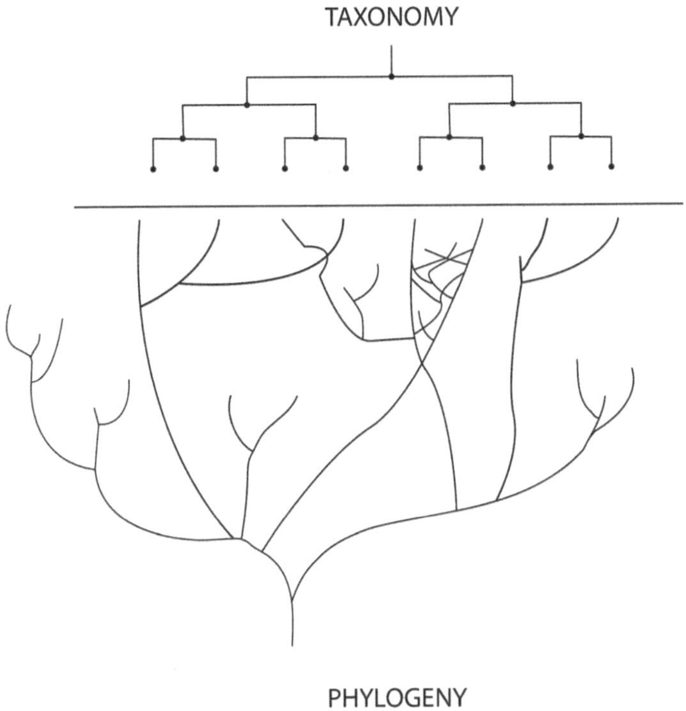

Fig. 6.4 A greater level of realism still introduces in evolution reticulate evolution, polyphyletism, full parallelism or convergent evolution (analogies). Only a commitment to the *strong contingency thesis* can make sense of such complexities, which are denied by Darwin. Every major threat to Darwin's theory is encountered here: the blurring of affiliations under common reproductive networks; selective forces that forge similarities among forms that are not necessarily closely related; and the profound evolutionary changes that generate apparently isolated or ancestorless groups

who seems to have thought he had a fairly clear view at the evolutionary past,[6] evolution's picture is closer to layers upon layers of evolutionary groups that lived in contemporaneity and succeeded each other in an inextricable complexity, with no clear phylogenetic connections between them. Currently existing forms constitute merely a very small sample of forms that have existed in the history of life, and a very unrepresentative sample at that. To attempt proposing phylogenetic hypotheses on the basis of extant forms exclusively can only lead to a very partial and distorted view.

Just as taxonomy in Figs. 6.1, 6.2, 6.3, 6.4, and 6.5 encompasses the same information, the introduction of ever more realistic cases ultimately reveal the gradual dissolution of the conceptual link between *taxonomy* and *phylogeny*.

[6]See Darwin (1859: 417, 425, 426, 1872: 367, 370, 372, 373, 381).

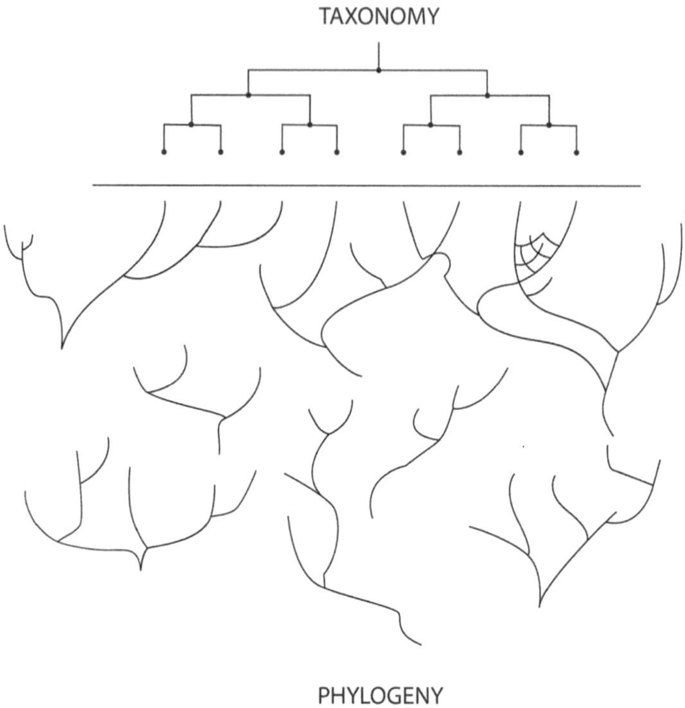

Fig. 6.5 The last step towards a truly realistic picture of evolution. In contradistinction to Darwin, who seems to have thought he had a fairly clear view of the evolutionary past, evolution's picture is organized around layers upon layers of evolutionary groups, hidden from the modern observer, in addition to having no clear phylogenetic connections between them. Currently existing forms constitute only a very small sample of all the forms that have existed in the history of life, and a very unrepresentative sample at that

Unfortunately for Darwin, and contrary to what he claimed, his Diagram is less a segment of tree of life than a genealogy of extant forms. No surprise, then, that he did not realize that the present was largely blind to the evolutionary past.[7]

It is no accident that Chapter 13 of the *Origins* opens with the principles of classification (systematics) before moving on to more specific fields like morphology, embryology, and comparative anatomy (Darwin 1859: 411–434, 1872: 363–381). After all, for Darwin evolutionary systematics constitutes the common thread uniting these other fields, "descent" being the conceptual intermediary through which knowledge is mediated. Speaking of the field of embryology, for instance, he writes: "We can see why characters derived from the embryo should be of equal importance with those derived from the adult, for our classification of course include all ages of each species" (1859: 418; see also 1872: 368). What is said here of embryology also applies to morphology and comparative anatomy. It is

[7]This theme has been more fully developed in Chap. 2 of this book.

reasonable to suppose, then, that the limitations associated with Darwin's use of systematics must be transmitted to these other fields. This having been established, we will highlight the argumentative strategy exploited by Darwin in Chapter 13:

> Whereas the first section on systematics deploys the full conceptual panoply accompanying Darwin's theory centered around the principle of divergence, the following sections devoted to morphology, embryology, and comparative anatomy are exclusively concerned with supporting the notion of common ancestry.

As we know, "divergence" and "common ancestry" are not necessarily equivalent: (a) forms derived from a common ancestor may well eventually converge toward each other; (b) forms diverging from one another bearing similarities may be taken to arise from an immediate common ancestor when, in reality, they arose from separate ancestors.

Apparently, Darwin took for granted that the arguments presented in the section on systematics were transferable to those about morphology, embryology, and comparative anatomy, thus relieving him of the effort of showing the full application of his theory in these other contexts. It is easy to miss the argumentative transition found in Chapter 13 and to be persuaded by Darwin's rhetoric. However, we know that the notion of common ancestry—and its accompanying concepts like a unique center of dispersal and monophyletism—cannot do justice to Darwin's complete theory. His theory is much more ambitious and also concerns the pattern and process of evolution, as seen in the theory's dual core comprising the principles of divergence and gradualism-gradation. To put it differently, the sections on morphology, embryology, and comparative anatomy support the notion of common ancestry but are of little help in supporting other crucial explanatory components contained in the theory's core. The explanatory weight of Chapter 13, therefore, largely rests on the section devoted to systematics. Three complementary considerations buttressed Darwin's assumption that sections exclusively devoted to common ancestry also supported the rest of his theory:

1. Again and again in the section on systematics, Darwin explains that evolution is about evolutionary divergence, using expressions like "divergence of character" or "radiating" to support the claim that the most divergent forms are usually more likely to persist than less divergent ones.[8] This theoretical commitment is further reinforced by Darwin's (1859: 412, 413, 1872: 364) invitation to consult his Diagram. In this section, Darwin unpacked in a rather compact and implicit fashion the explanatory components contained in his theory's core: the principle of divergence, common ancestry, unique center of dispersal, monophyletism, homologies, and gradualism-gradation.
2. The section on systematics contains another crucial source of information about Darwin's conception, one that specified that related lineages derived from hypothetical common ancestors would remain separate during their entire evolutionary history while simultaneously maintaining their position relative to one another (Darwin 1859: 420–422, 1872: 369, 370). In other words, no matter how much

[8]See Darwin (1859: 412, 430, 431, 433, 434, 1872: 363, 364, 379, 381).

related lineages might have changed under evolutionary divergence, taxonomy today would have registered all key evolutionary moments.
3. Although everything required to understand Darwin's view is already contained in points (1) and (2), the issue can be made clearer still by invoking another piece of information given by Darwin in Chapter 10 of the *Origin*: divergence is at its maximum today, since all past forms are either aligned with or fall in-between currently living forms (Darwin 1859: 329, 1872: 301). This assumption is carefully reflected in the Diagram.

Putting together the three explanatory components just outlined, it becomes perfectly clear why Darwin equated the "principle of divergence" and "common ancestry." For him, evolution since the beginning of life had no other route but to diverge gradually up to the maximum point observed today (and perhaps to be continued in the future), with no major past evolutionary events being hidden or lost. Darwin locked up the evolutionary past under a rigid and predictable conception. Nothing truly new or unexpected could have happened according to a theory that assumed the present always reveals more than the past.[9] It follows from Darwin's assumptions that the simple notion of "common ancestry" *implied* all the other notions contained in his theory's core. Sections on morphology, embryology, and comparative anatomy of Chapter 13 were, in Darwin's mind, logical extensions of the principle of divergence. From the critical viewpoint adopted in this book, we know that these assumptions are unfounded. More importantly, we now have a better understanding of the argumentative structure adopted by Darwin in Chapter 13.

6.3 The Limitations of Morphology, Embryology, and Comparative Anatomy

In light of our analysis above the rest of this chapter could seem anticlimactic, knowing that sections on morphology, embryology, and comparative anatomy are largely devoted merely to supporting the notion of common ancestry. We would argue, however, that the rest of Chapter 13—which is seen in the historiography as contributing to the application of Darwin's theoretical core to a host of fields—also comes with drawbacks, thus throwing more epistemological light on Darwin's theory. The application of a theory across various fields has been called a "consilience of inductions" by Darwin's contemporary William Whewell, a notion also promoted under a different guise by another contemporary, John Herschel (Ruse 1975, 1979, 2000; Hodge 1977, 1989; Laudan 1981: 163–180). Darwin knew both authors personally and consulted their main publications, in addition to having fairly

[9]This issue is treated more fully in Chap. 2 of this book. Interested readers may especially consult the section entitled "A Closed, Permanent and Segmented Tree of Life."

explicitly referred to the explanatory notions of vera causa and consilience of inductions in *The Variation of Animals and Plants Under Domestication* (1868: 8, 9).[10] The epistemological benefit of a "consilience of inductions" consists in increasing the robustness of a theory by widening the basis of its support through a number of scientific areas. Independent areas not only receive an explanation with a new common theoretical basis but also find in it a previously unexpected unity. However, just as an expanded application of a theory to a host of fields may contribute to increase its robustness, it should not be overlooked that tensions, contradictions, or inconsistencies may accompany such an expansion. Not only would the latter case weaken the value of such a consilience of inductions, but it must also be remembered that its value has already been reduced once before, as seen earlier in the fact that several sections of Chapter 13 are not concerned with supporting the core of Darwin's theory in its entirety.

The field of morphology has been useful to Darwin for setting the ultimate horizon of his evolutionary thinking: the class. We have seen in Chap. 2 of this book that Darwin's so-called tree of life is a closed and segmented one, organized around distinct classes, with evolutionary motion impacting entities at lower taxonomic levels within each class (orders, families, genera, species). For him, the notion of common ancestry is de facto incarnated in a class:

> We have seen that the members of the same class, independently of their habits of life, resemble each other in the general plan of their organisation. This resemblance is often expressed by the term 'unity of type;' or by saying that the several parts and organs in the different species of the class are homologous... [T]he parts may change to almost any extent in form and size, and yet they always remain connected together in the same order... Hence the same names can be given to homologous bones in widely different animals... The explanation is manifest on the theory of the natural selection of successive slight modifications... If we suppose that the ancient progenitor, the archetype as it may be called, of all mammals, had its limbs constructed on the existing general pattern, for whatever purpose they served, we can at once perceive the plain signification of the homologous construction of the limbs throughout the whole class. (Darwin 1859: 434, 435; see also 1872: 382, 383).

In the last edition of the *Origin*, Darwin specified that the archetype described in this last quote included not only the mammals but also the birds and reptiles. Regrettably for Darwin's equation between the principle of divergence and the notion of common ancestry, what he gained in favor of the latter he lost to the former. The field of morphology is too general—being naturally applied at high taxonomic levels—to serve the purpose of tracking the pattern-process of evolution with any precision. The study of the modification of fairly superficial features (homologies) grounded in a rock-solid common body plan (archetype) lacks in resolution for the kind of theory he is trying to build. Indeed, to claim that mammals, birds, reptiles, and fish share a common body plan is of little help in establishing the phylogenetic connections and transitions binding them (Darwin 1859: 338, 431, 1872: 310, 380). The same applies to Darwin's claim that the vertebrates of today are

[10]See also our analysis in Chap. 8 of this book.

built around a series of internal vertebrae bearing certain processes and appendages, a feature assumed to be present in an unknown progenitor of all vertebrates (1859: 437, 1872: 384, 385). It seems that the power of natural selection—when viewed from the perspective of currently living forms—is powerful enough to alter the vertebrate archetype to the extent that it blurs the connections and transitions between mammals, birds, reptiles, and fish. Furthermore, even the fossil record itself, continues Darwin, is of limited value in terms of connecting distinct families and orders:

> In nature the case will be far more complicated than is represented in the diagram; for the groups will have been more numerous, they will have endured for extremely unequal lengths of time, and will have been modified in various degrees. As we possess only the last volume of the geological record, and that in a very broken condition, we have no right to expect, except in a very rare cases, to fill up wide intervals in the natural system, and thus unite distinct families or orders. All that we have a right to expect, is that those groups, which have within known geological periods undergone much modification, should in the older formations make some slight approach to each other; so that the older members should differ less from each other in some of their characters than do the existing members of the same groups.... (Darwin 1859: 333; see also 1872: 304, 305)

Turning to embryology, Darwin (1859: 439, 1872: 387) gathers more support for common ancestry within classes: "The embryos, also, of distinct animals within the same class are often strikingly similar." So much so, Darwin insists that in the case of untagged embryos, Agassiz (1859: 439) or Von Baer (1872: 387, 388) were unable to identify the appropriate taxonomic groups, whether mammal, bird, or reptile. The fact that the section on embryology is more than twice the length of the morphology section in Chapter 13 may be indicative of the hope Darwin had invested in the former field, although it cannot be denied that it remains fairly short.

If embryos at this early age are relieved from struggling for survival—either because they are passively involved in life or sheltered for a time by parents—it could be argued that natural selection would not impact their development as much as it would in adult morphology. The traceability of descent, therefore, is increased in embryology. This is exactly what Darwin suggests in several places: "The point of structure, in which the embryos of widely different animals of the same class resemble each other, often have no direct relation to their conditions of existence" (1859: 439, 440; see also 1872: 388). Or again: "For the welfare of a very young animal, as long as it remains in its mother's womb, or in the egg, or as long as it is nourished and protected by its parents, it must be quite unimportant whether most of its characters are fully acquired a little earlier or later in life" (1859: 443, 444; see also 1872: 391). The value of embryology is, therefore, obvious:

> On this view we can understand how it is that, in the eyes of most naturalists, the structure of the embryo is even more important for classification than that of the adult. For the embryo is the animal in its less modified state; and in so far it reveals the structure of its progenitor. In two groups of animal, however much they may at present differ from each other in structure and habits, if they pass through the same or similar embryonic stages, we may feel assured that they have both descended from the same or nearly similar parents, and are therefore in that degree closely related. Thus, community in embryonic structure reveals community of descent. (Darwin 1859: 449; see also 1872: 395, 396)

It should be noted how this statement implies an epistemological distinction between the fields of morphology and embryology, a distinction at odds with Darwin's desire to comply with John Herschel's and William Whewell's methodological ideal of a "consilience of inductions." Beyond the superficial notion that morphology, embryology, and comparative anatomy all contribute to expanding the application of Darwin's theory to various fields—thus increasing its robustness and likelihood—it should again be remembered that such a consilience of inductions may be weakened by tensions between fields. By claiming that embryology is a more powerful tool for investigating descent, Darwin is also implicitly recognizing that morphology is more susceptible to the forces of natural selection (analogies, deletion of affiliative traces). What is gained in one field, then, is lost in another. Not only is the resolution level of the field of morphology low—restricted to supporting the notion of common ancestry—we now learn that, for this task, it may not be the most reliable field. This may explain why Darwin dedicated only five pages to the morphology section. What follows is especially revealing of the kind of epistemological break that may exist between the fields of morphology (adult) and embryology (embryo). The example concerns two different orders of Cirripedia (crustaceans):

> In most cases, however, the larvae, though active, still obey more or less closely the law of common embryonic resemblance. Cirripedes afford a good instance of this: even the illustrious Cuvier did not perceive that a barnacle was, as it certainly is, a crustacean; but a glance at the larva shows this to be the case in an unmistakeable manner. So again the two main divisions of the cirripedes, the pedunculated and sessile, which differ widely in external appearance, have larvae in all their several stages barely distinguishable. (Darwin 1859: 440; see also 1872: 389)

This example is surely a triumph for the equation of "common ancestry" and "embryology," but certainly not for that between "common ancestry" and "morphology." As Darwin made clear in the last edition of the *Origin*, his ambitions for embryology went beyond merely establishing common ancestry based on general resemblances of embryos. Indeed, he went one step further by arguing that embryos today may look like adults of the past, providing us with a means to take a peek at actual progenitors: "As the embryo often shows us more or less plainly the structure of the less modified and ancient progenitor of the group, *we can see why ancient and extinct forms so often resemble in their adult state the embryos of existing species of the same class*" (Darwin 1872: 396) [my emphasis]. Darwin's ambition is commendable, but how would he have known about the conformation of adult extinct forms, especially given he explicitly recognized that paleontology may never be able to independently confirm this (Darwin 1859: 338, 450, 1872: 310, 396)?

After having established, in principle at least, that embryology should be more revelatory of descent than other fields, Darwin had to face the harsh reality of an evolutionary engine erasing its traces as it moves through time. Once again, at issue in the embryology section is the maintaining of a delicate balance between "descent" and "natural selection." It should be noted that Chapters 9–13 of the *Origin* were designed to impose the principle of divergence (convergence backward in time) upon the wild power of natural selection, a power that when in excess blurs and mixes the pattern-process of evolution as postulated in the *strong contingency thesis*, a thesis to which Darwin did not subscribe. Darwin described the impact of adaptive pressure on embryological development in these words:

> The case, however, is different when an animal during any part of its embryonic career is active, and has to provide for itself. The period of activity may come on earlier or later in life; but whenever it comes on, the adaptation of the larva to its conditions of life is just as perfect and as beautiful as in the adult animal. From such special adaptations, the similarity of the larvae or active embryos of allied animals is sometimes much obscured; and cases could be given of the larvae of two species, or of two groups of species, differing quite as much, or even more, from each other than do their adult parent. (Darwin 1859: 440; see also 1872: 388, 389)

In addition, it should be noted that Darwin's enthusiasm for embryology was not free of interpretative biases. For instance, he writes: "No one will suppose that the stripes on the whelp of a lion, or the spots on the young blackbird, are of any use to these animals, or are related to the conditions to which they are exposed" (1859: 440; see also 1872: 388). Although he wanted to push his argument for early ontogenetic stages as revelatory of ancestry, a counterargument could be mounted for this specific case in favor of adaptive differences between mature and immature organisms, wherein youngsters are clearly marked from other individuals of the same group in order to reduce the impact of intraspecific competition. Darwin himself was well aware of the reality of intraspecific competition, as exemplified in the context of sexual selection in Chapter 4 of the *Origin* (1859: 87–90, 1872: 69, 70). Furthermore, Darwin's view of embryology was not always consistent throughout the *Origin of Species*. In Chapter 13, he did his best to use it as an investigative tool for "descent." However, in a passage in Chapter 4 intended to show what natural selection can achieve, embryological development becomes just another source of variation upon which it acts without restriction, across the embryo-adult spectrum:

> …in a state of nature, natural selection will be enabled to act on and modify organic beings at any age, by the accumulation of profitable variations at that age, and by their inheritance at a corresponding age. If it profit a plant to have its seeds more and more widely disseminated by the wind, I can see no greater difficulty in this being effected through natural selection, than in the cotton-planter increasing and improving by selection the down in the pods on his cotton-trees. Natural selection may modify and adapt the larva of an insect to a score of contingencies, wholly different from those which concern the mature insect. These modifications will no doubt affect, through the laws of correlation, the structure of the adult; and probably in the case of those insects which live only for a few hours, and which never feed, a large part of their structure is merely the correlated result of successive changes in the structure of their larvae. So, conversely, modifications in the adult will probably often affect the structure of the larva; but in all cases natural selection will ensure that modifications consequent on other modifications at a different period of life, shall not be in the least degree injurious… Natural selection will modify the structure of the young in relation to the parent, and of the parent in relation to the young. (Darwin 1859: 86, 87; see also 1872: 67)

Ultimately, Darwin resorted to adopting a theoretical posture to impose his view of evolution. The field of embryology, he insisted, generally complied with two complementary conditions that guaranteed its value for phylogenetic analyses (1859: 443–450, 1872: 391–396). Although not yet proven, continues Darwin, these two conditions are said to be to some degree probable and hopefully to be eventually demonstrated (1859: 446, 449, 1872: 396):

1. New features or changes in the embryological development appear during later stages rather than earlier ones. This condition encapsulates the idea that natural selection does not impact embryological development until later stages, thus leaving the earlier stages pristine.
2. Changes in embryological development appear at the same or corresponding age throughout all generations of the same lineage. This other condition enforces the idea of a great stability in the embryological development, one that even permits the use of changes or adaptive features as a means to trace descent, since they present themselves in a clockwork fashion throughout the generations.

Although Darwin (1859: 450) was committed to these two conditions, he clearly realized that what he calls the "law of resemblance" would not be valid if natural selection had obliterated the traces of affiliation: "It can be proved true in those cases alone in which the ancient state, now supposed to be represented in many embryos, has not been obliterated, either by the successive variations in a long course of modification having supervened at a very early age, or by the variations having been inherited at an earlier period than that a which they first appeared" (1859: 449, 450; see also 1872: 396). The search for a delicate balance between "descent" and "natural selection" is what Darwin was seeking in the embryology section, ultimately hoping to keep the power of the latter under the control of the former. Regrettably for Darwin's embryological theory, counterfactual evidence was encountered. Of the first condition, he writes: "But in other cases it is quite possible that each successive modification, or most of them, may have appeared at an extremely early period [rather than at a later one]" (Darwin 1859: 444).[11] Similarly, when speaking of the second condition appealing to changes at a corresponding age, Darwin wrote: "I am far from meaning that this is invariably the case; and I could give a good many cases of variations... which have supervened at an earlier age in the child than in the parent" (1859: 444; see also 1872: 391).

Darwin's embryological theory is not always convincing in part because of his convoluted way of presenting his arguments. First, he seeks examples from domestic animals like dogs, horses, and pigeons (1859: 444–446, 1872: 392). In doing so, not only does he present counterfactual evidence at odds with his theory, he also turns to an elaborate method based on measurements, after having implicitly recognized that a qualitative assessment of cases was also counterfactual. Darwin then moved from domestic cases to what he calls "species in a state of nature" (1859: 446, 447, 1872: 393). But here, the reader is not presented with genuine factual evidence. Rather, Darwin turns to an entirely hypothetical scenario appealing to the evolution of birds and other vertebrates, one that uses Darwin's own embryological theory as its argumentative structure! To call this subterfuge "rhetorical" would be putting it mildly.

[11]This phrase was cut from the last edition of the *Origin of Species*.

Let us raise a question that might throw more light on Darwin's ontological view of the world. Why would he want to subscribe to the "law of resemblance" as enshrined in his two embryological conditions? At a superficial level, this question finds a perfectly satisfactory answer: an explanation had to be found to make sense of the embryological similarities uniting presumably related forms, as conceived by Darwin's theory (common ancestry). Or to reverse the perspective, since related forms have embryological developments that exhibit similarities, especially in their earlier stages, it becomes possible to make sense of this fact under the law of resemblance. So far, so good. At a more profound level, however, this question becomes more interesting. Was it truly necessary for Darwin to commit himself to the two conditions accompanying the law of resemblance knowing, as Darwin did, that there existed many exceptions? Why did Darwin not follow the same strategy he adopted in the morphology section of Chapter 13, which consisted in keeping his argument at the *descriptive level* (morphological similarities = common ancestry)? After all, the epistemological benefit would have been the same with the establishment of an equation between embryology and common ancestry. Instead, Darwin engaged embryology at the *nomological or causal level*, searching for the conditions behind a law he hoped to be as universal as possible. Apparently, Darwin turned his back on the descriptive level thinking he could gain something more from embryology than mere support for common ancestry. The perspective of that gain outweighed, in Darwin's mind, the potential risk of exposing his theory to refutation, for it should be realized that the more precise Darwin was about the conditions of application of his theory, the more it was exposed to precise criticisms.

In line with one of the main theses proposed throughout this book, we would suggest that Darwin approached embryology the way he did to compensate, in part at least, for the obvious limitations that accompanied his ontological view of the world. As a proponent of a fully constituted world, Darwin assumed that the present always reveals more than the past. This is not the place to recapitulate Darwin's reasons for believing this. With embryology, Darwin added one more argument to his list, but one that he thought more decisive than others: presumably, *embryology allows him a glimpse at the actual conformation of common progenitors*. If we may be allowed a psychological speculation here, we would submit that Darwin felt insecure about the validity of his methodological procedures in light of his ontological commitments. To tackle the reality of evolution in the 1830s, 1840s, and 1850s with the intellectual toolkit of the seventeenth and eighteenth centuries—a set of tools grounded in a static worldview—was not without it risks (whether or not Darwin was fully conscious of it). To assume, as Darwin does throughout the *Origin of Species*, that one finds in the present everything essential for phylogenetic analyses constitutes a gutsy move that must be appreciated at its full value. As argued in Chap. 2 of this book, even William Whewell foresaw as early as 1832 the epistemological limitations that came from investigating the past with the tenets of actualism and uniformitarianism.[12] In this intellectual context, the kind of science proposed by Charles Lyell in the *Principles of Geology* (1830–1833) and even more so by Charles Darwin in the *Origin of Species* (1859–1872) has the feel of an earlier time. No surprise, then, that Darwin had to apply thick layers of rhetoric throughout the *Origin* to give the impression that his science was of a new age.

On this view, it is perhaps unsurprising to see Darwin push the field of embryology beyond its reasonable utility for his theory. This may explain his eagerness to go beyond the mere descriptive level for embryology by entering nomological territory, thus exposing his theory to a greater risk of refutation. Here, we reach the heart of the matter: the two conditions imposed by Darwin on behalf of the "law of resemblance"—(1) changes come at later embryological stages rather than at earlier ones; (2) changes come at corresponding embryological ages within the evolutionary history of lineages—are remarkably rigid conditions. They put evolution in a straitjacket, depriving it of its full evolutionary flexibility. Are we to suppose that such conditions are consistent with the kind of evolutionary contingency claimed under Darwin's rhetoric to the effect that "there is no law of development"?

It seems this constitutes yet another attack from the part of Darwin on the *strong contingency thesis*. Indeed, evolution's contingent character requires embryological flexibility, for several reasons. For instance, as evolutionary lineages continue to accumulate adaptations over geological ages, the embryological development will need to either to be extended or accelerated. If the process is accelerated, the entire series of stages will be compressed or deleted, thus impacting even the earliest stages. If the process is extended, the life history of the species involved will be modified by increasing both the length of the gestation period and the amount of energy required before birth. However, there are limits to how far embryological development can be extended without either putting the survival of females at risk or requiring some level of embryological compression, which opens the door to organogenesis (the internal laws of biological organization). Moreover, those forms lower on the scale of complexity are hardly equipped to face that degree of change as compared to higher forms. Darwin's law of resemblance deprives evolution of flexibility on issues like the embryological condensation/extension spectrum, the impact of such embryological changes on the life history of species, the laws of organogenesis (poorly understood during Darwin's lifetime), and the way adaptive features are introduced, or fail to be, in embryological development.

Nobody can criticize Darwin for not knowing the answers to these embryological questions at such an early period in the history of that field. One should be surprised, however, at how Darwin confined evolution within rigid embryological conditions. Just as Darwin imposed a rigid way to view the evolutionary history of lineages and

[12]It will be argued in Chap. 8 of this book that Darwin's actualism is founded on uniformitarianism (a steady-state view).

the tree of life,[13] he imposed the same rigidity upon embryology. Apparently, Darwin's methodical approach to various fields responded to the same epistemological imperative: to keep the past under the control of the present. Darwin's law of resemblance surely allows for some degree of evolutionary flexibility, but one that is more consistent with the *weak contingency thesis* than the *strong contingency thesis*. If most of the past is to be reflected in today's taxonomy, one must be certain that the evolutionary engine will not erase all there is to know about affiliation. Again, we return to the inherent tension contained in Darwin's theory between "descent" and "natural selection." In order to keep the power of natural selection under the control of descent, Darwin had to ensure that the latter was sheltered from the excessive impact of the former. This, we argue, constitutes the precise epistemological function of the two conditions behind the law of resemblance. In the last paragraph of the embryological section, Darwin writes: "Thus, as it seems to me, the leading facts in embryology, which are second in importance to none in natural history, are explained on the principle of slight modifications not appearing, in the many descendants from some one ancient progenitor, at a very early period in the life of each, though perhaps caused at the earliest, and being inherited at a corresponding not early period" (1859: 450; see also 1872: 396). Differently stated, embryology will preserve from "natural selection" what needs to be preserved about "descent."

Irrespective of Darwin's theoretical posture, the fact remains that he often uses embryology in too general a way to serve the purpose of documenting with any precision the pattern-process of evolution. To make a truly significant contribution to his theory as a whole—and not merely to the notion of common ancestry—Darwin had to identify embryological stages late enough in age to be used for the discrimination of competing phylogenetic hypotheses in groups at fairly low taxonomic levels. Interestingly, Darwin moved in that direction for some cases:

> A trace of the law of embryonic resemblance, sometimes lasts till a rather late age: thus birds of the same genus, and of closely allied genera, often resemble each other in their first and second plumage; as we see in the spotted feathers in the thrush group. In the cat tribe, most of the species [when adult] are striped or spotted in lines; and stripes can be plainly distinguished in the whelp [a young animal] of the lion. We occasionally though rarely see something of this kind in plants: thus the embryonic leaves of the ulex or furze, and the first leaves of the phyllodineous acaceas, are pinnate or divided like the ordinary leaves of the leguminosae. (Darwin 1859: 439; see also 1872: 388)

To these cases, one must add the example involving two different orders of Cirripedia (crustaceans) quoted above. This being said, Darwin stopped short of exploiting the potential of these cases. Apparently, he underestimated the value of such examples for bolstering the core of his theory in its entirety, preferring instead to rely on embryological cases of general value only (common ancestry). As already alluded to, this reasoning constituted a mistake on his part. For a consilience of inductions to be fully effective, it must be demonstrated that the core of a theory

[13]This has been noted in Chap. 2 of this book for the field of paleontology and in this chapter for the field of systematics.

equally applies to a range of fields. Perhaps for lack of sufficient proof or lack of clarity of mind, Darwin only partially supported his theory by relying on the assumption that "common ancestry" = "divergence." And as if things we not muddled enough, the tensions already noted between the fields of morphology and embryology, as well as the inconsistencies raised within the field of embryology itself, further contributed to weakening the value of the consilience of inductions pursued by Darwin. It is perhaps no accident that Darwin concluded his embryology section on an ambiguous note: "Embryology rises greatly in interest, when we thus look at the embryo as a picture, *more or less obscured*, of the common parent-form of each great class of animals" (1859: 450; see also 1872: 396) [my emphasis]. Was that formula intended to build confidence in the value of embryology for phylogenetic analyses?

Let us conclude this discussion by mentioning that it is not clear which of the two main embryological theories of his time Darwin embraced:

1. The "law of parallelism," promoted by Johann Meckel, Étienne Serres, and Ernst Haeckel, which sees the embryological development as a linear series of stages organized along the hierarchy of extant animals leading to humans, wherein all embryos follow an identical path, although lower animals stop at an earlier stage than higher ones
2. The "law of specialization," held by Karl Ernst von Baer, which argues that vertebrate embryos all start from a similar stage only to branch off in separate directions as they each differentiate into different classes

Darwin scholars are not agreed about this question (Gould 1977: 69–74; Ospovat 1981: 146–169; Richards 1992; Nyhart 2009). Although Von Baer's theory seems more congenial to Darwin's theory, since both rely on the notion of divergence, the fact remains that the whole issue is of little consequence for our thesis here, if only because Darwin used embryology in a general way only and at a low-resolution level.

Darwin closes Chapter 13 of the *Origin* with a section devoted to comparative anatomy, whose exact title reads "Rudimentary, atrophied, or aborted organs." On Darwin's theory of biological evolution, it is predicted that the building blocks of life are to be continually modified, with some new features appearing under positive selection and some others being eliminated under negative selection. It is expected, Darwin continued, that this ongoing process will generate "leftovers," that is, traces of past material no longer useful for the survival of organisms but now seen under an atrophied state owing to their neutral selectivity:

> Organs or parts in this strange condition, bearing the stamp of inutility, are extremely common throughout nature... In tracing the homologies of the same part in different members of a class, nothing is more common, or more necessary, than the use and discovery of rudiments... As the presence of rudimentary organs is thus due to the tendency in every part of the organisation, which has long existed, to be inherited—we can understand, on the genealogical view of classification, how it is that systematists have found rudimentary parts as useful as, or even sometimes more useful than, parts of high physiological importance. Rudimentary organs may be compared with the letters in a word, still retained in the spelling, but become useless in the pronunciation, but which serve as a clue in seeking for its

derivation. On this view of descent with modification, we may conclude that the existence of organs in a rudimentary, imperfect, and useless condition, or quite aborted... might even have been anticipated, and can be accounted for by the laws of inheritance. (Darwin 1859: 450, 452, 455, 456; see also 1872: 397, 399, 402)

In other words, the adaptive engine erases its traces as it moves along, but some traces are retained in the form of rudimentary features. Once again, the main epistemological stake of this section of Chapter 13 consists in the traceability of descent, with Darwin trying to maintain a delicate balance between "descent" and "natural selection." Of course, Darwin's goal in promoting the equation between "rudimentary features" and "common ancestry" was ultimately intended to support the notion that common ancestry also logically implies the principle of divergence.

Having seen the principles for rudimentary features set out so carefully, the reader might reasonably expect factual evidence supporting them. As is so typical of Darwin, however, the reader is guided through a convoluted series of facts explicitly supporting or disproving them, in addition to being offered a list of cases whose exact explanatory role remains unspecified or implicit only. Let us try to introduce some order into this situation by raising several legitimate questions. Among cases of rudimentary organs capable of throwing light on past evolutionary stages—although these are merely presented descriptively without any further comments from Darwin—are the following (1859: 450, 451, 1872: 397): (a) bastard wings (alulae) in birds representing a digit; (b) teeth in fetal whales that disappear in adult organisms; (c) teeth in baby cows (calves) that never cut through the gums; (d) teeth in the beaks of some embryonic birds; (e) vestigial and useless wings in many insects of different groups or within the same species; (f) four developed and two rudimentary teats among forms of the genus *Bos* (cattle) (although in domestic cows the six teats are developed and functional). The reader is certainly prepared to grant Darwin the possibility that wings were originally a collection of digits, that some toothless forms are derived from ancestors that earlier had fully developed teeth or had more of them, that vestigial wings in insects testify to a fully winged ancestor, and that the ancestor of the genus *Bos* originally had six functional teats.

However, other cases of rudimentary organs also presented without further comment from Darwin raise legitimate skepticism. For instance, speaking of mammals, Darwin writes: "[R]udimentary mammae are very general in the males of mammals... Rudimentary organs sometimes retain their potentiality, and are merely not developed: this seems to be the case with the mammae of male mammals, for many instance are on record of these organs having become well developed in full-grown males, and having secreted milk" (1859: 450, 451; see also 1872: 397). What is Darwin trying to suggest here? That males and females among the mammals were once both involved in breastfeeding? Although the *Origin* remains silent on this question, Darwin speculated about such an ancestral condition in *The Descent of Man* (1871: 207–211). Although his answer is not without ambiguity, he was apparently quite open to the idea:

The mammary glands and nipples, as they exist in male mammals, can indeed hardly be called rudimentary; they are simply not fully developed and not functionally active... At birth they often secrete a few drops of milk; and they have been known occasionally in man

> and other mammals to become well developed, and to yield a fair supply of milk. Now if we suppose that during a former prolonged period male mammals aided the females in nursing their offspring, and that afterwards from some cause, as from a smaller number of young being produced, the males ceased giving this aid, disuse of the organs during maturity would lead to their becoming inactive; and from two well-known principles of inheritance this state of inactivity would probably be transmitted to the males at the corresponding age of maturity. But at all earlier ages these organs would be left unaffected, so that they would be equally well developed in the young of both sexes. (Darwin 1871: 210, 211)

Darwin's interesting speculation notwithstanding, the fact remains that the reader of the *Origin* is not obliged to follow him, adopting instead a more obvious line of argumentation: the laws of organogenesis working through the duplication of identical structures, should it be between males and females of the same species or within the same organism. In other words, no need for elaborate speculation about ancestral conditions. Rather, a nonhistorical explanation can be proposed, one appealing to the internal cohesion of the whole organism by which its organization is guided. It should be remembered that Darwin speaks again and again of the "correlation of growth" throughout the *Origin*, wherein the organism's parts are somehow interrelated and impact each other.[14] There is, in Darwin, an awareness of this holistic reality in the organism. In *The Descent of Man*, for instance, Darwin (1871: 209) alludes to the biological notion of duplication of identical structures across the sexes: "We seem, therefore, compelled to... conclude that the nipples were first developed in the females of some very early marsupial [mammal] form, and were then, in accordance with a common law of inheritance, transferred in a functionally imperfect condition to the males." The point we want to raise here is the following: ancestral conditions may explain some of the rudimentary features, but Darwin himself implicitly suggested to the careful reader of the *Origin* through the notion of correlation of growth that a number of so-called rudimentary features may be more directly explainable by organogenesis than by ancestry. Affiliative lines may therefore just be illusory under the then-unknown laws of organogenesis. In light of two possible explanations—ancestry versus organogenesis—the following quote of the *Origin* finds two different interpretations:

> Rudimentary organs in the individuals of the same species are very liable to vary in degree of development and in other respects. Moreover, in closely allied species, the degree to which the same organ has been rendered rudimentary occasionally differs much. This latter fact is well exemplified in the state of the wings of the female moths in certain groups. Rudimentary organs may be utterly aborted; and this implies, that we find in an animal or plant no trace of an organ, which analogy would lead us to expect to find, and which is occasionally found in monstrous individuals of the species. (Darwin 1859: 452; see also 1872: 399)

Does the scattered manifestations of such features variously developed constitute support for the notion of common ancestry or for the laws of organogenesis, which include teratology (abnormal development)? In at least one place in this section of Chapter 13, Darwin explicitly recognizes the nonhistorical reality of organogenesis.

---

[14]See Darwin (1859: 5, 7, 11, 12, 33, 43, 85, 131, 143–150, 146, 168, 196, 199, 199, 200, 435, 466, 473, 486).

In a critical and sarcastic passage, he writes: "An eminent physiologist accounts for the presence of rudimentary organs, by supposing that they serve to excrete matter in excess... When a man's fingers have been amputated, imperfect nails sometimes appear on the stumps: I could as soon believe that these vestiges of nails have appeared, not from *unknown laws of growth*, but in order to excrete horny matter..." (1859: 453, 454; see also 1872: 400) [my emphasis]. The value of rudimentary features in phylogenetic analyses is further undermined by Darwin himself, when he candidly announces that such features appear under domestication. If true, this would mean that rudimentary features have arisen within a limited number of generations, thus contributing to blurring the distinction between features that emerged quite recently and those that arose out of a deep past. The way Darwin tried to extract himself from this uncomfortable predicament is both interesting and revealing:

> On my view of descent with modification, the origin of rudimentary organs is simple. We have plenty of cases of rudimentary organs in our domestic productions,—as the stump of a tail in tailless breeds,—the vestige of an ear in earless breeds,—the reappearance of minute dangling horns in hornless breeds of cattle, more especially, according to Youatt, in young animals,—and the state of the whole flower in the cauliflower. We often see rudiments of various parts in monsters. But I doubt whether any of these cases throw light on the origin of rudimentary organs in a state of nature, further than by showing that rudiments can be produced; for I doubt whether species under nature ever undergo abrupt changes. I believe that disuse has been the main agency; that it has led in successive generations to the gradual reduction of various organs, until they have become rudimentary,—as in the case of the eyes of animals inhabiting dark caverns, and of the wings of birds inhabiting oceanic islands, which have seldom been forced to take flight, and have ultimately lost the power of flying. (Darwin 1859: 454; see also 1872: 400, 401).

This extraordinary passage highlights three issues. First, Darwin claims that artificial selection is useful when it comes to showing that rudimentary organs are a reality but at the same time denies that those generated by such means have any phylogenetic value. The least that can be said is that Darwin's analogy between artificial selection and natural selection is a very loose and flexible one, epistemologically speaking, selecting from cases under domestication whatever he wants before transferring them to cases in nature. In fact, even a perfunctory analysis of the relationship between artificial selection and natural selection throughout the *Origin* shows Darwin's inconsistency in his use of this analogy, with artificial selection sometimes depicted as either more powerful or less powerful than natural selection.[15] It seems that Darwin subscribed to another double epistemological standard: when it suits him he buys it, when it does not he leaves it.

Second, by listing the rudimentary organs that appeared under domestication—and by dismissing them—Darwin introduces a general doubt in the mind of the reader about the significance of all the other cases mentioned in this section of Chapter 13. Which cases are truly revealing of deep ancestry and which are not?

[15]Darwin (1859: 29–31, 38–39, 61, 82–84, 109, 213, 269, 454).

Third, more light is thrown here on the value Darwin attaches to one of the two pillars at the core of his theory: gradualism-gradation. In the quote above, we see Darwin dismissing rudimentary organs produced under domestication on the grounds that evolution in nature is a slow process. He is even clearer about this in the last edition of the *Origin* (1872: 400): "species under nature do not undergo great and abrupt changes." In other words, any evidence showing that evolution is a rapid or sudden process should be dismissed. This blind commitment to gradualism-gradation is a proof of its irrefutability in Darwin's mind, having precedence over explanatory considerations emerging from the all-important analogy between artificial selection and natural selection in the *Origin of Species*.

Darwin's contradictions and inconsistencies in this section of Chapter 13 do little to establish the phylogenetic utility of rudimentary features. As if things were not already confused enough, Darwin ultimately turned to his embryological theory—and its two conditions of application[16]—in order to ground the legitimacy of such features:

> At whatever period of life disuse or selection reduces an organ, and this will generally be when the being has come to maturity and to its full powers of action, the principle of inheritance at corresponding ages will reproduce the organ in its reduced state at the same age, and consequently will seldom affect or reduce it in the embryo. Thus we can understand the greater relative size of rudimentary organs in the embryo, and their lesser relative size in the adult. But if each of the process of reduction were to be inherited, not at the corresponding age, but at an extremely early period of life (as we have good reason to believe to be possible) the rudimentary part would tend to be wholly lost, and we should have a case of complete abortion. (Darwin 1859: 455; see also 1872: 401)

We have already seen the ways in which Darwin's embryological theory confines the embryological development in a fairly rigid explanatory framework that is at odds with the *strong contingency thesis*. As for any other morphological features, this preceding quotation also reveals that Darwin is perfectly aware that rudimentary organs could be entirely erased under selective pressure, thus blurring or deleting the traces of affiliation.

The section devoted to rudimentary organs constitutes no great contribution to Darwin's attempt to work toward a consilience of inductions. Instead of reinforcing the sections already devoted to the fields of morphology and embryology, this section encounters its own set of problems:

1. Difficulties in distinguishing between ancestral features and features modulated under the laws of organogenesis
2. The limitations associated with Darwin's rigid embryological theory now transferred to the study of rudimentary organs

[16]The two conditions supporting Darwin's embryological theory are the following: (1) changes in the embryological development appear during later stages rather than earlier stages, thus preserving the state of earlier ones, and (2) changes in the embryological development appear at the same or corresponding age throughout all the generations of the same lineage.

3. The ad hoc explanations introduced to distinguish between rudimentary organs produced under domestication and those in a state of nature, the argument as a whole leading to confusion with respect to legitimate and illegitimate features
4. The recognition that rudimentary organs could also be erased under the action of natural selection, losing their value for phylogenetic analyses
5. Darwin's lack of ambition with respect to using rudimentary organs as a tool for discriminating between competing phylogenetic hypotheses, preferring instead to remain at a very general level of argumentation

It cannot be denied that Darwin's study of rudimentary organs somewhat strengthened the general idea of biological evolution and the specific notion of common ancestry. But other key components of Darwin's theoretical core, such as gradualism-gradation and the principle of divergence, were not directly supported by their study.

To conclude this analysis, a brief overview of Chapter 13 of the *Origin* may not be out of place. Darwin hoped to use the field of taxonomy as an explanatory backbone around which to organize auxiliary fields like morphology, embryology, and comparative anatomy. Clusters of forms regrouped within ever-more inclusive clusters along the taxonomic scale were interpreted as registering a pan-divergent view of evolution. The notion of convergence—the principle of divergence applied backward in time—was exploited as a conceptual instrument, organizing a unified and all-encompassing evolutionary area, with surrounding fields acting as independent yet complementary classes of facts responding to the ideal of a science based on a consilience of inductions. Such were Darwin's intentions. He was confronted with two main difficulties.

First, he had to combat the problem of an evolutionary engine erasing its traces as its moves forward in time, a by-product of natural selection's dual and extreme faces as seen in the rise of complete novelties (ancestorless forms or features) and perfect imitations (i.e., analogies, parallelism, evolutionary convergence). Darwin confronted the latter case when discussing the phylogenetic relations of marsupials and placental mammals (rodents), with limited success. Imposing order upon the complexity of evolution, he adopted simplifying assumptions that took him away from the strong contingency thesis by putting evolution in a conceptual straitjacket:

1. Evolution today is believed to be at its maximum state of divergence, implying that all past forms are either aligned with or falling in-between extant forms.
2. Generally speaking, the selective process is thought to retain the most divergent forms with less divergent ones being driven to extinction.
3. The taxonomic *arrangement* among extant forms is assumed to have registered all the main events by maintaining itself through the entirety of the evolutionary history.

Second, the sought-after explanatory unity was jeopardized by the nature of the relationship Darwin established between fields. Taking for granted that "common ancestry" equals "divergence," Darwin managed to gather support for the former concept when reviewing fields like morphology, embryology, and comparative anatomy, if only because his analyses were couched only in general terms, leaving aside considerations about the actual process and pattern of evolution. In addition, numerous contradictions and inconsistencies were introduced and remained unanswered throughout Chapter 13. In the case of rudimentary organs, for instance, Darwin had difficulties distinguishing between truly ancestral features and those generated by the laws of organogenesis. Even worse, he resorted to ad hoc explanations when attempting to distinguish rudimentary organs generated either under domestication or in a state of nature. Far from using the auxiliary fields to bolster the case for a full evolutionary contingency—contrary to his claim that "there is no law of development"—Darwin took advantage of them to confine evolution inside a rigid interpretation. His embryological theory constitutes an obvious case of this by being based on the two assumptions that changes appear at later embryological stages (not at earlier ones) and that such changes appear at corresponding embryological ages within lineages. Taken collectively, Chapter 13 of the *Origin* is far from complying with an ideal science founded on a consilience of inductions: the chapter's sections do not all independently support Darwin's theory, while the robustness of such an overarching theory is undermined by contradictions and inconsistencies stemming from the various fields. It is no surprise that William Whewell was not impressed by Darwin's achievements in the *Origin of Species*, notwithstanding his own opposition to the idea of biological evolution.[17]

6.4 Conclusion of Part II

Darwin aimed at offering a unified and wide-ranging evolutionary theory complying with the best epistemological canons of his time. That is why he divided the *Origin of Species* in two main parts:

Chapters 1–5: A part describing the process of divergence forward in time under the drive of a selective mechanism responsible for the ascent of the taxonomic levels (vera causa)

Chapters 9–13: A part reversing the viewpoint by projecting backward in time the pattern of convergence across taxonomic levels with the assistance of a host of scientific disciplines (consilience of inductions)

Darwin's overall picture of evolution is encapsulated in his Diagram (see Fig. 2.1). The visual representation depicts evolution as being perfectly identical at whichever taxonomic level the observer is placed. It must be recognized that this theoretical construct is impressive in its explanatory unity and symmetry. But how

[17]See a letter written by William Whewell in Todhunter, I. (1876), *William Whewell, D.D. Master of Trinity College Cambridge* (London: Macmillan), vol. 2, pp. 433–434. Part of this letter is reproduced in Hull, D.L. (1989), *The Metaphysics of Evolution* (Albany: State University of New York Press), p. 30. Whewell's reaction to the publication of the *Origin of Species* will be discussed in a different context in Chap. 8 of this book.

much could Darwin really gather through the lens of a theory ultimately founded on the ontological assumption that the world reveals most of itself in the present time? The neat and pan-divergent view Darwin imagined was, at low taxonomic levels, blurred by reticulate evolution and polyphyletism, and at higher levels, obscured by deleted traces of affiliation and parallel evolution (analogies). That is why the *Origin* contains from one end to the other implicit or explicit references to these issues. It seems, therefore, entirely legitimate to make a distinction between Darwin's ambitions and his actual achievements. This we have done through a formal separation between the rhetorical Darwin and a critical reading of his *magnum opus*. When looked at closely, the *Origin* offers to the critical reader a profoundly dislocated explanatory structure, and a series of competing evolutionary pictures, several of which are incompatible with one another:

Chapters 1–5, 8: A confused evolutionary pattern-process caught up in common reproductive networks at low taxonomic levels. This picture emerges from Darwin's commitment to the tradition of natural theology as applied to the life sciences, wherein life forms are characterized by complex interrelationships.

Chapters 6 and 7: The wild power of natural selection overrunning the pattern-process of divergence through a directional ascent of taxonomic levels and evolutionary grades, thus blurring the homology-analogy distinction and stressing "natural selection" over "descent." This picture seems an overreaction by Darwin to the difficulties he encountered in Chapters 1–5. His overall theory in the rest of the *Origin* is antithetical to this series of independent and vertical ascents. The cohabitation of these two theories within the same book undermines one another.

Chapters 9 and 10: The paleontological annals are seen as largely lacking in novelties, with past forms falling entirely within the range of biological variability or disparity observed among extant forms. This kind of paleontology appeals more to "shadows" than "ancestors." This picture of evolution applies with full explanatory force the notion of a world believed to be fully constituted, with nothing to be gained by looking into the past or the future. This is another intellectual by-product of natural theology.

Chapters 11 and 12: An ideal model is applied for which explanatory variables are calibrated along symmetrical proportions: amount of evolutionary change (taxonomic level) is proportional to the time elapsed since the separation of two forms from their common ancestor, a fact also proportional to the geographical distance traveled and the level of selective pressure. This equilibrium model is yet another manifestation of the tradition of natural theology. As a default position, this ideal model is applied unless the facts cannot be arranged to fit the theory. When this is not the case, core variables (divergence and gradation) are treated as irrefutable and unmovable, thus transferring the explanatory stress to other auxiliary variables (such as geological timeframe, taxonomic level, geographical distance, and selective pressure) which are thought to be somewhat flexible. Important clues about the nature of Darwin's theory are revealed when he moves beyond the default position, as this forces him (1) to negate the strong contingency thesis in favor of the weak contingency thesis and (2) to treat natural selection as a mere auxiliary variable that can be adjusted to protect the theory's core, thus increasing or reducing its power at will in order to match the theory's predictions.

Chapter 13: The application of the strong contingency thesis is also thwarted through the use of taxonomy as a conceptual tool to impose a rigid pan-divergent outlook, and enclosing evolution within the biological range of currently existing forms through convergence backward in time (the principle of divergence). The action of natural selection is thus tamed by preventing it from blurring evolutionary connections (such as analogies, ancestorless features). In addition, an attempted synthesis between the disciplines of taxonomy, morphology, embryology, and comparative anatomy leads to internal inconsistencies and contradictions, resulting in a permanent confusion between the distinct notions of "common ancestry" and "divergence," as well as weakening the general case in favor of Darwin's theory.

Taken as a whole, Darwin's self-proclaimed unified and all-encompassing theory of evolution—"this volume is one long argument" (1859: 459, 1872: 404)—constitutes quite a challenge to a critical reader. The following passage from a review of the *Origin of Species* (1859) by T. H. Huxley, a contemporary supporter of Darwin, takes on an interesting twist in light of our analysis:

> But this very superabundance of matter must have been embarrassing to a writer who, for the present, can only put forward an abstract of his views; and thence it arises, perhaps, that notwithstanding the clearness of the style, those who attempt fairly to digest the book find much of it a sort of intellectual pemmican—a mass of facts crushed and pounded into shape, rather than held together by the ordinary medium of an obvious *logical bond*: due attention will, without doubt, discover this bond, but it is often hard to find. (Huxley 1860: 542) [my emphasis]

Was there ever such a "logical bond"? We think not, at least not in light of our modern understanding of evolutionary biology.[18] Darwin's Diagram seems to serve the purpose of masking the tensions of a theory rather than of revealing its internal consistency. Jean Gayon (2009: 282) had this say of Darwin's Diagram:

> More than anything else, it is Darwin's tree-of-life diagram that summarized his proposal that the entire history of life can be representing as a general phenomenon of gradual modification, splitting, divergence and extinction of species... It had an almost immediate effect upon the entirety of the biological community. Within a remarkably short period of time, it became the paradigmatic representation of organic evolution and its status as an established fact... Such a theory looks like a descriptive generalization, but it is not: *it is a heuristic device, a plausible bet* about the general form and pace of the big classes of phenomena that constitute organic evolution. [my emphasis]

Although Darwin's overall metaphor has since been challenged from different perspectives (see Morrison 2014), we should like to add here that the first to have challenged Darwin's "heuristic device and plausible bet" was Darwin himself, in the

[18]The concluding chapter of this book will raise the question of the conditions for restoring unity to the *Origin of Species*.

form of the several competing views he enshrined in the *Origin of Species*. In our view, Darwin's Diagram is best envisioned as a propaganda machine at the service of his rhetoric.

References

Archibald JD (2012) Darwin's two competing phylogenetic trees: marsupials as ancestors or sister taxa? Arch Nat Hist 39(2):217–233

Bowler PJ (1990) Charles Darwin: the man and his influence. Basil Blackwell, Oxford

Darwin C (1859) On the origin of species. John Murray, London

Darwin C (1868) The variation of animals and plants under domestication, vol 1. John Murray, London

Darwin C (1871) The descent of man, and selection in relation to sex. John Murray, London

Darwin C (1872) The origin of species, 6th edn, with additions and corrections [1876]. John Murray, London

Feldhamer GA, Drickamer LC, Vessey SH, Merritt JF, Krajewski C (2015) Mammalogy: adaptation, diversity, ecology, 4th edn. Johns Hopkins University Press, Baltimore

Gayon J (2009) From Darwin to today in evolutionary biology. In: Hodge MJS, Radick G (eds) The Cambridge companion to Darwin, 2nd edn. Cambridge University Press, Cambridge, pp 277–301

Gould SJ (1977) Ontogeny and phylogeny. Belknap Press, Cambridge

Hodge MJS (1977) The structure and strategy of Darwin's 'long argument'. Br J Hist Sci 10:237–246

Hodge MJS (1989) Darwin's theory and Darwin's arguments. In: Ruse M (ed) What the philosophy of biology is. Kluwer, Dordrecht, pp 163–182

Hull DL (1989) The metaphysics of evolution. State University of New York Press, Albany

Huxley TH (1860) Darwin on the Origin of Species. Westminster Rev 17:541–570

Kemp TS (2005) The origin and evolution of mammals. Oxford University Press, Oxford

Laudan L (1981) Science and hypothesis: historical essays on scientific methodology. D. Reidel, Dordrecht

Mivart SG (1871) On the genesis of species. Macmillan, London

Morrison DA (2014) Is the tree of life the best metaphor, model, or heuristic for phylogenetics? Syst Biol 63:628–638

Nyhart LK (2009) Embryology and morphology. In: Ruse M, Richards RJ (eds) The Cambridge companion to the 'Origin of Species'. Cambridge University Press, Cambridge, pp 194–215

Ospovat D (1981) The development of Darwin's theory: natural history, natural theology, and natural selection, 1838–1859. Cambridge University Press, Cambridge

Richards RJ (1992) The meaning of evolution. University of Chicago Press, Chicago

Ruse M (1975) Darwin's debt to philosophy: an examination of the influence of the philosophical ideas of John F.W. Herschel and William Whewell on the development of Charles Darwin's theory of evolution. Stud Hist Phil Sci 6:159–181

Ruse M (1979) The Darwinian revolution. University of Chicago Press, Chicago

Ruse M (2000) Darwin and the philosophers: epistemological factors in the development and reception of the theory of the Origin of Species. In: Creath R, Maienschein J (eds) Biology and epistemology. Cambridge University Press, Cambridge, pp 3–26

Todhunter I (1876) William Whewell, D.D. master of Trinity College Cambridge, vol 2. Macmillan, London

Part III
Evolutionary Dynamics

Chapter 7
Cyclicity, Evolutionary Equilibrium, and Biological Progress

Abstract Charles Darwin believed in biological progress. Yet many modern scholars puzzled over how this belief could be compatible with his theory which, presumably, was based on "evolutionary opportunism." We hold that the puzzle is even more problematic for the opposite reason: whereas evolutionary opportunism might eventually have found the ways to progress, Darwin ultimately deprived the evolutionary process of flexibility by confining it within the straitjacket of a rigid and pre-established pan-divergent pattern. In addition, having constrained evolution within nearly stable and permanent classes/phyla, Darwin restricted evolutionary motion to lower taxonomic entities, envisioning them as falling under endless cycles of rising and declining forms. At a very fundamental explanatory level, Darwin's view appealed to an equilibrium system built around devices such as (1) demographic expansion performed at the proportional expense of demographic decline; (2) an evolutionary drive involving related forms that exhausts itself in reverse proportion to their taxonomic ascent; and (3) an evolutionary stalemate maintained between forms interlocked in a mutual adaptive deadlock. The biological progress recognized by Darwin does not ensue from the application of his theory; rather, it is an empirical fact that his theory cannot explain.

Keywords Biological progress · Evolutionary opportunism · Pan-divergent view · Equilibrium system · Proportional devices · Demographic expansion · Demographic decline · Fully stocked world · Not fully stocked word · Adaptive equilibrium · Species question · Realism · Nominalism

We have seen that Darwin confined the complexity of evolution within a pan-divergent view. The manifestation of alternative patterns and processes such as polyphyletism, reticulate evolution, parallel evolution, convergent evolution, and the deletion of affiliative signs under the action of natural selection were all severely constrained by its narrow purview. In addition, with the exception of Chapters 6 and 7 of the *Origin*, which present an entirely different theory of evolution founded on linear and independent ascent of evolutionary lines across taxonomic levels and evolutionary grades, Darwin refused the idea of directional or linear evolution.

R. G. Delisle, *Charles Darwin's Incomplete Revolution*, Evolutionary Biology – New Perspectives on Its Development 1, https://doi.org/10.1007/978-3-030-17203-9_7

Darwin's epistemological relation to the idea of linear evolution presents itself, therefore, as another potentially fruitful investigative topic, allowing us to probe the explanatory structure of the *Origin*. In this chapter, we will explore this through its complementary idea: biological progress.

7.1 Darwin and Biological Progress

Darwin's relation with the idea of biological progress is anything but straightforward. This explains why Darwin scholars have been divided on the issue, with a range of overlapping interpretations.[1] Before tackling our main topic, it will be useful to look closely at the recent historiography by laying out five characteristic interpretations.

1. Darwin's theory is entirely about opportunistic lineages diverging from one another. As evolution is the outcome of the encounter of a series of unpredictable conditions—such as the variation available at a precise geological time, the number of lineages involved in the competition process, and changing abiotic conditions—no sustained evolutionary directionality is possible. The history of life is merely the sum of separate and independent evolutionary histories of forms adapted only to local conditions. As formulated by Peter J. Bowler (1986: 41, 150): "Progressionism owed very little to the Darwinian theory of evolution by natural selection. In many respects, Darwin's mechanism challenged the most fundamental values of the Victorian era, by making natural development an essentially haphazard and undirected process... Darwin himself made a notable effort to break away from the progressionist assumption." It seems that until the early 1980s, Bowler's view was widely shared. As Michael Ruse (1988: 97) declared, Darwin "sees evolution as a directionless process, going nowhere rather slowly."
2. Darwin's theory is entirely committed to opportunistic lineages diverging from one another, but he eventually gave in to the cultural pressures of his time by supporting the idea of biological progress. Stephen Jay Gould holds that a contradiction exists in Darwin's view as seen in his simultaneous support for evolutionary opportunism and biological progress, a paradox he tried to conceal the best he could:

> The basic theory of natural selection offers no statement about progress, and supplies no mechanism whereby overall advance might be expected... None of Darwin's *outré* ideas could have been more unacceptable than his denial of progress as a predictable outcome of the mechanisms of evolutionary change... I believe that Darwin's views contain an unresolved inconsistency. Darwin, the intellectual radical, knew what his own theory entailed and implied; but Darwin, the social conservative, could not undermine the

[1]Here, I largely follow Timothy Shanahan's useful review of the literature by rearranging his analysis. See Shanahan (2004: 173–195, 285–294).

> defining principle of a culture... to which he felt such loyalty... Darwin did, of course, supply an argument to bridge the two starkly contradictory claims... Why, then, did he bother to smuggle progress back in through the rear door of a complex and dubious ecological argument about the predominance of biotic competition [over abiotic condition] in a persistently full world [leaving no room for empty places]?... I have no special insight into Darwin's psyche, but I do feel that his strained and uncomfortable argument for progress arises from a conflict between two of his beings—the intellectual radical and the cultural conservative. (Gould 1996: 136, 137, 141, 142, 144) [italics original]

3. Darwin recognized that biological progress (complexification) could be one of the several possible evolutionary outcomes of a selection process, the others being stagnation and simplification. Whereas biological progress could be a reality of evolution, however, the history of life is anything but centrally organized around progress. Yet, in stating this position, Darwin was at best ambiguous, unable as he was in clearly making a good case for biological progress, thus perhaps indicating that this was merely a secondary issue for him. David L. Hull (1988: 30) writes: "What I find remarkable about Darwin was that at a time when a belief in progress was pandemic, he had so little to say about it and, when he did, expressed himself so equivocally. In all 490 pages of the *Origin of Species*, he mentions progress only a dozen times, and in half of these cases he means only change." Hull holds that Darwin expressed in his few pronouncements on that topic doubts about the two key issues implicated in biological progress: the empirical support for it and its theoretical justification. About the first, Hull emphasizes that Darwin recognized that within the same class, the lower crustaceans have beaten the higher molluscs in the struggle for existence. About the second, Hull notes Darwin's own admission that he could not test his own supposition that, under similar conditions, later life forms should be able to outcompete earlier ones.

   Similarly, Robert C. Richardson and Thomas C. Kane hold that Darwin's ambiguities seem to be compatible with a recapitulationist doctrine that supports a strong directional evolution but that, ultimately, Darwin's dominant view settled for a nonrecapitulationist one based on the weaker notions of differentiations of parts and functional specializations (themselves underlying the principle of divergence). They write: "...Darwin continually insisted that natural selection does not *necessarily* include progressive development, and must be understood as an opportunistic procedure... Progressive improvement might be accommodated by the Darwinian theory, but is not required by it. The mechanism of natural selection [is] consistent with progress, stability, or retrogression" (Richardson and Kane 1988: 151) [italics original].
4. Darwin gave a significant place to the idea of biological progress, as he believed that the selective process would in the long run lead to many life forms climbing the scale of complexity. In the race to find new places in the economy of nature, diverging forms can create specialized solutions that increase the level of complexity. As Ernst Mayr (1982: 531, 532) explains:

   > The Darwinians had considerable difficulty in making it clear to their opponents that to deny the existence of an internal perfecting principle did not mean a denial of observed evolutionary progression... Darwin, fully aware of the unpredictable and opportunistic

> aspects of evolution, merely denied the existence of a law-like progression... [He] referred in the *Origin* frequently to evolutionary progress... This was necessary... to counter a newly developed school which denied any difference in perfection between the simplest and the most complex organisms... A careful analysis of everything Darwin wrote on evolutionary progress shows that he did not contradict himself. Where Darwin encounters improvements in the course of evolution, he finds that they can be easily explained as the a posteriori results of variation and natural selection... The main objection raised by the anti-Darwinians has always been that the living world is full of progressive trends but that it is inconceivable they should be caused by random variation and natural selection. The Darwinian answers, Why not? After all, any improvement, any new structural, physiological, or behavioral invention made in any gene pool can lead to evolutionary success and thus to progress as it is traditionally defined. This Darwin had already seen quite clearly.

In a somewhat similar vein, Timothy Shanahan argues that what seems a contradiction between the idea of specialized forms adapted to local conditions only and the idea of a complexification in life's history cohere, in reality, in Darwin (albeit not without ambiguities):

> Darwin had plenty to say about evolutionary progress; his views on the issue developed throughout his writings, especially in the successive editions of the *Origin*; and when considered carefully, a clear, unequivocal thesis emerges. To talk of divergence and specialization *rather than* progress as the motif of the *Origin* is to anachronistically draw a false contrast. For Darwin, an important form of evolutionary progress consists precisely in advancement in the organization of living things, where the latter is marked by increasing specialization of parts and division of labors... Darwin's vision of the evolution of life as an ascent driven by replacement of less fit organisms by superior forms seemed to reinforce the traditional view of the hierarchy of nature [chain of being], while his conception of the history of life as a branching treelike structure... seemed to call it into question. Darwin's own writings embody this dual conception. Despite Darwin's endorsement of a qualified form of evolutionary progress, he never felt entirely comfortable with the idea, calling it 'vague' and 'ill-defined' on more than one occasion. (Shanahan 2004: 194, 195) [italics original]

For Mayr, as for Shanahan, biological progress is a consequence of Darwin's conception of the evolutionary process and not an inevitable outcome.

5. Darwin was a progressionist in the fullest sense of the expression, in so far as his view comes with a strong evolutionary directionality. He believed in an organizing principle that structured life around an ascending scale of complexity. This thesis comes in two main versions. Dov Ospovat defends a weaker version, holding that early on in his intellectual development, Darwin was committed to progressionism, supporting it with the embryological theory of recapitulation (the law of parallelism), which states that higher forms pass through the organization levels of the lower forms. When Darwin gave up on the recapitulation theory in order to embrace the law of specialization, which states that higher forms merely pass through embryological stages similar to, and more general than, lower ones, he nonetheless continued to hold on to progressionism. The explanatory power lost by giving up on the recapitulation theory was shifted in Darwin's mind by turning to the dual notions of divergence and specialization, notions supporting

the idea of an inevitable biological progress on the basis that specialization constitutes an advantage that will be positively selected for:

> [T]he effect of Darwin's adopting the principle of divergence was to introduce into the theory of natural selection a developmental tendency—not a Lamarckian internal organic force, but a tendency for the large majority of lines of descent to show increasing specialization and complexity... Once he worked out the principle of divergence... he could argue for the inevitability of progress... [P]rogress was for him a necessary outcome of the evolutionary process. (Ospovat 1981: 210, 211, 225)

Ospovat (1981: 226) hints that theological considerations in favor of the creation of higher beings may be at play in Darwin's understanding of the evolutionary process, values widespread in mid-Victorian England: "Darwin's theory of progress went far, though perhaps not all the way, toward making a moral and intellectual being an inevitable—and therefore arguably designed—rather than a chance result of evolution." Robert J. Richards takes the step of proposing that such a cultural moral explanatory agent was actually internalized early in Darwin's theory by making it an inherent part of the workings of what will eventually be known as natural selection, giving progress a nearly teleological tone. Richards formulates a somewhat stronger version of the progressionist thesis in these words:

> In his early notebooks, [Darwin] portrayed progress as the inevitable outcome of the logic of the evolutionary process. Progress was the result... of an external dynamic *pulling* them to perfection. He supposed that the environment against which a creature would be selected would be the living environment of other creatures, so that each increase of competitive efficiency, each augmentation of specialization, each new trait evolved to meet a new challenge—that all these alterations of one individual would call forth reciprocal development in others. The evolutionary situation had a *built-in* progressive dynamic.... (Richards 1988: 131)[2] [my emphasis]

Two points should be made here about these five interpretations. First, they all capture something of Darwin's views. In the *Origin of Species*, one finds a mixed bag of arguments each with a different epistemological status (1859: 314, 336, 337, 345, 490, 1872: 97–100, 201): (1) naturalists are not agreed on how to define biological progress, seeing the notion as ill-defined, which explains their contradictory assessments; (2) anthropocentrism could be a problem when evaluating progress (an argument more implicit than explicit in Darwin); (3) there is no way of testing whether more recent forms are more progressive than more ancient ones; (4) since there is no fixed law of development, some lineages will change somewhat, others not at all; (5) some places in the economy of nature will be filled by forms going through simplification and by others remaining simple for indefinite ages; (6) successive forms in the history of life that have outcompeted their predecessors are higher in the scale of organization; (7) differentiation and specialization will be seized upon by natural selection, thus leading to higher organizations; (8) the selective process will generally lead to advancement; (9) the production of higher animals is the outcome of the struggle for survival.

---

[2]See also Richards (1992).

The second point is that there seems to be a historiographical trend following more detailed analyses of Darwin's publications, notebooks, correspondence, and marginalia, indicating that the notion of biological progress was more important to him than first evaluated. Scholars like Michael Ruse and Peter Bowler, for instance, have re-evaluated their initial assessments, the former even writing: "People who deny that Darwin was a progressionist—and I was one of them... are just plain wrong" (Ruse 1988: 104).[3]

Darwin did believe in biological progress of one kind or another. Taking this for granted, this chapter will contribute to the ongoing debate by asking how biological progress fits with Darwin's ontological commitment toward a world believed to be fully constituted, two seemingly opposed notions. This takes us to the central thesis of this chapter: *Darwin's support for biological progress is largely part of his rhetorical apparatus, as evidenced by the fact that he himself provided many explanatory devices that thwarted the expression of progress or genuine complexification. However, these devices are not only applicable to Darwin's special brand of evolutionary opportunism under the weak contingency thesis, but also extend to the epistemological implications accompanying his ontological commitment to a fully constituted world.*

As noted by Ernst Mayr (1982: 531), Darwin needed the notion of biological progress in order to avoid the possible refutation of the doctrine of evolutionism: "Denial of a progression from the infusorians to the angiosperms and vertebrates might well imply a rejection of evolution altogether." The fact that in the *Origin* Darwin repeatedly opposed the idea of *descent with modification* to the notion of an *independent act of creation* reminds us that the fight for the evolutionist doctrine was still believed to be important at the time.[4] However, Darwin's main theoretical apparatus was ill-designed to support biological progress, constraining him to resort to grandiloquent declarations such as the one closing the *Origin of Species*: "Thus, from the war of nature, from famine and death, the most exalted object which we are capable of conceiving, namely, the production of the higher animals, directly follows" (1859: 490; see also 1872: 429). As recognized by Darwin scholars, there exists a fundamental tension in Darwin's view. Ruse even complexifies the picture by describing layers of tensions between: (1) a theory of evolution based on opportunism, adaptedness (design), and branching; (2) the notion of relative progress whereby forms find ways to better their rivals; (3) the notion of absolute evolutionary progress culminating in humankind; and (4) the doctrine of progressionism as a common cultural value of the time (Ruse 1996: 136–177). Ruse's nuanced analysis reads this way:

> As far as progress was concerned, Darwin himself was certainly a cultural progressivist and saw evolution itself as progressive. However, he also saw that the link between progress and

---

[3]See also Bowler (2003: 12, 146, 164, 169).

[4]Darwin (1859: 3, 6, 44, 55, 59, 129, 133, 138, 139, 152, 155, 159, 162, 167, 185, 194, 203, 355, 372, 389, 390, 393, 394, 396, 398, 406, 413, 414, 420, 434, 435, 437, 456, 469, 470, 471, 473, 475, 478, 482, 483, 486, 487, 488).

> evolution was something that brought down the status of the latter, and he realized that natural selection is a mechanism that denies the inevitability of biological progress. In other words, conceptually Darwin broke the link between the two, although, having done this, he argued that natural selection could lead to progress as the end result. . .. (Ruse 2009: 23, 24)

The historiography thus far has mainly focused on Darwin's opposition between *biological progress* and *evolutionary opportunism*. These two notions may not necessarily be entirely incompatible with one another, as seen in Darwin's intellectual contortions in arguing that differentiation of parts could lead to sustained biological progress, a view given credence in the interpretations of Mayr and Shanahan. We would like to argue, however, that a more fundamental opposition exists in Darwin's reasoning: the utter incompatibility of the idea of biological progress and the notion of a *pre-established pattern-process of divergence as conceived from the present and projected backward in time*. The latter notion lacks in conceptual flexibility and does not allow lineages to easily find the road to biological progress. Furthermore, it lacks in explanatory power with respect to justifying the rise of genuine evolutionary novelties. Apparently, the idea of biological progress constitutes, for Darwin, a superficial addition superimposed upon a more fundamental explanatory apparatus, further enlarging the gulf between his evolutionary theory and his belief in progress. We will now present the underlying premises of our thesis.

7.2 Thwarting Biological Progress

Darwin asks us to read his Diagram (see Fig. 2.1) from the bottom up, following the historical deployment and diversification of lineages across geological times and taxonomic levels. He holds that forms showing more differentiation and specialization of parts will find new places in the economy, opening up ways for greater complexity and higher organizations. In reality, Darwin's ontology and its accompanying methodology (see also Chap. 8 of this book) require of us that, epistemologically speaking, we read the Diagram from the top down, that is, from the present to the past. Darwin's own approach to the history of life in the *Origin of Species* blocked the evolutionary process by imposing upon it a series of pre-established notions thwarting the expression of biological progress and evolutionary novelties. As these explanatory notions have already been exposed in previous chapters of this book, it will suffice to present them in a compact fashion only. Here, it will be useful to consider two complementary assumptions of Darwin.

According to Darwin's first assumption, the present always reveals more of the evolutionary history than the past itself, a natural assumption for someone who believes in a fully constituted world. This explains why he argues that biological diversity or disparity entirely falls under the purview of the present: variability in the past is either aligned with or falls in-between currently existing forms. This is ensured by Darwin's peculiar use of the principle of divergence as applied backward in time (convergence) (1859: 329, 330, 1872: 301–303). Darwin, then, redirects his reader to the Diagram, where it can be seen that virtually all the more ancient forms fall between more recent ones. Darwin assumes that the distinct classes and phyla of today are at their maximum divergence relative to their own past. This is consistent with his claim that the taxonomic *arrangement* was maintained throughout the entire evolutionary history, allowing us to keep track of the main events of the past as these are registered in today's taxonomy (1859: 420–422, 1872: 369, 370).

The epistemological implications of Darwin's first assumption must be appreciated for what it is: the evolutionary pattern-process is predetermined, foreclosing any other alternative unless it can be accommodated within the framework provided by the *weak contingency thesis* (which excludes the *strong contingency thesis*). It is no surprise, then, that Darwin (1859: 118, 119, 412, 1872: 91, 92, 364) confidently holds that it is the most divergent forms that are usually selected during the evolutionary process, leaving the others on the side road of extinction. As already mentioned, this is reflected in the Diagram, which shows that about two-thirds of the surviving lineages are among the most divergent ones.

Darwin wants the reader of the *Origin* to believe he has arrived at this conclusion by reflecting upon how diversity is produced, hence his conceptualization of the principle of divergence (associated with the notions of differentiation and specialization of parts). In light of Darwin's ontology, however, the reasoning must be inverted: because Darwin assumes that the deployment of each separate class/phylum is at its maximum today, he has no other logical option but to make the dual claim that the most divergent forms are usually selected and that differentiation-specialization of parts constitutes the main evolutionary road taken.

It is surely not unreasonable for Darwin to hold that two extant forms sharing many similarities are probably bound in common ancestry (setting aside the problems of interbreeding and analogies). But the principle of divergence is, for Darwin, much more than a pattern-process commonly observed in the history of life. Rather, it is a conceptual tool encased in a broader explanatory apparatus, designed to ensure that the present time constitutes the privileged approach to the evolutionary past. In other words, Darwin's principle of divergence is not *conceptually neutral*, contrary to what is often believed among evolutionists today. That is why Darwin coupled it with its twin notion—gradation—the latter playing the explanatory role of tying up the fabric of life under the tight control of the present, leaving no room for major morphological gaps or discontinuities. Simply put, Darwin's explanatory apparatus uses the principle of divergence as a way for the present to bracket the diversity of the past, while the notion of gradation becomes a privileged tool to probe that past using the gauge of extant forms. We have already seen how Darwin managed to protect the immovable and irrefutable core of his theory—divergence and gradation—by displacing the explanatory stress to auxiliary components believed to be adjustable enough to save the theory (taxonomic level, geological timeframe, selective pressure, geographical distance, or a host of other speculative and ad hoc explanations).[5]

---

[5]See our analysis in Chap. 5 of this book.

Our analysis raises the following question: how can biological progress and complexity of organization be generated out of a theory founded on a pre-established pattern-process of evolution depicting a rigidified tree of life? Had Darwin let the process of evolution flow freely, evolutionary opportunism might have succeeded in finding ways to progress, including through the differentiation of parts and divergence. But Darwin ultimately denied this.

A second and complementary assumption is found on Darwin's intellectual road to biological progress. By bracketing all of the past under the purview of the present, he leaves little conceptual room for thinking about the rise of genuine evolutionary novelties. Under Darwin's approach, the present devours the past in an epistemological process of assimilation, to the point where past forms merely serve the purpose of providing ancestors for living ones, with the former having little independent existence. This explains why imagined progenitors remarkably resemble miniatures of currently existing forms under Darwin's hands (1859: 437, 1872: 384, 385). Combatting discontinuities, Darwin probes the past with the help of the principle of divergence (convergence backward in time) in search of common ancestors. In so doing, ever more ancient forms are assimilated to extant ones.[6]

7.3 Motion in a Closed System: Recycling Mechanical Devices

We have seen that a number of Darwin scholars have struggled to unify in a single coherent whole Darwin's mechanism of evolution and biological progress, trying to match the relative success of forms adapted to local conditions only and absolute advance in organization (see also Ruse 2004: 18). Since Darwin assumes progress to be a fact, something must be profoundly modified in the course of evolution. As Michael Ruse puts it (1996: 169): "Darwin's picture was one of hard, slow, physical grind—of a real groping gradually through much sweat and toil, often unsuccessfully, toward a better state." The key topic to be investigated in the rest of this chapter concerns "evolutionary dynamics," that is, the explanatory conditions required to generate evolutionary motion, general advancement, and linear evolution. As mentioned in the "Intellectual Grid" presented in the Introduction chapter of this book, our analysis can ultimately be organized around two key topics: (1) examining how the world's entities come about or what kind of entities are being generated; (2) investigating the nature of such entities either by looking at their inherent structure or at how they are organized externally. In short, the matter confines itself to the *entities* themselves and their *context*. Let us begin with the latter.

Stephen Jay Gould (1996: 142–144) holds that Darwin skewed his theory to accommodate a notion of biological progress. He would have done so by overstressing biotic competition over abiotic ones, and by claiming that the world

[6]See Darwin (1859: 303–306, 313, 489, 1872: 283–285, 428).

is always full to capacity, thus neglecting episodes of relative emptiness. In Gould's (1996: 143) words: "if ecologies are always chockfull of species, and losers have no place to go, then the victors in biotic competition will truly eliminate the vanquished—and the buildup of these successive eliminations might produce a trend to general progress." According to Gould, then, Darwin created the conditions for biological progress by placing evolving entities within the competitive context of a living world permanently full to capacity. Robert J. Richards puts this in a slightly different way, although he too stresses the context of overpopulation. In his view, an upward spiral toward complexification is de facto established the moment it is realized that simplified forms derived from higher forms are always more advanced (and therefore at an advantage) than simplified forms descended from less highly developed forms:

> Darwin did recognize. . . that some species seemed to have fallen back, to have been chiseled to a simpler form. . . But even this partial slippage was, he thought, to be expected—under the assumption that where it occurred, the underlying simpler animals had been destroyed. In which case, the apparent anomaly disappear: natural selection inevitably would shave down some complex creatures and shove them into the gap. When the necessary *backfilling* was completed, however, natural selection would continue to *stack* species progressively higher in perfection. (Richards 1988: 138, 139) [my emphasis]

We will argue below that on closer analysis, Darwin deprives his theory of the proper evolutionary *context* to sustain biological progress by moving away from the evolutionary impetus created by overpopulation.

Turning now to the notion of the evolving *entities* themselves, some light is thrown on this topic by briefly revisiting John C. Greene's (1957, 1959, 1971, 1981, 1986) thesis. As mentioned in the earlier chapters of this book, Greene holds that the *Origin of Species* is to be found at the end of the road of the atomistic tradition going back to Robert Boyle's *Origin of Forms and Qualities* (1666), and which includes Descartes, Buffon, Kant, and Lamarck, among others. This tradition largely envisions the world as a dynamic, causal, non-teleological, and time-oriented law-bound system of matter in motion. Nature is broken down through the doctrine of corpuscularism, its constituent parts in the biological world (traits, varieties, species) being set in motion and channelled by the principle of competition along a progressive path. For Darwin, it seems, living atoms (parts) flow in the evolutionary current. As we saw in Chap. 4 of this book, Darwin moved away from a pan-divergent view in Chapters 6 and 7 of the *Origin* (1859) by presenting an overall picture that looked like a series of parallel evolutionary lines racing forward independently, seeking to acquire similar features (analogies) across the taxonomic spectrum, a process under the unstoppable impetus of natural selection.

It is worth noting that the tradition of atomism or corpuscularism was revived in the sixteenth and seventeenth centuries (Dijksterhuis 1961: 418–444; Hall 1981: 216–243; Sargent 1995: 98–108; Shapiro 2002). Relating Darwin to that tradition requires that we recognize he was dealing with the constituted realities of living forms and not those of inanimate bodies (Hodge 1985, 2014; Hodge and Radick 2009; Sloan 1985). Once this is acknowledged, Darwin's reasoning can be profitably linked to Newtonian dynamical systems—widely construed—through such abstract notions as forces and pressures (selection); constituent parts of a system (traits, organisms, species); closed systems (places in the economy of nature); states of equilibrium (increasing versus decreasing populations); and motion and paths (evolutionary change and divergence) (Schweber 1985; Depew and Weber 1995: 3–9, 71–98, 115–136). Although in the *Origin*, Darwin insisted that organisms often differ from one another—hence his support for what we today call "population thinking"—his thinking became much more abstract as he embraced the "big picture" of evolutionary pressure responsible for evolutionary change. At such an abstract level, living atoms were conceived as identical entities whose role consisted in putting evolutionary pressure on neighboring entities.

In what follows, we argue that Greene's thesis has retained its value but needs to be significantly weakened. Darwin can clearly be positioned within the atomistic tradition, since this is one of the many intellectual strains that influenced him. But as recognized by Greene himself, the atomistic worldview, which focused on a law-bound system of matter in motion, coexisted with the static worldview concerned with designed and contrived entities. It is this other worldview, we hold, that predominates in Darwin. Whereas Chapters 6 and 7 of the *Origin of Species* (1859) offer a picture of the wide taxonomic distribution of features floating freely in an evolutionary stream, we have already seen that the rest of the *Origin* tries to impose a strict order upon the living world by taming the action of natural selection through the principle of divergence. Because we are here concerned with the notions of evolutionary dynamics and biological progress, we will push our thesis one step further: *Darwin's commitment to a designed and contrived world was so profound that his own conception of the evolutionary dynamics thwarted the expression of both biological progress and the process-pattern of divergence.* Apparently, Darwin could not evade his foundational choice of erecting his evolutionism upon the conceptual tools inherited from the tradition of natural theology, a tradition that rested on a static and fully constituted world and which sees motion as confined to cycles and states of equilibrium. Let us begin by revisiting Darwin's overall view of the tree of life.

7.4 Rotating Shafts: Fixed Taxonomic Categories and Cycling Taxonomic Categories

Confronted with classes (fish, reptiles, mammals) already distinct from one another for a very long time, it becomes difficult for the reader of the *Origin* to shake off the impression that evolution is largely at a standstill (1859: 338, 483, 484, 1872: 310, 424). It is no surprise, then, that Darwin borrows another concept from earlier centuries, one associated with a fully constituted world, that of *eternity*. In a world assumed to have always been identical to itself, time has an intimation of eternity. Of course, Darwin had to adapt the concept to the evolutionary context; after all, he needed time for classes to deploy themselves in full. But it is surely interesting to note that he required massive amounts of it even to go from one ancestral species to several distinct orders within the same class. Darwin estimated that the time period extending from the Silurian or Cambrian to the present represented about half the amount of time provided for the entire history of life (1859: 307, 1872: 286). That much had been required for classes and phyla to reach their current state. Clearly, Darwin is a gradualist of the purest kind. The notions of "gradualism" and "uniformity" fuse together in Darwin, allowing us to define the principle of uniformity as the quality or state of being uniform, regular, unvarying, or as applied to laws and processes now in operation and assumed to be the same everywhere and for all eternity (in the past, present, and future).

In implicitly adopting this definition of uniformity, Darwin followed in the footsteps of Lyell, who himself had followed Hutton. While Hutton had conceived the geological Earth to be a steady-state system perpetually renewing itself with no beginning and no end in sight, Lyell had superimposed on it a steady-state system of life forms in a state of perpetual renewal: lost species were replaced by new ones, although the main types of living forms have always been maintained from an unknown and unknowable beginning.[7] Lyell's uniformitarian/gradualistic geologico-biological system required an extensive amount of time to produce its effects (Rudwick 1970: 11). Similarly, Darwin's own special brand of evolutionism retained from it a nearly boundless time horizon:

> . . .it may be objected, that time will not have sufficed for so great an amount of organic change, all changes having been effected very slowly through natural selection. It is hardly possible for me even to recall to the reader. . . the facts leading the mind feebly to comprehend the lapse of time. He who can read Sir Charles Lyell's grand work on the Principles of Geology. . . yet does not admit how incomprehensibly vast have been the past periods of time, may at once close this volume (1859: 282) . . .What an infinite number of generations, which the mind cannot grasp, must have succeeded each other in the long roll of years! (1859: 287) . . .The whole history of the world, as at present known, although of a length quite incomprehensible by us, will hereafter be recognised as a mere fragment of time, compared with the ages which have elapsed since the first creature, the progenitor of innumerable extinct and living descendants, was created. (1859: 488)

For Darwin, geological time really has a strong eternal flavor, as do extant classes and phyla whose independent ancestries are lost in the unfathomable depths of time and whose deployment has been complete as far back as, perhaps, the Silurian or Cambrian age. Having contemplated this near-permanence of classes and phyla, all that was left for Darwin's theory to explain was evolution at lower taxonomic levels. Differently stated, since classes and phyla were apparently unmovable entities profoundly entrenched in nature, Darwin's theory of biological evolution was forced to put all the explanatory weight on changes at lower taxonomic levels. To use a mechanical analogy, Darwin's view might be compared to a series of parallel running shafts of locomotives—i.e., the independent and parallel classes and phyla in the history of life—each being perfectly stable and permanent and distributing power or motion to the various pieces of the machinery, the latter being the varieties,

---

[7]This theme has been treated more fully in Chap. 2 of this book.

species, genera, families, and orders. This analogy is meant to express the distinction between stable and permanent parts and moving and cycling parts.

Again, it is useful to turn to T.H. Huxley for comparison. Reflecting upon the value of the animal fossil record in support of evolutionism, Huxley had this to say in his 1862 Address:

> The positive change in passing from the recent to the ancient animal world is greater, but still singularly small. No fossil animal is so distinct from those now living as to require to be arranged even in a separate class from those which contain existing forms. It is only when we come to the orders, which may be roughly estimated at about a hundred and thirty, that we meet with fossil animals so distinct from those now living as to require orders for themselves; and these do not amount, on the most liberal estimate, to more than about ten per cent of the whole. (Huxley 1862: xlviii)

It is worth introducing here a short digression about Huxley. Michael Bartholomew (1975: 529) had this interesting comment: "[T]he point that I wish to make here is that when Huxley decided to become Darwin's champion, he carried along with him a lot of the trappings of his old views: with hindsight, we can see that they were not very well suited to a defence of the *Origin, even though Huxley himself thought that they were tailor-made for the job*" [my emphasis]. Although Huxley (1862) was less than impressed at first by the facts garnered in favor of an evolutionary view supporting change going from a generalized state to a less generalized one, he showed more openness to this idea by 1870 (Ruse 1979: 220, 221). By the time Darwin published the last and 6th edition of the *Origin of Species* in 1872, he and Huxley shared at least one key idea about the history of life: higher taxonomic entities are nearly stable and permanent; evolution occurs at lower levels through specialization with an understanding that it is extant forms that offer the key to classify past forms. Equipped with his focus on archetypes and his desire to seek deep geological roots for extant types (Lyons 1995; Bowler 1996: 263–267, 331–335), perhaps Huxley saw beyond Darwin's rhetoric about universal evolutionism. This may throw a different light on their intellectual relationship, independently of the fact that Huxley's support for natural selection was rather superficial insofar as it was essentially confined to its naturalistic aspect.[8]

The intellectual relationship between Darwin and Richard Owen should also be considered. Although Owen wrote a very critical review of the *Origin of Species* in 1860—one that strongly opposed his own view to Darwin—Owen's own brand of evolutionism is also based on an interpretation of the annals of life similar to Darwin's, irrespective of the fact that Owen falls for Darwin's rhetoric about life being ultimately derived from a single common ancestor. Owen (1860a: 515) writes:

> Accordingly we find that every grade of structure, from the lowest to the highest, from the most simple to the most complex, is now in being,—a result which it is impossible to

[8]It is instructive to contrast what Huxley had to say about his own work and about the work of Darwin. See Huxley, T.H. (1893), *Darwiniana: Collected Essays, Vol. 2* (London: Macmillan) and Huxley, T.H. (1898–1902), *The Scientific Memoirs of Thomas Henry Huxley*, 4 Vols, edited by M. Foster and E. Ray Lankester (London: Macmillan). See also Lyons (1999: 245–250) and Di Gregorio (1984: 57–68).

> reconcile with the Darwinian hypothesis of the one and once only created primordial form, the parent of all subsequent living things. The changes which our planet has undergone in the course of geological time have been accompanied by the loss of many *minor* links which connected together the existing evidences of gradational structure; but the general laws regulating the progress and diversity of organic forms, having been the same throughout all time, so it happens, according to the testimony of the most experienced palaeontologists, that. . .[9] [my emphasis]

Then, Owen goes on by directly quoting from his own book, *Palaeontology* (1860b: 18), with the difference that the original passage is no longer presented as only applying to protozoans and invertebrates, but also to all life forms: "Every known fossil belongs to some one or other of the existing classes, and that the organic remains of the most ancient fossiliferous strata do not indicate or suggest that any earlier and different class of being remains to be discovered, or has been irretrievably lost in the universal metamorphism of the oldest rocks" (Owen 1860a: 515). Not opposed to evolutionism, Owen (1860a: 515) adds: "that their [forms] limitation in time may, in some instances, or in some degree, be due to constitutional changes, accumulating by slow degrees in the long course of generation. . .."

The picture of the triumvirate comprising Darwin-Huxley-Owen is a complex one. On the one hand, Huxley may have perceived some similarities between his view of the annals of life and Darwin's, something Darwin apparently largely overlooked. Indeed, responding to Huxley's 1862 Address, Darwin writes to him: "I cannot tell why, but when I finished your address I felt convinced that many would infer that you were dead against change of species, but I clearly saw you were not."[10] On the other hand, Owen overlooked the similarities uniting him with Darwin on the same topic, entirely buying into the rhetoric of the latter about a single primitive common ancestor for all life forms. It is tempting to say that Darwin, too, fell for his own rhetoric, explaining his original reaction to the 1862 Address of Huxley. As if the picture was not already confused enough, one must add the well-known oppositions between Owen and Huxley over scientific, institutional, and personal matters (Di Gregorio 1984: 35–43; Lyons 1999: 206–209; Rupke 2009:180–224). In this context of friendship and rivalry, of different evolutionary theories, of second-guessing each other's work, and of rhetorical discourse, it is no surprise that a fundamental agreement between the three was lost: they were all dedicated to the notion of a fully constituted world but now reinterpreted against the backdrop of a rising evolutionism.[11]

It is in this context that the "manufactory of species" spoken of by Darwin is relevant (1859: 53–59, 1872: 43–47). Since variability is key to survival, life forms

[9]Owen's text is reproduced in Hull (1973: 175–213) with the quote on pp. 195–196.

[10]Letter from Darwin to Huxley, May 10th 1862, in F. Darwin and A Seward (eds.), *More Letters of Charles Darwin*, vol. 1, p. 82 (London: John Murray, 1903). Reproduced in Bartholomew (1975: 529).

[11]These brief considerations over the Darwin-Huxley-Owen triumvirate would require a much deeper analysis than can be provided here.

offering more variability to natural selection will, in general, have a better chance of survival. Darwin thus conceived an evolutionary dynamics in which more variable forms would become ever more dominant, at the expense of other groups that would gradually be pushed to marginality and eventually extinction. Just as large, widely diffused species often contain more varieties than small species, and large genera more species than small genera, evolution is driven by a diversification process of ever more dominant forms climbing the taxonomic scale:

> We have seen that it is the common, the widely-diffused and widely-ranging species, belonging to the larger genera, which vary most; and these will tend to transmit to their modified offspring that superiority which now makes them dominant in their own countries. Natural selection... leads to divergence of character and to much extinction of the less improved and intermediate forms of life. On these principles, I believe, the nature of the affinities of all organic beings may be explained... [A]ll animals and all plants throughout all time and space [are]... related to each other in group subordinate to group... namely, varieties of the same species most closely related together, species of the same genus less closely and unequally related together, forming sections and sub-genera, species of distinct genera much less closely related, and genera related in different degrees, forming sub-families, families, orders, sub-classes, and classes. The several subordinate groups in any class cannot be ranked in a single file, but seem rather to be clustered *round points*, and these *round other points*, and so on in almost *endless cycles*... [This] is explained through inheritance and the complex action of natural selection, entailing extinction and divergence of character, as we have seen illustrated in the diagram. (1859: 128, 129; see also 1872: 103, 104) [my emphasis]

Darwin's utilization of the expressions "round points" and "endless cycles" to describe taxonomic groups nested within ever more inclusive groups, along the taxonomic scale, is most interesting. Since taxonomy itself is a static classificatory system, there would seem to be little reason to make use of such a "dynamic" concept. Darwin's slip, of course, makes sense considering that he has in mind an evolutionary process that sees a diversification of ever-more dominant forms that climb the taxonomic scale. How long, then, will such a turnover dynamics last? As Darwin sees it, "[l]ooking to the future, we can predict that the groups of organic beings which are now large and triumphant... will for a long period continue to increase" (1859: 126; see also 1872: 96). Then, metaphorically comparing affiliated forms within a single class to a "tree of life," Darwin writes: "From the first growth of the tree, many a limb and branch has decayed and dropped off; and these lost branches of various sizes may represent those whole orders, families, and genera which have now no living representatives, and which are known to us only from having been found in a fossil state" (1859: 129, 130; see also 1872: 104). From this we learn that Darwin recognizes that groups belonging to the order level and below have gone extinct during the evolutionary history of the distinct classes. Stepping back to look broadly at the history of life, Darwin explains the relative rarity of higher taxonomy groups, as compared to lower ones, by the process of ever more dominant groups driving others to extinction: "Thus we can account for the fact that all organisms, recent and extinct, are included under a few great orders, under still fewer classes" (1859: 428, 429; see also 1872: 378). Or, again: "we can understand how it is that there exist but very few classes in each main division of the animal and vegetable kingdoms" (1859: 126; see also 1872: 97).

Even though Darwin holds that classes required large amounts of time to deploy in full, we have seen above that he also thought that the different classes of the vertebrate phylum (such as the fishes, reptiles, and mammals), for instance, have been sharply differentiated from one another for a very long time. Indeed, he argues that the history of life has long been stocked with groups of many different taxonomic levels: "at the most remote geological period, the earth may have been as well-populated with many species of many genera, families, orders, and classes, as at the present day" (1859: 126; see also 1872: 97). It is no surprise, then, that Darwin's interpretation of the fossil record in the paleontological chapters of the *Origin* leaves the reader with a sense of a world largely completed. Darwin's Cambrian or Silurian world looks strangely like ours.

What kind of evolutionary room does this leave to accommodate "biological progress"? Factually speaking, very little, which explains why the notion of progress in Darwin does not stem from the core of his theory, but rather presents itself like a rhetorical veneer that must be peeled off. The ongoing evolutionary dynamics of ever-more dominant groups replacing less successful ones seems to have been ongoing since very ancient times, busy going nowhere but recycling the same variations seen today within the confines of permanent and independent classes. Looking back at the long history of life since the pre-Silurian or pre-Cambrian period, Darwin projects himself into the future: "Hence we may look with some confidence to a secure future of equally inappreciable length" (1859: 489; see also 1872: 428). Not only has Darwin's world long been nearly fully constituted, it is also likely to maintain itself for quite some time. Apparently, extant classes and phyla are here to stay for nearly an eternity.

7.5 Opening and Closing Devices

Building upon the discussion of the influence of the tradition of natural theology in Chap. 3 of this book, let us ask a simple question: Why was Darwin not a straightforward progressionist like Spencer was? Why did Darwin take such a convoluted pathway to express his belief in biological progress, making the work of historians so difficult? The immediate and obvious answer is that Darwin's theory relied heavily on the notion of real species living in a real world: from the very outset species need to be adapted (designed) to places in the economy of nature to even get the chance of reproducing and perhaps eventually evolving. Instead of assuming, as Lamarck does, that life is endowed with an "innate and inevitable" drive to change (Darwin 1872: 98), Darwin elected to go down the difficult road of evolutionism through the tradition of natural theology. Recalling our earlier discussion, we note that this tradition was built around several key ideas, especially the following: (1) only the present reveals in full what there is to know about the world; (2) the living world is organized around an economy of nature wherein forms are interconnected in complex ways, thus maintaining the equilibrium or balance of nature; (3) the living world is largely deprived of voids, since it is filled with contiguous forms in contact with each other; (4) living forms occupy places in the economy of nature; (5) goal-directed ends (teleologies) and designed forms are key features of this world; and (6) motion exists in this world but as expressed in cyclicity or deployment of pre-existing forms (preformationism), change being a real but superficial aspect of the world's structure in place since a very long time.

In short, the tradition of natural theology enshrines the key notions of order, completeness, equilibrium, harmony, symmetry, interconnection, and contrivance. Let us postpone for now the considerations of how Darwin reinterpreted, or added to, these notions by introducing more motion than Linnaeus had done before him. Instead, let us review the explanatory devices Darwin carried from the tradition of natural theology into the deep explanatory structure of the *Origin of Species*. Three overlapping devices are especially revealing of Darwin's difficulties in extracting himself from the static worldview and taking the full evolutionistic turn. Taken collectively, these devices act to thwart evolutionary change and biological progress.

Increasing Population Versus Decreasing Population Darwin promoted an evolutionary dynamics based on a symmetrical yet inverted relationship wherein the increase of population A^1 in space and time is done at the expense of the decrease of population A^2, as seen in this visual representation: $\backslash a^1/a^2\backslash$. Darwin writes:

> Hence, as more individuals are produced than can possibly survive, there must in every case be a struggle for existence, either one individual with another of the same species, or with the individuals of distinct species, or with the physical conditions of life. It is the doctrine of Malthus applied with manifold force to the whole animal and vegetable kingdoms... There is no exception to the rule that every organic being naturally increases at so high a rate, that if not destroyed, the earth would soon be covered by the progeny of a single pair... [I]t follows that as each selected and favoured form increases in number, so will the less favoured forms decrease and become rare... As natural selection acts solely by the preservation of profitable modifications, each new form will tend in a fully-stocked country to take the place of, and finally to exterminate, its own less improved parent or other less-favoured forms with which it comes into competition. Thus extinction and natural selection will, as we have seen, go hand in hand. (Darwin 1859: 63, 64, 109, 172; see also 1872: 50, 51, 134)

In reality, Darwin's statement constitutes a conceptual knot. The contiguity of forms in contact is expressed in the fact that increase of A^1 is possible only if accompanied by decrease of A^2. If the latter holds its ground in the economy of nature, the argument implies, the former will be blocked in its expansion. The picture that emerges from this view of nature is one of a fairly rigid mechanical equilibrium: while the evolutionary gate of A^1 opens, the one of A^2 closes. The order of nature is preserved through this inverted yet remarkably symmetrical relationship when viewed from the long geological perspective:

> Thus the appearance of new forms and the disappearance of old forms... are bound together. In certain flourishing groups, the number of new specific forms which have been produced within a given time is probably greater than that of the old forms which have been exterminated; but we know that the number of species has not gone on indefinitely increasing, at least during the later geological periods, so that looking to later times we

> may believe that the production of new forms has caused the extinction of about the same number of old forms. (Darwin 1859: 320; see also 1872: 296)

Reading Darwin's Diagram under this light, what seem to be evolutionary lineages diverging from one another in relatively open space and time should, in reality, be read by filling in the blanks to capacity (plenum), a visual form Darwin could not adopt without significantly blurring the clarity of the diagram. Only a world filled to capacity can create the necessary demographic pressure to instigate evolutionary change. Evolution takes place in the context of a "pressure cooker," a pressure strong enough to expel a number of forms forward along a divergent path while extinguishing others: "Within the same large group, the later and more highly perfected sub-groups, from branching out and seizing on many new places in the polity of Nature, will constantly tend to supplant and destroy the earlier and less improved sub-groups" (1859: 125, 126; see also 1872: 96). For Darwin, the rise of forms and the extinction of others are two sides of the same coin (1859: 320–322, 1872: 295–297).

So much, then, for the exposition of the general principles. A closer look at the *Origin* reveals, however, an overall picture that is much less straightforward. First, we know that Darwin treated classes/phyla as permanent and stable entities profoundly entrenched in nature. As such, the evolutionary dynamics characterized by $\backslash a^1/a^2\backslash$ merely constitutes a means for generating cyclical motion in the lower taxonomic entities through their perpetual rise and fall. Nothing truly new is thus created. Second, Darwin's postulated evolutionary dynamics holds that the world is a plenum without which an evolutionary drive would be lacking.[12] More than that, the world is assumed to be a plenum composed of atomistic entities with perfectly smooth surfaces, on which nothing could get a grip. On this abstract view, the separate and neighboring entities merely serve the purpose of putting demographic pressure on the next entity, pushing one another through action-by-contact, so to speak. Let us take a closer look at how Darwin challenged, implicitly or explicitly, both the plenitude of the world and the atomistic nature of living entities.

A World Fully Stocked Versus a World Less Fully Stocked One difficulty for Darwin is that his "pressure cooker" is not air-tight and loses the evolutionary steam necessary to institute the kind of change he has in mind. In other words, Darwin does not always hold that the world is full to capacity. Of course, Darwin is not the Presocratic thinker Parmenides who argues, through logic and reasoning only, that the world is constituted like a single, solid, and unmovable monolithic block (Guthrie 1965: 26–43; Barnes 1998: 38–41). Darwin does provide room for motion by allowing the various parts of his world to articulate and move relative to one another. For instance, room is created for a time when forms can become more specialized through the exploitation of specific niches in the economy of nature: a higher biomass density can be sustained when the same plot of land is inhabited by

[12]For two contrasting approaches to this question, compare Gould (1996: 142–143) and Radick (2000).

distinct specialized forms rather than by more generalized ones (Darwin 1859: 113–116, 1872: 87–90). Or room opens up when an evolving lineage finds a new adaptive trick that places it at an adaptive advantage relative to other potential competitors: evolutionary space is created for a time, until the niche is filled again either by the descendants of the same successful lineage or by other forms that find the doorway to the same niche. When Darwin repeatedly claims *Natura non facit saltum*, he does not intend to mean that the world is packed to the point where no room exists for evolutionary change.[13] Within evolutionary motion as a whole, fairly small and relatively empty places are created here and there for limited amounts of time or exist under a potential state pending future evolutionary responses.[14] That much is granted.

More problematic for Darwin are other cases the reader encounters in the *Origin* that appeal to larger "empty places" in nature, some lasting extensive amounts of time. These create a disruption in the driving tension needed to sustain the evolutionary dynamics. In one case, the concept of the competitive exclusion principle is not consistently applied in the context of a world sufficiently filled to institute divergence all the way up to the highest taxonomic levels. As seen in Chap. 3 of this book, Darwin holds that competition is at its most severe among allied forms, considering they fill the same place in the economy of nature (1859: 75, 76, 320, 321, 1872: 58, 59, 295, 296). Since they are identical, they come to exploit nearly the same resources for their survival, putting them in direct competition with one another under the demographic pressure of populations increasing in number. In this context of high density, the stress accompanying forms pressing tightly against each other is, according to Darwin, dissipated or released inside an evolutionary dynamic that pushes them away through divergence. As explicitly recognized by Darwin, however, species belonging to the same genus are under greater pressure than allied species belonging to distinct genera, since the former share more similarities than the latter (Darwin 1859: 75, 76, 321, 1872: 58, 59, 296).

The careful reader of the *Origin* understands that demographic pressure for such forms is gradually reduced in reverse proportions to the ascent of the taxonomic scale, tending toward "zero" or the exhaustion of the evolutionary drive (a logical issue upon which Darwin remains strangely silent). The more allied forms diverge away, the less the competitive exclusion principle constitutes a valid active principle. What was at first a world filled to capacity by allied forms at low taxonomic levels turns into, under that principle, an ever-more empty world at higher levels. Although Darwin tried to force his theoretical way, his competitive exclusion principle could not account for an ongoing divergence process: the evolutionary pattern of high level entities should look more like a random walk or evolutionary stagnation. We should keep in mind that Darwin's theory relied on divergence as a way to sustain biological progress. Unfortunately, the exhaustion of the evolutionary drive at high taxonomic levels was not to his explanatory advantage.

---

[13]Darwin (1859: 171, 194, 206, 210, 243, 460, 471).

[14]See, for instance, Darwin (1859: 77, 81–82, 102–111, 119, 350, 388, 401, 472).

In a second case, we again see Darwin facing difficulties, this time at lower taxonomic levels. Darwin wavered between two different models of speciation: the sympatric model and the allopatric model (Vorzimmer 1970; Sulloway 1979; Mayr 1982: 411–417; 1991: 31–34; Bulmer 2004). In the first model, the splitting of two lineages is conceived in the context of forms living in open space pressing against each other, whereas in the second, forms are geographically isolated from one another (1859: 102–108, 1872: 80–84). Since we have already discussed this at some length in Chap. 3 of this book, we will here simply reconsider the question from the perspective of a spectrum ranging from an entirely filled world to a totally empty one. Unsurprisingly, Darwin inclined toward the sympatric model as the more important of the two for instigating speciations (1859: 105, 106, 1872: 82, 83). This is perfectly consistent with a theory founded on the principle of divergence: allied forms are more effectively repelled from one another in the context of a crowded world (competitive exclusion principle). At the same time, Darwin also made significant use of the allopatric model in hopes of taking advantage of both speciation models, alternating between the two. Apparently, Darwin hoped that sympatry would allow for the build-up of demographic pressure in preparation for divergence, whereas allopatry would allow for the completion of the process through the isolation of a lineage placed on an irreversible path.

Let us pause to raise a question: Are we to assume that such an isolated lineage will continue to diverge from its allied forms? Having been deprived of the influence of the competitive exclusion principle for long periods of time (Darwin 1859: 108, 109, 1872: 84, 85), would not a random evolutionary walk or stagnation be the most likely evolutionary outcome for isolated lineages? If we follow Darwin's own logic on selective pressure, evolution of allied forms in the context of a filled world or in the context of an empty world (geographical isolation) cannot be accompanied by the same evolutionary outcome. Nevertheless, Darwin seems determined to force his theoretical way by applying the principle of divergence at low taxonomic levels (see his Diagram), irrespective of changing selective contexts:

> To sum up the circumstances favourable and unfavourable to natural selection... I conclude, looking to the future, that for terrestrial productions a large continental area, which will probably undergo many oscillations of level, and which consequently will exist for long periods in a broken condition, will be the most favourable for the production of many new forms of life, likely to endure long and to spread widely. For the area will first have existed as a continent, and the inhabitants, at this period numerous in individuals and kinds, will have been subjected to very severe competition. When converted by subsidence into large separate islands, there will still exist many individuals of the same species on each island: intercrossing on the confines of the range of each species will thus be checked: after physical changes of any kind, immigration will be prevented, *so that new places in the polity of each island will have to be filled up by modification of the old inhabitants*; and time will be allowed for the varieties in each to become well modified and perfected. When, by renewed elevation, the islands shall be re-converted into a continental area, there will again be severe competition: the most favoured or improved varieties will be enabled to spread : there will be much extinction of the less improved forms, and the relative proportional numbers of the various inhabitants of the renewed continent will again be changed; and again there will be a fair field for natural selection to improve still further the inhabitants, and thus produce new species. (Darwin 1859: 107, 108; see also 1872: 84) [my emphasis]

In the context of affiliated forms isolated from one another, each lineage must face its own selective constraints, owing to the unique variations carried by its members, the competition of different non-allied forms, and the specificity of the physical milieu. Let us try to imagine the independent evolutionary outcomes for a whole set of affiliated forms long isolated from each other. The evolutionary picture cannot be what Darwin has in mind. Darwin scholars have long been aware of his difficulties with the notion of speciation. This issue is of no concern to us here. The point, rather, is an epistemological one: by modulating the selective contexts of his evolutionary examples—by moving back and forth along the filled world/empty world spectrum of related forms—Darwin offered the careful reader contradictory pictures of evolution. The theory he tries to promote is ill-suited to face significant and long-lasting empty places in nature, as these disturb the march of divergence and, secondarily, of biological progress.

It is interesting to note that the long passage quoted above couch life's history against the background of a cyclical geology, as seen in oscillations of sea levels and continental subsidence-elevation. Although Darwin probably thought he had only borrowed Charles Lyell's neo-Huttonian geology while substituting for biology a directional view of life, one can argue with some justice that Darwin himself seriously compromised that latter view. Again, it is tempting to argue that, empirically speaking, life manages to find its own way to biological progress precisely because evolution does not comply with Darwin's rigid and pre-established pattern-process of divergence.

Thus far we have analyzed the filled world/empty world question through the organism-to-organism perspective, according to which population density and competition level constitute the engine behind the evolutionary dynamics. However, there is in Darwin another dimension accompanying the same question: the relationship between organisms and the physical environment. This will be our third case to consider. Darwin made clear he believes that selective pressure is less affected by the physical milieu than by the competition of forms living in the same niche (1859: 351, 487, 1872: 84). Why that choice? For him, biotic competition is more susceptible to creating the condition for high adaptive pressure, especially under the competitive exclusion principle, which then creates the condition of a perfectly filled world. On the other hand, the adaptive pressure is more easily relaxed through the life-physical milieu relationship, since many territories offer a whole range of resources for survival (excluding extreme environments like deserts and polar regions). That issue does not arise in the context of an organism-to-organism plenum.

More specifically, a shift in climate affecting a large region of the Earth can easily open up a new vacant niche to a form already pre-adapted to these conditions, the climate shift itself disposing of potential competitors. In this context of a victory without a fight, little selective incentive exists for evolutionary change. From Darwin's viewpoint, therefore, the life-physical milieu relationship is yet another context for which an empty world can deprive him of an evolutionary drive: "...forms have retained nearly the same character from an enormously remote geological period, so certain species have migrated over vast spaces, and have not become greatly or at all modified" (1872: 319; see also 1859: 351). Let us provide a case presented by Darwin in which identical yet widely scattered forms are encountered:

> I do not pretend to indicate the exact lines and means of migration, or the reason why certain species and not others have migrated; why certain species have been modified and have given rise to new groups of forms, and others have remained unaltered... I have said that many difficulties remain to be solved... I will only say that as far as regards the occurrence of identical species at points so enormously remote as Kerguelen Land, New Zealand, and Fuegia, I believe that towards the close of the Glacial period, icebergs, as suggested by Lyell, have been largely concerned in their dispersal. (Darwin 1859: 380, 381; see also 1872: 341)

Independently of whether or not the dispersal mode postulated here is convincing, the reader gathers from this passage that the principle of divergence is entirely disabled in some cases, thus obstructing the march of evolution as Darwin understands it.

We know that Darwin believed in biological progress. He needed something to be profoundly modified during evolution, whether this is thought of as "specialization of parts" or something else. For that purpose, the evolutionary process had to be kept under tension at all times, hence a theory of biological evolution founded on a world filled to capacity. Unfortunately, Darwin could not evade the fact that his pressure cooker was leaking in a number of important ways: the gradual deactivation of the competitive exclusion principle during taxonomic ascent, the loss of selective pressure during the allopatric phase of speciation at low taxonomic levels, and the relative lack of adaptive pressure for pre-adapted forms during migration or climate shift.

Atomism Versus Connectedness Darwin had to face yet another set of difficulties in his defence of biological progress. His pressure cooker model rested not only on a filled world but, significantly also, on the atomistic assumption that evolutionary entities are merely pressurizing each other without grabbing on to one another, so to speak. Indeed, competing forms are, for Darwin, often envisioned like atoms with perfectly smooth surfaces to which nothing could attach itself, atoms moving freely in a plenum along a rigid and pre-established divergent pattern. This model, based on the notion of "action-by-contact" borrowed from physics, was translated in biology through selective pressure becoming, by analogy, the agent responsible for increasing or decreasing the number of forms. However, Darwin challenged this model in a number of different cases, explicitly or implicitly holding instead that evolutionary entities are connected to each other in intimate ways, to the point of preventing them from either evolving along the expected path or from evolving at all.

The first and most obvious case comes from reproduction, that is, from entities caught up in tight and complex reproductive networks at low taxonomic levels and from which it is difficult to break free. Although Darwin tried to impose his own view upon this inextricable complexity, the *Origin of Species* (1859) is replete with counterfactual evidence. The most salient features of life resisting change and, ultimately, segregation are the following: (a) life resisting change for a number of generations (p. 7); (b) life reversing toward the parental condition (pp. 14, 15, 25, 26, 152–154, 159); (c) life being organized around non-monolithic taxonomic entities like "strains" and "sub-breeds (pp. 31, 32, 96, 267); (d) nature favoring interbreeding rather than close inbreeding (pp. 70, 71, 96, 104–106, 248–250, 253); (e) life being characterized by various levels of fertility below the family level (pp. 22, 23, 248, 253, 255–257, 261, 267–272); (f) breeders being able to create forms separated at the generic level yet still fully fertile with one another (pp. 22, 23, 26, 445); (g) polyphyletism as a common genealogical arrangement among domesticated forms (pp. 17, 28, 29, 40); (h) the impossibility for natural selection to pick out a negative feature like sterility, the latter being a mere by-product of other features themselves positively selected (pp. 245, 260, 269); and (i) the imperfect relationship between a "phenotype" and "fertility" since dissimilar forms can sometimes reproduce together while similar ones cannot (p. 269), thus muddling the action of natural selection during the segregation process, for on exactly what basis is natural selection working? As if this list is not sufficiently damning, it should be remembered that Darwin never produced a clear process of speciation and wavered between sympatric and allopatric models (pp. 102–108) with the former requiring a close physical contact between organisms. To this we must add Darwin's support for an heredity based on blending (p. 108).

The second case involves the two main selective contexts imagined by Darwin when thinking about propelling evolution forward. In his search for a sustained evolutionary dynamic, Darwin always preferred to work in the context of a world filled to near capacity. As already seen, the competitive exclusion principle as applied to closely related forms meets this conceptual requirement. By the same demographic logic, this selective context was extended in its application to unrelated forms living in the "entangled bank" (Darwin 1859: 74, 489, 1872: 58, 429), that is, forms not held together by genealogical connections but bound both by sharing a common place in the economy of nature and by various kinds of interdependence toward each other. Reading this part of the *Origin* (1859: 71–79, 1872: 55–61) is strongly reminiscent of Carl von Linnaeus's *Oratio de Telluris Habitabilis Incremento* (1744), *Oeconomia Naturae* (1749), and *Politia Naturae* (1760). In one instance, Darwin presents this interdependence in these words (to quote this passage again):

> [I]n several parts of the world insects determine the existence of cattle. Perhaps Paraguay offers the most curious instance of this; for here neither cattle not horses nor dogs have ever run wild, though they swarm southward and northward in a feral state; and Azara and Rengger have shown that this is caused by the great number in Paraguay of a certain fly, which lays its eggs in the navels of these animals when first born. The increase of these flies, numerous as they are, must be habitually checked by some means, probably by birds. Hence, if certain insectivorous (whose numbers are probably regulated by hawks or beasts of prey) were to increase in Paraguay, the flies would decrease—then cattle and horses would become feral, and this would certainly greatly alter (as indeed I have observed in parts of South America) the vegetation : this again would largely affect the insects; and this, as we have just seen in Staffordshire, the insectivorous birds, and so onwards in ever-increasing circles of complexity. We began this series by insectivorous birds, and we have ended with them. Not that in nature the relations can ever be as simple as this. (Darwin 1859: 72, 73; see also 1872: 56, 57)

What Darwin seeks to preserve at all costs through the notion of the entangled bank is the conceptual benefit of forms harshly pressing against each other, forms presumably left with only two possible evolutionary outcomes: extinction or evolutionary change along a divergent path. Regrettably, Darwin's entangled bank constitutes a mixed bag of examples that connect forms through different kinds of relationships, for instance, parasitism as seen in the configuration of the legs and claws of a parasite clinging to the hair of tigers (1859: 77, 1872: 60); contrivance or a designed structure for fertilizing orchids (1859: 73, 193, 203, 1872: 57, 154, 155, 200)[15]; and a predator-prey relationship as illustrated between cats and mice (1859: 74, 1872: 58). This raises two obvious issues. First, evolutionary change may not necessarily ensue from these relationships, as a high level of interdependence may constitute a stabilizing factor by-passing the risk of change. This was eventually explicitly contemplated by Darwin. In the sentence immediately following the passage quoted above, he adds: "Battle within battle must ever be recurring with varying success; and yet in the long-run the forces are so nicely balanced, that the face of nature remains uniform for long periods of time..." (1859: 73; see also 1872: 57). Second, why should we assume that evolutionary change generated among *unrelated forms* living in the entangled bank leads to an ongoing divergence process among *closely related forms*? After all, Darwin argues that the stalemate of mutually dependent forms stuck in a stabilizing relationship will eventually be broken. The same passage just quoted continues: "...the face of nature remains uniform for long periods of time, though assuredly the merest trifle would often give the victory to one organic being over another" (1859: 73; see also 1872: 57). And we know only too well that for Darwin, victory or evolutionary change usually means "divergence." However, would a random evolutionary walk not be more plausible in the context of the entangled bank? Yet, Darwin moved back and forth indiscriminately between the two different selective contexts—the competitive exclusion principle and the entangled bank—as if the two were equivalent in their evolutionary outcome (1859: 71–79, 1872: 55–61).

In the third case, life forms are not dependent on one another but are rather engaged in a sort of "adaptive equilibrium" wherein living forms sharing the same place in the economy of nature stand their ground relative to one another—each being entrenched in its own niche—natural selection having adapted them so perfectly to each other that evolutionary stagnation or equilibrium is reached and maintained for extremely long geological periods. From this viewpoint, the world is filled to capacity, but the evolutionary dynamics founded on a symmetrical yet inverted relationship of increasing/decreasing populations responsible for propelling life forward is replaced by another kind of equilibrium that appeals, this time, to selective forces interlocking competing forms in an evolutionary stalemate. Perhaps the most concise exposition of this notion can be found in the following passage:

[15]See also Darwin, C. (1862), *On the Various Contrivances by which British and Foreign Orchids Are Fertilized by Insects, and on the Good Effects of Intercrossing* (London: John Murray).

> If, for instance, a number of species, which stand in direct competition with each other, migrate in a body into a new and afterwards isolated country, they will be little liable to modification; for neither migration nor isolation in themselves can do anything. These principles come into play only by bringing organisms into new relations with each other, and in a lesser degree with the surrounding physical conditions. As we have seen in the last chapter that some forms have retained nearly the same character from an enormously remote geological period, so certain species have migrated over vast spaces, and have not become greatly modified. (Darwin 1859: 351; see also 1872: 319)

This evolutionary phenomenon is not restricted to *unrelated forms* living in the entangled bank but applies also to *closely related forms*, thus impinging upon Darwin's cardinal concept of the competitive exclusion principle as a means to initiate and sustain evolutionary divergence:

> But we often take, I think, an erroneous view of the probability of closely allied species invading each other's territory, when put into free intercommunication. Undoubtedly if one species has any advantage whatever over another, it will in a very brief time wholly or in part supplant it; but if both are equally well fitted for their own places in nature, both probably will hold their own places and keep separate for almost any length of time. (Darwin 1859: 401; see also 1872: 356)

The notion of "adaptive equilibrium" is no minor component in the *Origin of Species*, contrary to Darwin's rhetoric downplaying its role.[16] Indeed, Darwin's Diagram shows that out of eleven lineages, nine are in a state of stagnation, irrespective of the taxonomic level considered, in addition to the fact that Darwin returns to it on numerous occasions.[17]

Let us pause to summarize our analysis. Darwin believed in biological progress. For this he relied on his analogy of the pressure cooker to propel life forward: forms tightly pressed against each other are condemned either to go extinct or climb the scale of complexity. The pattern-process of divergence is the privileged evolutionary path to survival and, ultimately, to persistence. Regrettably for Darwin, his intellectual connections with the tradition of natural theology—through the various expressions of the concepts of equilibrium and interdependence—put him at a disadvantage when it came to tackling the issue of biological progress. The overall explanatory structure he proposed is constructed around conceptual elements that disrupt the evolutionary drive, either by leading to evolutionary stagnation or by means of a random walk. Among the disruptive factors found in the *Origin of Species,* one finds the following: (1) increasing populations versus decreasing populations, (2) sympatric speciation versus allopatric speciation, (3) organism-to-organism adaptive context versus organism-to-the-physical-milieu adaptive context, and (4) competition between affiliated forms versus competition between non-related forms. Like Charles Lyell, who, in the *Principles of Geology* (1830–1833), imagined life forms as continually challenged by cyclical physical

---

[16]Darwin (1859: 281, 314–315, 351, 429, 1872: 265–266, 291–292, 319).

[17]Darwin (1859: 52, 73, 313, 331–332, 351, 368, 390, 400–402, 421, 429, 488, 1872: 57, 290–292, 304, 332, 319, 332, 348, 355–356, 370, 428).

changes, Darwin created a list of different contexts that put pressure on life forms and, hopefully, initiate motion. Underlying such competitive contexts, however, is the cardinal notion of a spectrum that ranges from a world filled to capacity to one that is mostly empty.

Darwin's demonstration encounters two main sets of difficulties. First, there is his leaky pressure cooker: important biotic voids are created in a world depleted of competition. The evolutionary drive exhausts itself in the following cases: (a) under the competitive exclusion principle during taxonomic ascent, (b) in the allopatric phase of speciation, and (c) when a changing climate brings about the migration of pre-adapted forms. Second, Darwin's atomistic and particulate assumption is challenged by life forms that are interconnected in various ways. This leads to a whole range of possible evolutionary outcomes, ranging from a random walk (divergence, parallelism, and convergence) to stagnation. This is illustrated in such cases as (a) interbreeding among affiliated forms at low taxonomic levels, (b) owing to interdependence among unrelated forms in the entangled bank (parasitism, contrivance, predator-prey relationship), and (c) during evolutionary equilibrium when forms are interlocked in a mutual adaptive entrenchment.

We refer to Darwin's explanations listed above as "opening and closing devices" since the evolutionary directionality he hoped to insufflate into the history of life is thwarted. It seems that for each possible evolutionary step forward, Darwin provides a counterbalance, a sort of contrary force:

1. The demographic expansion of a form is produced at the proportional expense of a declining one.
2. The speciation process is presented as occurring under maximum pressure (sympatric) or minimum pressure (allopatric).
3. Competition is increased among affiliated forms at low taxonomic levels but decreases, proportionally, to the taxonomic ascent. In other words, closely related forms first encounter high pressure (the competitive exclusion principle), but as they move away from each other, this pressure is relaxed in the context of non-related forms (entangled bank).
4. A state of evolutionary equilibrium exists between forms interlocked in a mutual adaptive entrenchment in which maximum adaptive pressure is perfectly maintained to the point of depriving this stalemate of evolutionary efficiency. A world filled to capacity, on this view, is no guarantee of an evolutionary drive.
5. Life forms are treated as sufficiently stable entities occupying real places in the economy of nature (realism), or life forms are envisioned as inherently unstable and transitional evolutionary entities (nominalism). To this latter issue we now turn.

Darwin scholars have been engaged in a subtle debate about Darwin's position on the species question. The intricacies of it shall not concern us here (Ghiselin 1969: 78–102; Mayr 1982: 265–269; Beatty 1985; Stamos 2007; Sloan 2009; Richards 2010: 78–96). It will suffice to mention that at least two distinct views are found in the *Origin of Species*, both of which flow naturally from the tension we have already documented in Darwin, who simultaneously promotes a world deprived of real change and a world profoundly changing. It is indeed expected to read under

Darwin's pen—the same that supports biological progress—a nominalistic declaration to the effect that life is inherently fleeting: "Certainly no clear line of demarcation has as yet been drawn between species and sub-species... or, again, between sub-species and well-marked varieties, or between lesser varieties and individual differences. These differences blend into each other in an insensible series; and a series impresses the mind with the idea of an actual passage" (1859: 51; see also 1872: 41). Interestingly, in the passage immediately following the one just quoted, Darwin hints at some varieties being more permanent than others, implying that entities become ever more permanent and solidly entrenched in the economy of nature as they ascend the taxonomic scale:

> Hence I look at individual differences... as of high importance for us, as being the first step towards such slight varieties as are barely thought worth recording in works on natural history. And I look at varieties which are in any degree more distinct and permanent, as steps leading to more strongly marked and more permanent varieties; and at these latter, as leading to sub-species, and to species. (Darwin 1859: 51, 52; see also 1872: 41, 42)

It is unsurprising that it is in this explanatory context that Darwin presents his realistic view, a context that assumes life to be characterized by designed entities entrenched in nature:

> To sum up, I believe that species come to be tolerably well-defined objects, and do not at any one period present an inextricable chaos of varying and intermediate links: firstly, because new varieties are very slowly formed, for variation is a very slow process, and natural selection can do nothing until favourable variations chance to occur, and until a place in the natural polity of the country can be better filled by some modification of some one or more of its inhabitants. (Darwin 1859: 177, 178; see also 1872: 137)

This statement is nearly a truism in light of the attention devoted by Darwin to evolutionary stagnation. From our viewpoint, the nominalism/realism issue constitutes only one more manifestation of the "opening and closing" explanatory devices enshrined by Darwin within his theoretical structure. It must be noted, however, that the more Darwin climbs the taxonomic scale, the more entities seem to be solidly and permanently entrenched in nature, thus tipping the conceptual balance in favor of the realistic outlook. It must be remembered that, at the highest possible levels of classes and phyla, entities are, for Darwin, nearly perfectly stable and permanent.

The idea of a world built around a fundamental equilibrium pervades Darwin's thinking well beyond what is usually identified by Darwin scholars under the expression "checks and balances." This also includes our previous discussion of Chapters 11 and 12 of the *Origin*, in particular Darwin's *ideal model* in which explanatory variables are interrelated in a symmetrical or proportional manner—taxonomic level, geographical space, geological time, amount of selective pressure—all arranged to protect the theory's dual core composed of divergence-gradation at the service of the *weak contingency thesis*.[18] This is all perfectly consistent

[18]This topic has been analyzed in great depth in Chap. 5 of this book.

with Darwin's commitment to a fully constituted world, for nothing truly new is expected to occur within that ordered world.

7.6 Conclusion of Part III

We have seen that a number of Darwin scholars have struggled to reconcile Darwin's theory of biological evolution and his support for biological progress. Our own account provides additional reasons for such a justified puzzlement. Having fixed the overall picture of evolution within the straitjacket of stable classes/phyla, Darwin conjured instances wherein evolutionary motion could manifest itself at lower taxonomic levels:

1. Biological diversity observed among extant forms is recycled by being projected backward in time.
2. The evolution of taxonomic entities at or below the order level presents itself as an endless cycle of rising and declining entities.
3. The expansion of forms comes at the proportional expense of others.
4. The exhaustion of the evolutionary drive supporting the divergence of two related forms occurs in reverse proportion to their taxonomic ascent.

The intellectual road taking us from Linnaeus to Darwin (documented in Chap. 3 of this book) is shorter than is often believed. During his attempted transition to evolutionism, Darwin borrowed many conceptual tools from the seventeenth and eighteenth centuries (see Egerton 1973; La Vergata 1988; Worster 1994). We saw that Linnaeus's view is one of an expanding life through historical times (Egerton 2007: 80), wherein nothing entirely new is introduced, since the same proportions between species are maintained and regulated beforehand to avoid encroaching species driving others to extinction (Linnaeus 1744: 48, 49). A pre-established equilibrium between species is thus maintained throughout an ongoing demographic expansion (Limoges 1972: 10, 11). Motion exists in this world but only a circular motion with no sign of a beginning or ending (Linnaeus 1760: 104).

To what extent was Darwin able to move beyond this view of the world in order to create more evolutionary motion? First, it is clear that Darwin dissolved the one-to-one relationship between a given particular species and a single specific physical place in nature. Second, by accepting the reality of extinction, Darwin saw that the removal of extinct forms opened up opportunities for remaining forms. However, this recognition did not apply to higher taxonomic entities like classes and phyla. Third, the balance of nature under Darwin's hand was released or freed but *only partially*. Despite the fact that he released it from its original constraints founded on pre-established proportions between species, Darwin replaced them with the numerous constraining explanatory devices listed in this chapter.

A classic exercise of Darwin studies consists in contrasting the views of Charles Darwin and Jean-Baptiste Lamarck. Whereas Lamarck subscribed to the idea of an *internal* evolutionary drive, inevitably pushing life forward up the scale of complexity, so the story goes, Darwin embraced the notion of evolution under an *external* pressure adapting species to places in the economy of nature, thus instigating diversification and perhaps progress. It will be remembered that Darwin wrote: "I believe in no fixed law of development, causing all the inhabitants of a country to change abruptly, or simultaneously, or to an equal degree" (1859: 314). Attacking Mivart this time rather than Lamarck, Darwin wrote more explicitly:

> Mr. Mivart believes that species change through 'an internal force or tendency,' about which it is not pretended that anything is known. That species have a capacity for change will be admitted by all evolutionists; but there is no need... to invoke any internal force beyond the tendency to ordinary variability, which through the aid of selection by man has given rise to many well-adapted domestic races, and which through the aid of natural selection would equally well give rise by graduated steps to natural races and species. The final result will generally have been... an advance, but in some few cases a retrogression, in organisation. (Darwin 1872: 201)[19]

At stake here, apparently, is the issue of an internal drive leading to a *predictable* outcome (complexity) versus an external evolutionary pressure leading to *opportunistic* possibilities only (stability, degeneration, or complexity). We now know that Darwin's so-called "opportunistic" view of evolution was much more channelled and constrained than is usually recognized in the historiography, thus reducing the intellectual gap between Darwin and Lamarck on that count.

Darwin's commitment to the idea of the balance of nature was such an integral part of his theoretical structure that he could not discuss the issue of evolutionary change without presenting it under the dual aspect of "progression/retrogression." Apparently, Darwin could not escape the conclusion that his theory could indifferently generate progressive evolution or degenerative evolution:

> [B]earing in mind that all organic beings are striving to increase at a high ratio and to seize on every unoccupied or less well occupied place in the economy of nature, that it is quite possible for natural selection gradually to fit a being to a situation in which several organs would be superfluous or useless: in such cases there would be retrogression in the scale of organization... On our theory the continued existence of lowly organisms offers no difficulty; for natural selection... does not necessarily include progressive development—it only takes advantage of such variations as arise and are beneficial to each creature under its complex relations of life. (Darwin 1872: 98)

This passage is ensconced in a section where Darwin argues that evolution is, in general, progressive, as seen in the overall increase of organization over time (1872: 97–100). We have argued in this chapter that Darwin's theory was ineffective when it came to supporting biological progress, forcing him to resort to rhetoric. Darwin is certainly right when he says: "But to suppose that most of the many now existing low forms have not in the least advanced since the first dawn of life would be extremely rash" (1872: 99). What Darwin recognizes here, however, is not a consequence of the implementation of his theory but rather an empirical fact (progress) his theory cannot explain. Having made this distinction, we are now in a better position to appreciate that Darwin's theory contains yet another manifestation of the idea of

[19]For a direct reference to Lamarck's support of an internal drive, see Darwin (1872: 98).

equilibrium: the progression/retrogression duality. Darwin's theory is truly one of continual tension between a world entirely finished and one in a state of perpetual construction.

References

Barnes J (1998) Les penseurs préplatoniciens. In: Canto-Sperber M (ed) Philosophie grecque, 2e edn. Presses Universitaires de France, Paris, pp 3–88
Bartholomew M (1975) Huxley's defence of Darwin. Ann Sci 32:525–535
Beatty J (1985) Speaking of species: Darwin's strategy. In: Kohn D (ed) The Darwinian heritage. Princeton University Press, New Jersey, pp 265–281
Bowler PJ (1986) Theories of human evolution: a century of debate, 1844–1944. Johns Hopkins University Press, Baltimore
Bowler PJ (1996) Life's splendid drama: evolutionary biology and the reconstruction of life's ancestry, 1860–1940. University of Chicago Press, Chicago
Bowler PJ (2003) Evolution: the history of an idea, 3rd edn. University of California Press, Berkeley
Bulmer M (2004) Did Jenkin's swamping argument invalidate Darwin's theory of natural selection? Br J Hist Sci 37:281–297
Darwin C (1859) On the origin of species. John Murray, London
Darwin C (1862) On the various contrivances by which British and foreign orchids are fertilized by insects, and on the good effects of Intercrossing. John Murray, London
Darwin C (1872) The origin of species, 6th edn, with additions and corrections [1876]. John Murray, London
Darwin F, Seward A (eds) (1903) More letters of Charles Darwin, vol 1. John Murray, London
Depew DJ, Weber BH (1995) Darwinism evolving: systems dynamics and the genealogy of natural selection. MIT Press, Cambridge
Di Gregorio M (1984) T.H. Huxley's place in natural science. Yale University Press, New Haven
Dijksterhuis EJ (1961) The mechanization of the world picture. Clarendon Press, Oxford
Egerton F (1973) Changing concepts of the balance of nature. Q Rev Biol 48:322–350
Egerton F (2007) A history of the ecological sciences, part 23: Linnaeus and the economy of nature. Bull Ecol Soc Am 88:72–88
Ghiselin MT (1969) The triumph of the Darwinian method. University of California Press, Berkeley
Gould SJ (1996) Full House: the spread of excellence from Plato to Darwin. Belknap Press, Cambridge
Greene JC (1957) Objectives and methods in intellectual history. Miss Val Hist Rev 44:58–74
Greene JC (1959) The death of Adam: evolution and its impact on Western thought. Iowa State University Press, Iowa
Greene JC (1971) The Kuhnian paradigm and the Darwinian Revolution in natural history. In: Roller D (ed) Perspectives in the history of science and technology. University of Oklahoma Press, Norman, pp 3–25
Greene JC (1981) Science, ideology, and world view. University of California Press, Berkeley
Greene JC (1986) The history of ideas revisited. Rev Synth (3):201–227
Guthrie WKC (1965) A history of Greek philosophy: the Presocratic tradition from Parmenides to Democritus. Cambridge University Press, Cambridge
Hall AR (1981) From Galileo to Newton. Dover Publications, New York
Hodge MJS (1985) Darwin as a lifelong generation theorist. In: Kohn D (ed) The Darwinian heritage. Princeton University Press, Princeton, pp 207–243
Hodge MJS (2014) On Darwin's science and its contexts. Endeavour 38:169–178

Hodge MJS, Radick G (2009) The place of Darwin's theories in the intellectual long run. In: Hodge MJS, Radick G (eds) The Cambridge companion to Darwin, 2nd edn. Cambridge University Press, Cambridge, pp 246–273
Hull DL (1973) Darwin and his critics: the reception of Darwin's theory of evolution by the scientific community. Harvard University Press, Cambridge
Hull DL (1988) Progress in ideas of progress. In: Nitecki MH (ed) Evolutionary progress. University of Chicago Press, Chicago, pp 27–48
Huxley TH (1862) The anniversary address. Q J Geol Soc Lond 18:xl–liv
Huxley TH (1893) Darwiniana: collected essays, vol 2. Macmillan, London
Huxley TH (1898–1902) In: Foster M, Lankester ER (eds) The scientific memoirs of Thomas Henry Huxley, 4 vols. Macmillan, London
La Vergata A (1988) Theodicy and nature's economy. Nuncius 3:139–152
Limoges C (1972) Introduction. In: Linné C (ed) L'équilibre de la nature. J. Vrin, Paris, pp 7–22
Linnaeus C (1744) Oratio de Telluris habitabilis incremento. In: Linné C (ed) L'équilibre de la nature (translated from Latin to French). J. Vrin, Paris, pp 29–49
Linnaeus C (1760) Politia naturae. In: Linné C (ed) L'équilibre de la nature (translated from Latin to French). J. Vrin, Paris, pp 103–121
Lyons SL (1995) The origins of T.H. Huxley's saltationism: history in Darwin's shadow. J Hist Biol 28:463–494
Lyons SL (1999) Thomas Henry Huxley: the evolution of a scientist. Prometheus Books, Amherst
Mayr E (1982) The growth of biological thought. Belknap Press, Cambridge
Mayr E (1991) One long argument: Charles Darwin and the genesis of modern evolutionary thought. Harvard University Press, Cambridge
Ospovat D (1981) The development of Darwin's theory: natural history, natural theology, and natural selection, 1838–1859. Cambridge University Press, Cambridge
Owen R (1860a) [Darwin on the Origin of Species]. Edinb Rev 111:487–532
Owen R (1860b) Palaeontology or a systematic summary of extinct animals and their geological relations. Adam and Charles Black, Edinburgh
Radick G (2000) Two explanations of evolutionary progress. Biol Philos 15:475–491
Richards RJ (1988) The moral foundation of the idea of evolutionary progress: Darwin, Spencer, and the neo-Darwinians. In: Nitecki MH (ed) Evolutionary progress. University of Chicago Press, Chicago, pp 129–148
Richards RJ (1992) The meaning of evolution: the morphological construction and ideological reconstruction of Darwin's theory. University of Chicago Press, Chicago
Richards RA (2010) The species problem: a philosophical analysis. Cambridge University Press, Cambridge, pp 78–96
Richardson RC, Kane TC (1988) Orthogenesis and evolution in the 19th century: the idea of progress in American neo-Lamarckism. In: Nitecki MH (ed) Evolutionary progress. University of Chicago Press, Chicago, pp 149–167
Rudwick MJS (1970) The strategy of Lyell's Principles of Geology. Isis 61:5–33
Rupke N (2009) Richard Owen: biology without Darwin. University of Chicago Press, Chicago
Ruse M (1979) The Darwinian Revolution. University of Chicago Press, Chicago
Ruse M (1988) Molecules to men: evolutionary biology and thoughts of progress. In: Nitecki MH (ed) Evolutionary progress. University of Chicago Press, Chicago, pp 97–126
Ruse M (1996) Monad to man: the concept of progress in evolutionary biology. Harvard University Press, Cambridge
Ruse M (2004) The Romantic conception of Robert J. Richards. J Hist Biol 37:3–23
Ruse M (2009) The history of evolutionary thought. In: Ruse M, Travis J (eds) Evolution: the first four billion years. Belknap Press, Cambridge, pp 1–48
Sargent R-M (1995) The diffident naturalist: Robert Boyle and the philosophy of experiment. University of Chicago Press, Chicago
Schweber SS (1985) The wider British context in Darwin's theorizing. In: Kohn D (ed) The Darwinian heritage. Princeton University Press, Princeton, pp 35–69

Shanahan T (2004) The evolution of Darwinism: selection, adaptation, and progress in evolutionary biology. Cambridge University Press, Cambridge

Shapiro AE (2002) Newton's optics and atomism. In: Cohen IB, Smith GE (eds) The Cambridge companion to Newton. Cambridge University Press, Cambridge, pp 227–255

Sloan PR (1985) Darwin's invertebrate program, 1826–1836: preconditions for transformism. In: Kohn D (ed) The Darwinian heritage. Princeton University Press, Princeton, pp 71–120

Sloan PR (2009) Originating species: Darwin on the species problem. In: Ruse M, Richards RJ (eds) The Cambridge companion to the 'Origin of Species'. Cambridge University Press, Cambridge, pp 67–86

Stamos DN (2007) Darwin and the nature of species. State University of New York, Albany

Sulloway FJ (1979) Geographic isolation in Darwin's thinking: the vicissitudes of a crucial idea. Stud Hist Biol 3:23–65

Vorzimmer PJ (1970) Charles Darwin: the years of controversy. Temple University Press, Philadelphia

Worster D (1994) Nature's economy: a history of ecological ideas, 2nd edn. Cambridge University Press, Cambridge

Part IV
A Question of Methods

Chapter 8
Methodologies for a World Fully Revealed

Abstract We argue that Charles Darwin's methodological commitments in the *Origin of Species* which appeal to *vera causa*, uniformitarianism, gradualism, actualism, and extrapolationism constitute a conceptual nexus founded on an ontological commitment to a fully constituted world, that is, a homogeneous world unchanging over time, with the past and the present conceived as identical (the "steady-state view"). Ultimately, this ontology prevented Darwin from successfully completing the intellectual transition to full and modern evolutionism. In addition, because a steady-state world is best investigated through empirico-inductivism, this approach condemned Darwin to integrating too many facts that could not be explained under his so-called all-encompassing theory of evolution. Confronted with this ill-integrated pluralism, the reader of the *Origin* is left with an array of competing evolutionary pictures and theories.

Keywords Uniformitarianism · Steady-statism · Ontology · Methodology · *Vera causa* · Method of analogy · Actualism · Inductivism · Empiricism · Gradualism · Gradation · Extrapolationism · Interpolationism · Isaac Newton · Thomas Reid · Charles Lyell

This chapter will investigate one final epistemological matter arising in the *Origin of Species*, Darwin's methodology. Relevant comments about this aspect of his thought can be found throughout the previous chapters of this book. They will be reiterated here and presented in a more synthetic and consistent fashion. Darwin's methodology presents itself as a tangle of complementary and mutually reinforcing methodologies: a science of real cause (*vera causa*), the method of analogy, inductivism-empiricism, actualism-uniformitarianism, gradualism-gradation, and extrapolationism-interpolationism. This conceptual knot does not exclude Darwin's use of other methodologies, but we argue that Darwin's methodological core is caught up in this tangle because it is tailor-made to fit a fully constituted world, that is, a world entirely revealed to us. In other words, we hold that Darwin's *methodological* core is intimately connected to his *ontological* worldview. In so arguing, we follow the view of science promoted by Thomas S. Kuhn in *The Structure of Scientific Revolutions* (1970: 41–42) and Larry Laudan in *Progress*

R. G. Delisle, *Charles Darwin's Incomplete Revolution*, Evolutionary Biology – New Perspectives on Its Development 1, https://doi.org/10.1007/978-3-030-17203-9_8

and Its Problems (1977: 78–81). Laudan provides an explicitation of the nature of the ontologico-methodological connection in an area directly relevant to our own thesis:

> Although it is vital to distinguish between the ontological and methodological components of a research tradition, the two are often intimately related, and for a very natural reason: namely, that one's views about the appropriate *methods* of inquiry are generally compatible with one's views about the *objects* of inquiry. When, for instance, Charles Lyell defined the 'uniformitarian' research tradition in geology, his ontology was restricted to presently acting causes and his methodology insisted that we should 'explain past effects in terms of presently acting causes.' Without a 'presentist' ontology, his uniformitarian methodology would have been inappropriate; and without the latter, the presentist ontology would not have allowed Lyell to explain the geological past (Laudan 1977: 80–81). [italics original]

Laudan makes clear that the area of application of ontology is not exclusively restricted to *processes* (acting causes) but also includes assumptions about existing *entities*. It is in this broader sense that Darwin's ontology was conceived in the previous chapters of this book. The main thesis of this chapter can thus be formulated: *Darwin's commitment to a world that has, in its essentials, always been complete was consistent enough to also extend to his methodology, making the separation of his ontology and methodology arbitrary.*

8.1 Behind a Science of Real Cause (*Vera Causa*)

It is well known that Darwin wanted somehow to comply with the methodological prescriptions of a science based on real causes (*verae causae*). This is revealed, among other cases, in his use of an analogy binding the work of breeders on domesticated life forms (artificial selection) and the cause of evolutionary change in nature (natural selection). Through this analogy, Darwin tried to show that natural selection was truly an evolutionary cause or mechanism that really existed by being empirically supported, and not merely a conjectural cause lacking either in existence or in evolutionary effectiveness in producing the postulated effects.[1]

We have seen that Darwin was not very successful in complying with this ideal of science, as he himself offered abundant implicit and explicit counterfactual evidence counting against the effectiveness of both artificial selection *and* natural selection: life resisting change for several generations; the reversion to the parental condition; the effect of blending inheritance; the imposition of the monophyletic agenda over the polyphyletic view of breeders; the inconsistent calibration of the power of artificial selection and natural selection throughout the *Origin*; the lack of clear sterility barrier below the family level; and so on.

---

[1]The debate over Darwin's use of some kind of "analogy" between artificial selection and natural selection has reached a sophisticated level among Darwin scholars. See Lloyd (1983), Evans (1984), Richards (1997), Sterrett (2002), and Gildenhuys (2004).

Darwin's failure is of no concern to us here. Rather, we would like to profit from Darwin's rhetoric about the assumed effectiveness of the artificial/natural selection analogy in order to probe into his fundamental epistemological commitments. Was the *vera causa* ideal merely driven by methodological considerations? Whereas Darwin scholars have often treated the methodological question as being cut off from its ontological underpinnings, we wish to revisit the question with that explicit aim in mind. To this end, it will be useful to turn our attention to Isaac Newton, Thomas Reid, and Charles Lyell. As already argued, the intellectual road taking us from Newton to Darwin is not a direct one,[2] but Newtonianism was nonetheless a very influential aspect of science during Darwin's time (Bellon 2014).

Not content with formulating the theory of universal gravitation, and extending it to phenomena as different as planetary motion, falling bodies on Earth, and cycles of tides in his *Mathematical Principles of Natural Philosophy* (1687 for the first edition), Newton also enumerates under "Rules of Reasoning in Philosophy" a number of basic epistemological rules for a sound practice of science. Although Newton gradually modified some epistemological prescriptions over the years (Koyré 1965: 261–272), one can find in the last and third edition of the *Principia* (1726[1952]: 270–271 [italics original]) the following four:

> Rule I: *We are to admit no more causes of natural things than such as are both true and sufficient to explain their appearances.* To this purpose the philosophers say that Nature does nothing in vain, and more is in vain when less will serve; for Nature is pleased with simplicity, and affects not the pomp of superfluous causes.
>
> Rule II: *Therefore to the same natural effects we must, as far as possible, assign the same causes.* As to respiration in a man and in a beast; the descent of stones in Europe and in America; the light of our culinary fire and of the sun; the reflection of light in the earth, and in the planets.
>
> Rule III: *The qualities of bodies, which admit neither intensification nor remission of degrees, and which are found to belong to all bodies within the reach of our experiments, are to be esteemed the universal qualities of all bodies whatsoever.* For since the qualities of bodies are only known to us by experiments, we are to hold for universal all such as universally agree with experiments; and such as are not liable to diminution can never be quite taken away. We are certainly not to relinquish the evidence of experiments for the sake of dreams and vain fictions of our own devising... We no other way know the extension of bodies than by our senses, nor do these reach it in all bodies; but because we perceive extension in all that are sensible, therefore we ascribe it universally to all others also... The extension, hardness, impenetrability, mobility, and inertia of the whole, result from the extension, hardness, impenetrability, mobility, and inertia of the parts; and hence we conclude that the least particles of all bodies to be also extended, and hard and impenetrable, and movable, and endowed with their proper inertia. And this is the foundation of all philosophy.
>
> Rule IV: *In experimental philosophy we are to look upon propositions inferred by general induction from phenomena as accurately or very nearly true, notwithstanding any contrary hypotheses that may be imagined, till such time as other phenomena occur, by which they may either be made accurate, or liable to exceptions.* This rule we must follow, that the argument of induction may not be evaded by hypotheses.

---

[2]The indirect intellectual road between Newton and Darwin has already been considered in Chap. 1 of this book.

What is Newton claiming in his Rules of Reasoning? That the world as currently revealed to us is totally knowable. This chapter will address two questions. First, *why* is nature knowable? Because nature is ontologically simple[3]: there is no need to multiply unduly the number of explanations, since the same effects are explained by the same causes (Rules I and II). This is also a way of supporting the uniformity of nature: laws are universal, precisely because the same causes explain the same effects (Rossi 1999, Chap. 17).[4] Furthermore, nature is knowable because microscopic manifestations that are hidden from us are reflected in an isomorphic manner at our macroscopic level (Rule III). Differently stated, nature is homogeneous, with entities being invariable, constant, and predictable (Rossi 1999: 310). Again, Newton's ontology is directly relevant to the uniformitarian principle, widely construed: whether hiding at lower microscopic levels or in past times, phenomena can nonetheless be grasped in our current time and space.

It will be remembered that Newton (1730: 375–376) wrote in his *Opticks*: "All these things being consider'd, it seems probable to me, that God in the beginning form'd Matter in solid, massy, hard, impenetrable, movable Particles, of such Sizes and Figures, and in such proportion to Space, as most conduced to the End for which he form'd them. . .." The multiplicity of world's entities is thus ultimately reducible to their atomic components. At this level, the world was created fully constituted: building blocks were permanent, immutable, and complete; nothing else could be expected out of such a full-fledged act of creation. The atomistic doctrine carried a threat, however. It could be argued that the world was complete at the atomic level but that at the macroscopic level, different things were being generated through the shifting of atomic combinations (Greene 1981: 13–14). While Descartes rushed into the breach by speculating about the evolution of the universe, as already alluded to in Chap. 1 of this book, scholars like John Ray (1701, preface) sealed the breach by adding to the world another layer of permanence and immutability: ". . .the Works created by God at first, and by him conserved to this Day in the same State and Condition in which they were first made." Similarly, Newton prevented the schism between the microscopic and the macroscopic levels by assuming an isomorphic relationship between the two, the latter being sufficient evidence of the nature of the

[3]Reijer Hooykaas explains the ontological component of the simplicity argument in the following way: "[T] he principle of economy of causes (or that of the simplicity of explanation): no more different causes should be assumed than is strictly necessary for explanation (or: explanatory systems should be as simple as possible). As soon as such a principle is transformed into the thesis that Nature *is* economic and simple. . . it has acquired an ontological instead of a purely methodological character." See Hooykaas (1975), quote on p. 356 footnote #206.

[4]Paolo Rossi's book was originally published in Italian in 1997 and translated in English in 2001 under *The Birth of Modern Science* (Oxford: Blackwell). I have used the French edition of 1999. For Newton, see Chap. 17.

former (*Principia*, Rule III).[5] As formulated by John C. Greene (1981: 130): "...the same God who had created the particles of matter and set them in motion... have arranged their motions from the beginning so that they would produce the harmonious system of mutually adapted structures constituting the present order of nature."

Considering that Newton's ontology carries implications for the application of the uniformitarian principle in the time dimension, some qualifications are in order here. Newton's support for a stable and static worldview neither excluded God's potential intervention in preventing it from decaying over very long periods of time, nor speculation about the cosmos being characterized by a series of different yet stable states in time, like a succession of different static worlds. The world today may show no traces of a beginning or ending, but this does not necessarily imply that the world was or will always be the same (Kubrin 1967). For Newton, then, the stability of our universe was a clear sign of contrived activity divinely planned (Brooke 1991: 135–151; Rossi 1999: 312, 325–328). In the General Scholium appended to the second edition of the *Principia* (1713), for instance, Newton expounds on the impossibility of explaining the current order of the world if it originated purely accidentally or out of chaos. Final causes or teleology must be appealed to in order to make sense of the observed ontological order of the world whose manifestation is seen in several instances:

1. In all the planets revolving around the Sun in the same direction along their respective curved paths at exactly the right distance and velocity, thus preventing them from crashing into each other under their mutual gravitational pull
2. In the stability of the distant and fixed stars that, other things being equal, should be attracted to each other but are not
3. In matter being lost in the tails of passing comets, thus allowing for the replenishment of matter gradually lost by the Sun and the planets

As explained by John H. Brooke (1991, 147): "Newton's position has been described as one in which God had taken every precaution to minimize the destabilizing forces and yet had willed a world in which His intervention would also be required." Once formed, continues Newton in Query 31 of his *Opticks*, the world as it currently stands may go on under the action of natural laws for many ages. In Newton's words, as far as the world's "negligible irregularities" are concerned—such as the slowdown effect produced by the mutual attraction of celestial bodies on one another—one could count on God to readjust the system when necessary.

Newton's worldview was uniformitarian through and through: in space, in time, and at the microscopic and macroscopic levels. As for space and time, I. Bernard Cohen (1985a: 182–183) writes: "Newton believed that there *was* a sense in which

[5]Nature's isomorphism or homogeneity was even extended to attractive forces in the *Queries* of Newton's *Opticks*: universal gravitation for planetary systems and attractive/repulsive forces for atomic particles. This allowed Newton to insist again that the world is simple. And, we would add, a simple world is more easily knowable than a complex one.

space and time were 'absolute' physical entities... [He] constructed a complete system based on a world view, encompassing ideas of space, time, and order... Not only could Newtonian science account for present and past phenomena; the principles could be applied to the prediction of future events" [italics original]. Perhaps nothing was more impressive when it came to binding the past, the present, and the future under the uniformitarian doctrine than the prediction made by Edmond Halley in 1705 concerning the projected reappearance of a comet in 1758 (now known as Halley's comet), a prediction based on Newton's physics and the study of human historical records.

When turning to the nature of the relationship between microscopic and macroscopic levels, it is tempting to say that Newton positioned himself between the two by being equipped with his "uniformitarian apple." On the one hand, the falling apple on Earth is seen as similar to the perpetual fall of the Moon toward the Earth. From there, the Earth is merely another apple perpetually falling toward the Sun (Westfall 1993: 171). On the other hand, an object as small as an apple was believed by Newton to be made up of tiny invisible particles held together by similar attractive forces, thus extending by analogy macroscopic phenomena to microscopic realities (Westfall 1993: 258). Perhaps the transition between the two levels was made easier for Newton by the modest yet significant breakthrough toward atomism in microscopic studies. Indeed, the belief in tiny invisible parts was no longer exclusively the province of the rationalistic imagination in Newton's day, as documented in *Micrographia* (1665) written by Robert Hooke. More importantly, the analogy between the microscopic and macroscopic levels was also taken for granted by scholars like Robert Boyle, Robert Hooke, and John Locke (Laudan 1981: 67–68). The apparent evidence of this analogy was most vividly illustrated by Boyle when he stated that to refuse it would be "as if a man should allow, that the laws of mechanism may take place in a town clock, but cannot in a pocket-watch."[6] Apparently, it was believed that a homogeneous or uniformitarian world was readily open to our scientific investigations. Again we see Newton supporting uniformitarianism by appealing to nature's simplicity, with the same effects being explained by the same causes. Speaking of Newton, Richard Westfall (1993: 258) writes, "[t]hus nature will be very comfortable to herself, he concluded, performing all her great motions by the attraction of gravity and her small ones by the forces between the particles."

This whole viewpoint, of course, is consistent with the nonhistorical character of physics and chemistry as conceived at the time: the ontology accompanying the steady-state worldview naturally extended to methodology, "uniformity" having a dual ontologico-methodological character. With the rise of "historical" disciplines in the nineteenth century, however, some qualifications were required (Hooykaas 1963: 140). Charles Lyell thought he had provided them in the *Principles of Geology* (1830–1833)

[6]Quote extracted from Laudan (1981: 68).

for both the fields of geology and biology.[7] By introducing more motion to the world's system, Lyell proposed an alternative view to the rise of a directionalist one. After having recounted with approval James Hutton's attempt at building a geology entirely devoid of "hypothetical causes" (Lyell 1830, vol. 1, p. 61)—the *vera causa* doctrine attributed to Newton (Camardi 1999; Ruse 1976)—Lyell explained how the world's uniformity was preserved through changes:

> There can be no doubt, that periods of disturbance and repose have followed each other in succession in every region of the globe, but it may be equally true, that the energy of the subterranean movements has been always uniform as regards the *whole earth*. The force of earthquakes may for a cycle of years have been invariably confined, as it is now, to large but determinate spaces, and may then have gradually shifted its position, so that another region, which had for ages been at rest, became in its turn the grand theatre of action (Lyell 1830, vol. 1, p. 64). [italics original]

It will be remembered that Lyell's neo-Huttonian solution for a steady-state world was based on the idea that while aqueous factors erode Earth's surface, igneous factors act as a counterbalance to restore it (Rudwick 1970: 12–18; Hodge 1982: 7). Lyell's uniformitarianism is not merely based on the actualistic methodological precepts that past geological events must be explained by the same causes now in operation, but rather on the more inclusive ontological conviction that empirical investigations show the past and the present to be isomorphically connected (Di Fate 2011). Newton's homogeneous view of the world with respect to both time and space is transposed to geology by Lyell: beyond superficial geological changes one ultimately discovers a stable world. Also applying his ontological commitment to biology, Lyell holds that life forms on Earth are definitely under the influence of cyclical geological changes, but not to the point of endangering the order of the living world, past, present, and future: "Instead of being astonished at the proofs there manifested of endless mutations in the animate world, they will appear to one who has thought profoundly on the fluctuations now in progress, to afford evidence in favour of the uniformity of the system..." (Lyell 1832, vol. 2, p. 157). We have seen in Chap. 2 of this book that Lyell was prepared to accept the extinction of life forms at low taxonomic levels but not at higher ones. Thus was preserved the stability of the world through time.

8.2 From Ontology to Methodology and Back

The second question we want to investigate in this chapter is *how* is nature knowable? By collecting the facts with the assistance of our senses. Newton's ontology is logically connected with his methodology. Against idealism and the sheer power of

[7]There is no need to repeat here what has already been said in Chap. 2 of this book about the strategic retreats in effect during eighteenth and nineteenth centuries designed to accommodate the discovery of deep geological times in order to save the notion of a fully constituted world or steady-state view.

reason, he argues that in experimental philosophy knowledge is gathered through our senses (*Principia*, Rule III). Knowledge should not be approached with elaborate and speculative ideas (hypotheses) in mind but rather with *verae causae* or real causes known to exist (Rule I).[8] As such, scientific generalizations are reached and corrected with the inductive method (Rule IV) (Rossi 1999: 310, 313). Experimentation is part of this widely construed empirico-inductivistic apparatus, which also includes a deductive component, which sooner or later will fall under the purview of induction. I. Bernard Cohen (1985b: 165–170) showed how Newton's astronomy was gradually elaborated through an ongoing confrontation between abstract mathematical models and the real world in what he calls "Newton's style" of doing science (see also Westfall 1993: 172). George E. Smith expanded upon Cohen's understanding of Newton's special brand of empirical investigation. Newton had to devise a way to solidly ground his work in the real world, confronted as he was with a complex world in which no two orbits of a single body are exactly identical under, for instance, the mutual effect of celestial body's gravitational field on each other[9]:

> [Newton's] mathematical theories of motion... [aim] to provide a basis for specifying experiments and observations by means of which the empirical world can provide answers to questions—this in contrast to conjecturing answers and then testing the implications of these conjectures. Newton is using mathematical theory in an effort to turn otherwise recalcitrant questions into empirically tractable questions (Smith 2002: 147).

Newton distrusted the hypothetico-deductive approach on the grounds that too many distinct hypotheses were compatible with the same empirical evidence. What Newton did was to render his theory firmer and firmer with each passing investigative stage: discrepancies with observations served the purpose of indicating how to go forward with an increasing empirical support. It is worth quoting Smith (2002: 156–157) at some length on these more technical matters:

> [I]n every case in which Newton deduces some feature of celestial gravitational forces, mathematical results established... allow him to identify specific conditions under which the phenomenon from which the deduction is made would hold... exactly. For instance, the orbiting body would sweep out equal areas in equal times if the only forces acting on it were centripedal, and its line of apsides would be stationary if the only forces acting on it were inverse-square centripedal forces... By imposing this restriction on the phenomena from which laws are deduced, Newton is assuring that these phenomena are not just arbitrary approximations to the true motions; at least according to the theory of the 'deduced' physical force, the true motions would be in exact accord with the phenomena were it not for specific complicating factors... For any deviation of the actual motions from a given approximation will then be physically meaningful, and not just a reflection of the particular mathematical scheme employed in achieving the approximation, as in curve fitting... [F]rom the point of view of the theory, then, any observed systematic pattern in the deviations from a given approximation would have the promise of being physically informative, and hence a promise of becoming telling evidence. In sum, judging from the details of Newton's 'deductions'

---

[8]What naturally comes to mind here is Newton's famous adage concerning the unclear nature of gravity, "*Hypotheses non fingo*" (I frame no hypotheses), found in the General Scholium of the *Principia* (1713).

[9]The following analysis is largely based on Smith (2002).

> from the phenomena, his approach to the complexities of real-world motions is to try to address them in a sequence of progressively more complex idealizations, with systematic deviations from the idealizations at any stage providing the 'phenomena' serving as evidence for the refinement achieved in the next [investigative stage].

These "successive approximations" allowed Newton to test his theory in such a stringent way as to eventually render unreasonable any opposition to the empirical well-foundedness of the manifestation of gravity. In considering the nature of the gravitational force, Newton declared: "I do not feign hypotheses" (*Hypotheses non fingo*). But when his successors like Clairaut, Euler, d'Alembert, Lagrange, and Laplace were done tying up the loose ends of Newton's incomplete theory by the early nineteenth century, the mathematical manifestation of the gravitation force or law—under quantities of motion—was just about to be seen as a physical and empirical one (Smith 2002: 150, 152). The extension of the mechanistic view originally based on action-by-contact to include also action-at-a-distance (gravity) was now complete. Newton's theory served as a powerful heuristic and predictive instrument for investigating apparently unrelated phenomena for which it was not originally designed, like the tide cycles, the precession of the equinoxes, and the trajectories of comets. This not only served the purpose of showing the universality of the law but also of broadening its evidential support. The fascination with Newtonian achievements during Darwin's time, therefore, comes as no surprise.

The reader may permit us a brief digression into methodological issues, since these are often at stake in the elaboration of science. In trying to avoid being criticized for using unreliable methods, scientists may succumb to rhetoric and intellectual fashion. In the twentieth century, for instance, scientists sometimes saw themselves as following the epistemologies of Karl Popper or Thomas Kuhn when these were in vogue in the 1960s and 1970s. A close examination of their methodologies, however, often revealed that these works were neither particularly Popperian nor Kuhnian in character. When reflecting on the methodologies between the seventeenth and the nineteenth centuries, one should keep the rhetorical dimension in mind.[10]

A key methodological shift occurring during this period concerned the disfavor for the "method of hypothesis" (sometimes called hypothetico-deductive) that proceeded *from* conjectures about the object of study *to* its verification. As summarized by Larry Laudan (1981: 9): "the method of hypothesis amounted to the claim that an hypothesis could be validated by ascertaining whether all of its examined consequences were true." This method faced the main objection of not being able to establish with certainty whether or not the postulated hypothesis is truly supported by its consequences, since several competing hypotheses could be compatible with them. It should be noted that the method of hypothesis relies significantly on the imagination (speculation) and reason (rationalism) of a scientist for its elaboration.

---

[10]For a concise review of the methodologies discussed here, see Rachel Laudan (1987: 10–14).

Several alternatives were often judged better to the method of hypothesis at the time. The most important of these, which gradually imposed itself in the eighteenth century—thanks to Francis Bacon's and Isaac Newton's reputations—was the method of induction based on a sensationist (senses) epistemology: secure knowledge should be sought through a sequence moving *from* direct observation and experimentation *to* a careful and gradual series of ever more inclusive generalizations, leaving as little room as possible for speculation and ad hoc arguments. The method of induction includes strategies like simple enumerative induction[11] and eliminative induction.[12] In short, the information flow should go from empirical facts to the elaboration of a theory (and not the reverse). In the method of hypothesis, empirical facts are exploited only to verify the anticipated consequences of an already formulated hypothesis.

Another alternative is the method of analogy. It consists of a reasoning process that moves from a known case 1 to an unknown case 2 on the presumption that the two are believed to be similar enough to warrant a meaningful rapprochement, thus extending to case 2 what is already known about case 1.

Finally, a third alternative is the one already reviewed under the name of *vera causa* and presented by Newton himself under Rule I of the *Principia*: "We are to admit no more causes of natural things than such as are both true and sufficient to explain their appearances." According to Newton, there is no need to multiply unduly the number of natural causes on the premise that nature is simple; to do otherwise is to venture into speculative (hypothetical) guesses deprived of a solid foundation.

Although there have always been proponents for the methods reviewed here, the methodological trends of the nineteenth century were more ambiguous with the rediscovery of the epistemological constraints imposed by the investigation of invisible and microscopic entities. After all, these entities can only be dimly grasped through their anticipated manifestations at the macroscopic level, which takes us back to the utility of the method of hypothesis. As summarized again by Laudan (1981: 23): "if we cannot establish our first truths by deriving them *from* experience, the only seeming alternative is to test them (via their consequences) *against* experience" [italics original].

Concluding this digression on methodologies will allow us to reflect upon the status of the inductive method (widely construed) between the seventeenth and nineteenth centuries. First, it is true that several methodologies tend to blend into one another on close inspection of the works of individual scientists. For instance, Francis Bacon and Isaac Newton were not exclusively committed to a narrow application of the inductive method (Malherbe 1996; Smith 2002). Second, it is also true

[11]Enumerative induction: an induction by enumeration of all the instances to which it applies—if one observes 100 swans and all of them are white, one can then formulate a generalization of the kind "all swans are white."

[12]Eliminative induction: an induction through the listing of all possible hypotheses about a specific issue followed by their gradual elimination under inspection of new evidence until only one remains.

that community pressure to comply with a specific methodology can lead to mere lip service being paid to a methodology. However, these caveats should not lead one to succumb to epistemological relativism, on the presumption that early modern science was envisioned by its practitioners the way we do today. Centuries of solid scientific work have taught us that knowledge is hard to come by, as testified by the intellectual movements that took us from positivism in the nineteenth and early twentieth centuries, to post-positivism around the mid-twentieth century, to the sociology of science of the post-modernist kind beginning in the 1970s (total relativism). The area of physics loomed large in the fall of our confidence in an all-powerful ability to understand nature. It must be recognized that humans are in a difficult place, stuck between subatomic realms (quantum mechanics) and super-macroscopic ones (general theory of relativity), a physical reality out of scale with our human condition, which can leave the impression that we are strangers in our own universe. Suddenly, it seems, a single universal reality has evaded us. Yet, it should be remembered that despite the recent discovery of new states of affairs, the transition was itself based entirely upon the scientific work of the previous centuries: our investigations first had to assume the complete uniformity of nature in order to eventually discover the limitations of this approach.

This is precisely the point to be made here regarding empirico-inductivism. For scientists between the seventeenth and nineteenth centuries, a new world presented itself to those willing to take a fresh and hard look at it, irrespective of the epistemological difficulties that confronted them. It is entirely legitimate to present the rise of early modern science as the "age of empiricism."[13] Turning their back on the passive approach to knowledge that traditionally consisted in reading the classical authors of Antiquity, scholars were then invested with a new confidence in our ability to gather knowledge. This new proactive attitude instilled a more interventionist approach to knowledge, experimentation being its most conspicuous aspect. Epistemologically speaking, observation and experimentation became twin pillars of scientific investigation as encapsulated in the doctrine of empiricism and the method of induction. This is not to say that everything about this period was new and that nothing was retained from more ancient traditions.[14] Similarly, this is not to hold that science was empirical and inductive through and through: the Scientific Revolution[15] was composed of distinct intellectual strains, several of them falling more on the side of rationalism (the method of hypothesis) and formalism (the language of nature is written in mathematics, claimed Galileo) (Cohen 1994: 21–150). This being said,

[13]In an otherwise excellent work, the philosopher of science Gaston Bachelard was only too happy to somewhat unfairly scorn eighteenth-century empiricism and inductivism in the name of rationalism, thus overlooking the significance of the intellectual trend that accompanied the "age of empiricism." See the 2002 English translation of his book originally published in French in 1938: *The Formation of the Scientific Mind* (Manchester: Clinamen Press), Chap. 3.

[14]See our "Sketch of an Intellectual Grid" in Chap. 1.

[15]It is recognized here that the label "Scientific Revolution" as a historiographical category may require some qualifications. These are provided in Chap. 1.

empiricism and induction were at the forefront of this revolution.[16] How could it have been otherwise in the age of the invention of the telescope and the microscope? It is but a small exaggeration to say that Galileo nearly single-handedly redrew the map of the sky equipped with a low-resolution telescope in his *Sidereus Nuncius* (1610). Similarly, how could it have been otherwise at a time of worldwide exploration and long sea voyages that rejuvenated the tradition of natural history (Levine 1983)? The recent historiography stressing the impact of mechanical and skilled people like engineers, chemists, metallurgists, and sailors—in short, practical people confronting the world—on scientific knowledge has much going for it (Hooykaas 1987; Conner 2005). The recent revival of the interest in empirico-inductivism in science under the appellation of "new experimentalism" constitutes a much needed counterweight to rationalism in our understanding of the natural sciences (Hacking 1983; Franklin 1986; Galison 1987; Gooding 1990; Mayo 1996; Chalmers 1999; Delisle 2012a).

Newton may not have been as exclusively committed to inductivism as sometimes claimed, whether by himself and by enthusiasts of the following centuries. But behind the mere rhetoric of empirico-inductivistic science, there existed solid knowledge erected upon a fresh and massive intake of empirical facts. In fact, it could well be argued that the rhetoric in favor of an empirico-inductivistic science associated with Bacon and Newton was widely voiced precisely because the importance of empirical facts was universally recognized at the time.

Having established some of the arguments in favor of the "empirical age" during the seventeenth to nineteenth century period, we now come to one of the main theses of this chapter: *not only was empirico-inductivism a major intellectual strain of the Scientific Revolution; it was a strain particularly well suited for scholars embracing the uniformitarian doctrine*. The rationale is simple enough: if the world is believed to be fully constituted by being uniform or homogeneous in time as in space, what better epistemology could there be than empirico-inductivism to explore it? At that precise junction, ontology naturally extends to methodology, generating a powerful and consistent ontologico-methodological approach. Put in other terms, a world with no hidden differences in terms of time and space is entirely open to our gaze, hence the choice of empiricism and inductionism.

On this understanding, Newton's uniformitarian stance about space, time, and the microscopic and macroscopic levels accords perfectly with his admonitions against speculation (hypotheses). The famous passage concerning the nature of gravity found in the General Scholium of the second edition of the *Principia* (1713) reads as follows:

> I have not as yet been able to deduce from phenomena the reason for these properties of gravity, and I do not feign hypotheses. For whatever is not deduced from the phenomena must be called a hypothesis; and hypotheses, whether metaphysical or physical, or based on occult qualities, or mechanical, have no place in experimental philosophy. In this

[16]For a confrontation between the mathematical and the experimental traditions, see Kuhn (1977).

> experimental philosophy, propositions are deduced from the phenomena and are made general by induction (Quote in Smith 2002: 139).

Newton was even more explicit about methodological prescriptions in another famous passage of the 1706 edition of his *Opticks*:

> This Analysis consists in making Experiments and Observations, and in drawing general Conclusions from them by Induction, and admitting of no Objections against the Conclusions, but such as are taken from Experiments, or other certain Truths. For hypotheses are not to be regarded in experimental Philosophy. And although the arguing from Experiments and Observations by Induction be no Demonstration of general Conclusions; yet is the best way of arguing which the Nature of the Thing admits of, and may be looked upon as so much the stronger, by how much the Induction is more general. And if no Exception occur from Phenomena, the Conclusion may be pronounced generally (Quote in Smith 2002: 165).

In addition, there was for Newton a theological justification for this epistemology. In the words of John H. Brooke (1991: 140): "If the workings of nature reflected the free agency of a divine will, then, the only way to uncover them was by empirical investigation. No armchair science, premised on how God *must* have organized things, was permissible. This meant that theological moves could be made in criticizing any science that seemed too conjectural" [italics original]. As the source of a contrived and orderly world, God is also the creator of a mechanical world whose activities are open to our scientific investigations. Why is it so? Because God does not play tricks on us by continually changing the rules of the game. As Gary Deason explains (1986: 187): "although God was sovereign over a world that He created and, in principle, could suspend or change natural laws to accomplish a special purpose, in practice He did not tamper with the laws of nature." This provided, Deason continues (1986: 187), the necessary "assurance of scientific explanation."

An important intellectual connection between Newton and nineteenth-century scholars like Lyell, Herschel, Whewell and Darwin is Thomas Reid.[17] Author of *An Inquiry into the Human Mind on the Principles of Common Sense* (1764) and of the *Essays on the Intellectual Powers of Man* (1785), Reid extracted what he thought was the essence of Newton's epistemology. It will be noted how he also subscribed to the epistemological nexus described above (Laudan 1981: 86–110):

1. A central place given to a science-based *verae causae*
2. A belief in a designed and contrived world too complex for scientists to outguess God's plan of the creation through the method of hypothesis (speculation)[18]
3. A belief that God's wisdom created a universe governed by fixed laws and in which uniformity prevails

[17]For intellectual connections between Newton and nineteenth-century scholars, consult Wilson (1974), Ruse (1975, 1976), Hodge (1977), L. Laudan (1981: 86–110, 163–180), and R. Laudan (1982).

[18]Thomas Reid noted Newton's infatuation with nature's simplicity, which he criticized. Reid replied that Newton's theories seem simple in retrospect only and were actually complex enough to match nature's complexities. See Laudan (1981: 91).

In light of Reid's ontologico-methodological commitments, it is with perfect coherence that he embraces the method of induction. To learn about the world, one must actually look and investigate how it works. As Larry Laudan (1981, 98) summarizes: "Basically, Reid's inductive method had three components: (1) observation of facts and experimentation; (2) 'reduction' of these facts to a general rule or law; and (3) the derivation of further conclusions from the set of general rules."

It should be stressed that in the background of the debate over whether or not evolutionism or directionalism is a reality (in cosmology, geology, biology, or human history), one finds a fundamental epistemological issue: if evolution is recognized as an ontological reality of our world, it becomes much more difficult for scientists to investigate it. Suddenly, the world no longer reveals itself in its fullness. To accept the inherent historicity of this world is to accept losing sight of most of it in the past and eventually the future. Or to reverse the perspective, a uniform and static world is much more easily investigated since it is presented to us in its entirety.

This obvious epistemological fact posed no problem to Newton and Reid, as both thought they lived in an age in which a steady-state world was a reality. Things presented themselves differently to Lyell in light of slowly accumulating empirical facts potentially supporting directionalism. One may even want to argue that Lyell's *Principles of Geology* (1830–1833) was a work of the past, an impossible defense of an uncompromising uniformitarianism. Be that as it may, the fact remains that behind this lost cause was also the refusal to see the investigative power of the empirico-inductive approach diminished. The following analysis by Rachel Laudan (1987: 206–207) is instructive:

> Lyell did not simplistically equate 'the present' and 'the past.' For him, the core dichotomy was between the observable and the unobservable, *not* between the present and the past. Lyell is often interpreted as being committed to 'actualism,' meaning that the past should be interpreted in terms of present causes... If this were the extent of his position, then he would be inconsistent in arguing from past to present, as he often did... However, this apparent inconsistency disappears if Lyell's basic commitment was to the *vera causa* principle, not to actualism.

The empirico-inductive approach constitutes the safest epistemological road to a *vera causa*. Uniformity, *vera causa*, analogy, empiricism, and inductivism were powerfully bound together in the minds of many scholars. When one looks more closely at the *Principles of Geology*, one finds a work in the British "empiricist-inductivist tradition" taking us from Newton, Thomas Reid, and Lyell, a work filled with examples presented in the style of enumerative induction (Gillispie 1951: 128; Laudan 1981: 96; Camardi 1999: 554; Di Fate 2011:145). As Michael Ruse puts it (1975: 168): "[Lyell] never achieved even the limited hypothetico-deductive success of Darwin, preferring rather to make his points with *strings of related examples*" [my emphasis]. Leaving aside Darwin, to which we will return presently, were there other serious options open to Lyell in light of his epistemological commitments? In the relatively new field of geology, the systematic survey and comparison of geological structures and processes was the most legitimate approach for someone who believed in the world's uniformity. When structures and processes looked similar enough, it was only natural for Lyell to imagine a relationship between them, whether they occurred in the same time horizon or across geological time. We would here stress the reality of the 1830s: the empirical diversity of geological manifestations. It was far from illegitimate for Lyell to think that it was premature to reject uniformitarianism when so much remained to be surveyed and compared. No doubt the minds of Lyell's contemporaries were moving more swiftly than his, but from Lyell's viewpoint, the haste with which some of them hid entities and causes in the past was certainly not deprived of objections.

We have already encountered William Whewell's (1832: 126) epistemological warning to Lyell in 1832 about the nature of historical objects of study. It is worth quoting him again:

> It seems to us somewhat rash to suppose, as the uniformitarian does, that the information which we at present possess concerning the course of physical occurrences, affecting the earth and its inhabitants, is sufficient to enable us to construct classifications, which shall include all that is past under the categories of the present. Limited as our knowledge is in time, in space, in kind, it would be very wonderful if it should have suggested to us all the laws and causes by which the natural history of the globe... is influenced—it would be strange, if it should not even have left us ignorant of some of the most important of the agents which, since the beginning of time, have been in action....

The warning, however, is a double-edged sword. Whereas Lyell risked being blind to novelties hiding in the womb of geological times, Lyell could reverse the argument by replying to Whewell that the value of scientific explanations derived from phenomena no longer directly accessible to us comes with its share of speculations and uncertainties. Lyell thought he avoided these risks altogether by defending the ideal of an empirically founded *vera causa* while simultaneously holding on to an ontologically complete or uniform world. His fight in the 1830s for a steady-state view may well have been on the losing side of history, but it was not yet a lost cause. One does not easily relinquish the epistemological safety of uniformity. This allows us to grasp better the profound implications accompanying the emergence of the Transformist Revolution. Apparently, Darwin wanted to embrace that revolution in full while retaining the epistemological safety of uniformity (see also Table 2.1).

8.3 The Touchstone of Darwin's Methodology: Integral Uniformitarianism

As we know, it has been argued by a number of Darwin scholars that the arguments in the *Origin of Species* were arranged to respond to the epistemological imperatives of the time: whereas its first chapters are devoted to demonstrating that natural selection is a *vera causa* through its analogy with artificial selection, the last chapters are concerned with a "consilience of inductions" and attempt to show that unrelated classes of facts in various fields are explained and united through this *vera causa*. It is interesting to note that this interpretation of Darwin's work gained momentum after the constitution of the so-called Evolutionary Synthesis between the 1930s and 1950s. That synthesis is often presented as organized around two central explanatory components (Mayr 1980: 1; Gould 1980: 119–120): (1) gradual evolution is explained by small genetic changes (variations) oriented by natural selection, thus leading to adaptation; (2) evolutionary trends and speciational events are macro-evolutionary phenomena explained by the extension of processes and mechanisms occurring at the previous microevolutionary level. Retrospectively, it seems, mid-twentieth-century developments threw new light on Darwin's work, with Darwin's arguments now being viewed as strongly structured and organized, forming part of the first unification phase of an even more comprehensive second phase called the Evolutionary Synthesis (Ghiselin 1969; Ruse 1973: 48–52; 1979: 198).

It is always surprising to see how much the present weighs on the understanding of the past. Be that as it may, it has recently been argued that the Evolutionary Synthesis may lack historical support (Levit et al. 2008; Delisle 2008, 2009a, b, 2012b; Largent 2009; Cain 2009). The issue is not whether that synthesis is comprehensive enough, should be expanded, or simply replaced, as scholars in the last decades have wondered. The question is more radical: Was there ever such an event as the Evolutionary Synthesis? Some historians and historically inclined philosophers argue that the event was more sociological than conceptual, with the rhetoric of "unification" cloaking an inherent pluralism of views that has existed since the inception of evolutionary biology.[19]

With a changing perspective on twentieth-century evolutionary biology, it was to be expected that Darwin studies would be revisited. The deflated discourse of a unified evolutionary biology was sure to lead to the rediscovery of the pluralism in Darwin's own work (Delisle 2014, 2017b). Among contemporaries of Darwin, for instance, some were struck by the lack of argumentative unity within the *Origin of Species*. Perhaps the most vivid exposition came in 1860, from Darwin's own friend and supporter, T.H. Huxley. Let us present this passage again:

> But this very superabundance of matter must have been embarrassing to a writer who, for the present, can only put forward an abstract of his views; and thence it arises, perhaps, that notwithstanding the clearness of the style, those who attempt fairly to digest the book find much of it a sort of intellectual pemmican—a mass of facts crushed and pounded into shape, rather than held together by the ordinary medium of an obvious logical bond: due attention will, without doubt, discover this bond, but it is often hard to find (Huxley 1860: 542).

Apparently, this abundance of matter also caught the attention of William Whewell. When speaking of Darwin's theory in a letter to a contemporary, Whewell writes that Darwin "has accumulated a vast array of hypotheses," before adding "that most of his hypotheses are quite unproved by fact."[20] It goes without saying that Whewell thought that Darwin had failed to achieve a "consilience of inductions"

[19]For a compact overview of this issue, including additional references, see Delisle (2017a).

[20]See Whewell's letter in Todhunter, I. (1876), *William Whewell, D.D. Master of Trinity College Cambridge* (London: Macmillan), Vol. 2, pp. 433–434. A part of this letter is reproduced in Hull (1989: 30).

whereby two or more classes of facts, each inductively generated, coincide in being explained by the same theory.[21] In another review of Darwin's *Origin of Species* (1859), Adam Sedgwick criticized Darwin in 1860 for not having succeeded in elaborating a theory emerging from the facts he himself presented. Underlying Sedgwick's inductivist critique, one implicitly detects the claim that a wealth of facts presented by Darwin remain unconnected under a unified theory: "But I must in the first place observe that Darwin's theory in not *inductive*,—not based on a series of acknowledged facts pointing to a *general conclusion*,—not a proposition evolved out of the facts, logically, and of course including them" (Sedgwick 1860: 334–335) [italics original].[22] In a similar vein, William Hopkins (1861: 86) notes in a review of the *Origin* that the numerous facts it presents do not fit the proposed theory: "It indicates a large acquaintance with the facts of natural history, and still larger stores of knowledge in reserve. In the statement of facts, the author is uniformly impartial... But when... he comes to the discussion of facts in their theoretical bearings, we recognise a want of strict adherence to philosophical and logical modes of thought and reasoning." Closer to our time, Charles C. Gillispie had this to say in 1958 when comparing the works of Darwin and Lamarck:

> More generally, it might be argued—indeed, I do argue—that in the relative cogency with which the two theories organize actual biological information, Lamarck's presentation in the great *Histoire naturelle des animaux sans vertèbres* is the more interesting and elegant. It is analytical and informs a systematic taxonomy, whereas Darwin simply amassed detail and pursued his argument through the accumulated observations in a naturalist's commonplace-book. To be single-minded and relentless is not necessarily to be systematic, and the merit of Darwin's approach must be sought elsewhere. (Gillispie 1958: 389)

Under the influence of an assumed Evolutionary Synthesis, the recent historiography has overstressed the theoretical coherence of the *Origin of Species*.

It is sometimes argued that John Herschel's *Preliminary Discourse on the Study of Natural History* (1830), William Whewell's *History of the Inductive Sciences* (1837) and *The Philosophy of the Inductive Sciences* (1840), and John Stuart Mill's *System of Logic* (1843) are the most important epistemological influences on Darwin.[23] Here, the work of David L. Hull (1989: 27–42, 62–75; 2000, 2009) will be useful in the next phase of our analysis. Hull shows the extent to which the epistemological issues of the time were anything but clearly defined, whether in

---

[21]Whewell explains the epistemological nature and implications of a consilience of inductions as follows: "But the evidence in favour of our induction is of a much higher and more forcible character when it enables us to explain and determine cases of a *kind different* from those which were contemplated in the formation of our hypothesis. The instances in which this has occurred, indeed, impress us with a conviction that the truth of our hypothesis is certain" [italics original]. See Whewell, W. (1840), *The Philosophy of the Inductive Sciences Founded Upon Their History*, Vol. 2 (London: J.W. Parker), p. 230.

[22]Sedgwick's 1860 paper is reproduced in Hull (1973: 159–166).

[23]I shall not try here to weigh their respective influence on Darwin's intellectual development between the 1830s and the 1860s. Among Darwin scholars, this has been debated by David Hull, Michael Ruse, and Jonathan Hodge.

the minds of these philosophers of science or in the mind of Darwin himself. Taken collectively, the epistemological objectives were rather blurred:

1. The philosophers and Darwin declared they were faithful to the ideal of an inductive science, thus following Bacon and Newton, even when their views departed from these authorities.
2. Herschel, Whewell, and Mill each had their own distinct assessments of the distinction between the method of hypothesis and the method of induction necessary to propose a generalization.
3. At the time, epistemology was caught up in a debate over fundamental issues: the difference and role of induction and deduction in the various scientific disciplines; the difference between the formation of concepts and the discovery of laws; the connection between discovery and justification; the distinction between theoretical entities and occult qualities; and the conflict between a naturalistic science and God's involvement.
4. Confusions and contradictions found in the works of philosophers were replicated in Darwin's considerations of the topic. Darwin held that one should avoid the tendency to excessive generalization; distrust deductive reasoning; work with true Baconian principles by avoiding collecting facts through a pre-established theory; object to scholars who refuse to theorize; and claim that observation can only be made within one theory or another.

It is therefore unsurprising to read Hull's assessment (1989: 29): "Perhaps Darwin had no clearer a conception of science and scientific method than the philosophers of his day [but] he was certainly no more confused." The evaluation of Darwin's position is not made any easier when one considers that he was perhaps often merely paying lip service to the epistemological views of his time, while simultaneously adopting other approaches albeit in a concealed manner. As Jonathan Hodge (2000: 42) writes:

> Finally, when one reviews the full sweep of Darwin's theorizing, whether in geology, biology, or psychology, there is one general conclusion that is surely difficult to resist. In his theorizing he was often, although not always, not merely markedly innovative but radical and subversive, and knowingly and deliberately so. But in his own eyes this theorizing did not require him to be correspondingly innovative about the epistemology and methodology of science. On the contrary, he seems anxious that his new theories be developed and presented within the familiar, reputable constraints satisfied by what his contemporaries count as the best science of the day.

Darwin therefore engaged in rhetorical strategies to defend his choices. The traditional historiography seems to assume that this was done to blunt criticism of his evolutionary views: Darwin needed breathing space as he attempted to break new ground in the area of evolutionary biology with its own historical epistemology based on historical narrative (the strong contingency thesis). However, we have seen in the previous chapters of this book how little Darwin was committed to this new epistemology, one that had to wait for the twentieth century for its full expression.

We are not denying that Darwin attempted some kind of structuration of his views in the *Origin of Species*. What we are contesting, rather, is the effectiveness of his attempt, irrespective of what Darwin claimed or thought he had achieved. Our thesis is the following: *regardless of criticisms of the time,*[24] *Darwin was sufficiently committed to empirico-inductivism and the "age of empiricism" to have been overwhelmed by a multitude of contradictory sources of information.*[25] His superficial attempt to impose order on that material through a compact theory guided by epistemological principles taken from Herschel, Whewell, or others largely failed. The world of biology was just too complex and multidimensional to fit into Darwin's narrow conceptual box. What is most interesting about Darwin is that he was honest enough to include many facts not easily accommodated by his theory but seems to have overlooked the difficulties accompanying such an approach. In other words, had Darwin been more selective in his choice of facts, he could have presented them in a more favorable light. He probably realized, however, that what would have been gained in explanatory coherence would have been lost to what had been left out of his theory. This would have exposed him to more damaging criticisms still. The *Origin* constitutes a premature attempt at unifying the field of evolutionary biology around a single compact explanatory apparatus. As admirable as this synthetic effort may have been, its weaknesses were all-too-obvious to many of Darwin's contemporaries, judging from the reception given to the *Origin of Species*.

Having presented these preliminary remarks, we will now tackle Darwin's epistemological choices head on. In line with Newton, Reid, Lyell, and others, one finds in the intellectual background of the *Origin of Species* a commitment to the epistemological nexus consisting of uniformity, *vera causa*, the method of analogy, empiricism, and inductivism. Let us organize this epistemological toolkit under three headings.

Empiricism and Inductivism The *Origin of Species* overwhelms the reader with a superabundance of material: "The modern reader frequently grows impatient with Darwin's method in the *Origin* of piling example on example..." as David Hull (1973: 32) puts it, while I. Bernard Cohen (1985b: 291) refers to "a mountain of observational evidence." Not unlike Lyell in geology, Darwin was living in the early stages of a newly emerging field in need of a systematic survey of entities and processes. It is unsurprising to see that such an empirical wealth of information is reflected, in part at least, in the *Origin*, irrespective of Darwin's commitments to other methods (especially the methods of hypothesis and deduction). It is because

[24]A number of Darwin's contemporaries criticized him either for not being inductive enough or for not having considered enough new facts, sometimes simultaneously. See Ellegard (1990[1958]: 174–197).

[25]I am grateful to Sydney Vickers for having helped me see more clearly the implications that come with Darwin's strengths as, in her words, a "strong collector." Her work has led me to concur that Darwin was simply overwhelmed by his own talent. Mrs. Vickers conducted an Independent Study under my supervision in the Spring 2018 under the title "Another View of Darwin" in which she compares the traditional view of Darwin presented in my Darwin course and the alternative view that emerges from my recent work.

Table 8.1 The *Origin of Species*: the many competing theories of evolution

1. Strong evolutionary contingency	5. Directional (progressive) and parallel evolution
2. Weak evolutionary contingency	6. A random evolutionary walk
3. A rigid pan-divergent view	7. Evolution at a standstill
4. Reticulate evolution	8. Cyclical and equilibrium states

The reader willing to embrace the *Origin* in all its complexity and ambiguity encounters a series of different pictures or theories of evolution, many of them being contradictory

Darwin is sufficiently committed to empirico-inductivism that one encounters competing and contradictory views based on a nearly infinite array of very specific topics: the breeding of pigeons, the growing patterns of Scotch firs, the behavior of bees, the propagation of Jerusalem artichoke, the reproduction in orchids, the taxonomic rank of ruminants and pachyderms, the preservation of seeds in water, geological marks left by glaciers in the Himalaya, the presence of a unique species of bats in New Zealand, among other examples. Darwin was just unable to synthesize this material under a single compact explanatory apparatus, as reflected in a host of explanatory tensions in the *Origin* (see Tables 8.1 and 9.1)[26]:

1. A tension between blurred evolutionary patterns among life forms united in common reproductive networks and a neat pan-divergent tree of life.
2. A systematic incoherence in the respective powers of artificial selection and natural selection: in some places, artificial selection is assumed to be either weak or strong; the same applies to natural selection.
3. The notion of life forms rushing vertically forward to climb evolutionary grades under an all-powerful selective drive at the cost of destroying both the pattern of divergence and the traces of affiliation.
4. While a pan-divergent view is promoted throughout the *Origin*, the reader systematically encounters facts easily interpreted as supporting evolutionary analogies rather than homologies. This suppressed background assumption keeps resurfacing.
5. The strange neglect of paleontological facts in favor of neontological ones.
6. The tension between the idea of an overall biological progress and the paleontological annals being interpreted as lacking in evolutionary novelties.
7. The opposition between living forms entrenched in a series of near static and isolated taxonomic classes/phyla and the assumption that all are derived from a few or even a single common ancestor.
8. A tension between an assumed evolutionary depth versus "ancestors" imagined by projecting backward in time variability observed among extant forms.

[26]The reader will be spared citations of the exact pages of the *Origin of Species* as these have been provided in the previous chapters of this book.

9. A discourse calling for an epistemology adapted to deal with historical narrative (the strong contingency thesis) versus an appeal to an evolutionary process trapped in a rigid pattern of divergence.
10. An open evolutionary dynamic closed in on itself under a generalized built-in equilibrium system organized around a series of stabilizing explanatory devices.

These tensions and contradictions reflect the dichotomy we have introduced between a rhetorical Darwin fully committed to modern evolutionism and a Darwin heavily indebted to the intellectual inheritance of the seventeenth and eighteenth centuries. The careful reader of the *Origin of Species* is implicitly or explicitly presented with facts supporting all these views. Wherever Darwin stands on epistemology, he certainly never neglects to take in significant amount of empirical facts, which ultimately leads him to suggest whole sets of contradictory views.

***The Method of Analogy and* Vera Causa** Like many scholars of his time, Darwin wanted to establish the credentials of his theory by basing it on a *vera causa*: on a real cause known to exist. Now, the epistemological status of *verae causae* was open to debate at the time. Because of its intangible character, Newton chose to provide gravity with an empirical foundation through its mathematical manifestations (quantity of motion). Those investigating the nature of light faced a greater challenge still. The wave or undulatory theory of light, for instance, came into being thanks to those who accepted its phenomenal reality in a series of indirect yet corroborative manifestations, including reflection, refraction, double refraction, interference, polarization, dipolarization, circular polarization, and the colors of thin plates (Laudan 1981: 111–140). The epistemological route to a *vera causa*, therefore, never reached a unanimous assessment, ranging between a direct empirical appraisal and an indirect, rationalistic one (Wilson 1974).[27] Fortunately for Darwin, who already had his hands full, the area of natural history had something more tangible than the nature of light. Coupling his commitments toward empirico-inductivism and *vera causa*, Darwin imposed a stringent demand upon the latter by providing it with what he thought was a strong empirical foundation. This is seen, among other instances, in his rapprochement between artificial selection and natural selection.[28] The argument is well known: we see breeders working on natural variations within populations in order to modify them; the process in nature is similar enough to accept the proposition that evolutionary change is instigated by a similar cause called natural selection. The fact that Darwin faced obvious objections from his contemporaries about the effectiveness of this connection is of no concern to us here,[29] since our concerns are purely epistemological.

---

[27]See also footnote #65 on p. 90 of Wilson (1974).

[28]For a list of other factual elements provided by Darwin in response to the ideal of a *vera causa* and of the empirico-inductivist tradition, see Ruse (1979: 174–180) and Hull (2009: 183–185).

[29]For his claim of a *vera causa* to be valid, Darwin had to convincingly show that breeders are actually able to modify forms across the species' barrier. However, this he could not do, as recognized by many scholars of the time, including Adam Sedgwick (1860: 334–335), William Whewell (in Todhunter 1876, Vol. 2, pp. 433–434), and Thomas Henry Huxley (1896: 6).

Darwin was trying to anchor his theory within the empirical tradition, encouraged by Thomas Reid and others along that intellectual path. Darwin scholars have debated whether, among his contemporaries, Darwin was closer to John Herschel or William Whewell on the issue of *verae causae*, the former being more an empiricist and the latter more a rationalist. This being said, Herschel himself was not always as inclined to empirico-inductivism as claimed, just as Whewell was not always so pulled toward rationalism. This returns us to the muddled epistemological context of the period (Wilson 1974).[30] It is suggested here that we should avoid focusing exclusively on Darwin's relation to Herschel or Whewell. Irrespective of whether or not Darwin operated a strategic or rhetorical retreat in favor of a rationalist *vera causa*—after criticisms following the publication of the *Origin of Species*[31]—what he wanted above all is to stay close to the empirico-inductivistic ideal. There is no need for us, then, to take a position with respect to Darwin's rhetorical or strategic retreat. Our claim is that Darwin was so overwhelmed by empirical facts that he could not succeed to any significant degree in imposing order on this material, whether or not he intended to do it through an empiricist or a rationalist *vera causa*.

To pursue our example, what was missing in Darwin at this stage was the establishment of a logical connection between artificial selection and natural selection. This he remedied with the method of analogy: if the cause of case 1 is factually known and the cause of case 2 is factually unknown, and if the two cases are believed to be similar enough, it becomes legitimate to explain the latter in terms of the

[30]See also David Hull (2009: 178) who writes about Herschel: "Two centuries later Herschel repeated Bacon's praise of induction in science, but then went on to remind the reader that 'the successful process of scientific enquiry demands continually the alternative use of both the *inductive* and *deductive* method'. In the study of nature, 'we must not, therefore, be scrupulous as to how we reach to a knowledge of such general facts: provided only we verify them carefully when once detected'"[italics original].

[31]A possible strategic or rhetorical retreat toward a rationalist *vera causa* may be found in the following statement of 1868 (Darwin, Vol. 1, pp. 8–9): "In scientific investigations it is permitted to invent any hypothesis, and if it explains various large and independent classes of facts it rises to the rank of a well-grounded theory. The undulations of the ether and even its existence are hypothetical, yet every one now admits the undulatory theory of light. The principle of natural selection *may* be looked at as a mere hypothesis, but rendered in some degree probable by what we positively know of the variability of organic beings in state of nature,—by what we positively know of the struggle for existence, and the consequent almost inevitable preservation of favourable variations,—and from the analogical formation of domestic races. Now this hypothesis may be tested,—and this seems to me the only fair and legitimate manner of considering the whole question,—by trying whether it explains several large and independent classes of facts; such as the geological succession of organic beings, their distribution in past and present times, and their mutual affinities and homologies. If the principle of natural selection does explain these and other large bodies of facts, it ought to be received" [my emphasis]. Jonathan Hodge (2000) and David Hull (2009) hold that this retreat is more rhetorical than anything else, since they argue that Darwin consistently maintains that his theory was cast within the empirico-inductive tradition. For his part, Michael Ruse (2000) argues that from the start Darwin wanted to comply with Herschel's empirical *vera causa* and Whewell's rationalist *vera causa*, requiring him to make a simple strategic retreat, stressing the rationalist part of his argument rather than its empiricist one.

former. Darwin would have found a definition of "analogy" in Herschel's *A Preliminary Discourse on the Study of Natural Philosophy* (1831: 149).[32]

Uniformity We come now to the last epistemological element at the core of Darwin's *Origin*. This piece is fundamental, since it gives full meaning to the other components: the method of analogy, *vera causa*, empiricism, and inductivism. In themselves, these last four components can be freely and independently used for other epistemological purposes. For Lyell as for Darwin, however, their full strength comes to the fore when collectively exploited under the uniformitarian doctrine. For Newton, Reid, and Lyell, the world presented itself in a homogenous manner. A world identical to itself in space, time, and phenomenal levels is always revealed to us in its entirety. This ontological assumption finds its natural extension in the methodological choices just enumerated. The question for Darwin was "direct observation." As explained by Larry Laudan (1981: 130), the older "methodological tradition of *verae causae* had rested upon a rigid distinction between directly observable entities and not-directly observable ones, endorsing the former and eschewing the latter." We have seen that Lyell was able to move back and forth between the present and the past because his uniformitarian assumption guaranteed he could directly observe processes and entities. Once the method of direct observation is justified by the ontological commitment to uniformity, the empirico-inductive approach acquires its full rationale. Within this epistemological package, the method of analogy becomes a natural extension of observation. We argue, therefore, that behind Darwin's methodological choices, one finds his ontological commitment to a fully constituted world. It is not necessary to recount again the explanatory components accompanying this ontological commitment. It will suffice to mention the following:

1. Extant classes/phyla exist as separate entities in their current stable state as far back as we can probe in geological times. It is as if each had independently arisen full-blown out of the Cambrian era.
2. As entities climb the taxonomic scale, the more deeply they become entrenched in the economy of nature to the point of becoming nearly permanent.
3. No classes/phyla ever went extinct.
4. The taxonomic arrangement seen today was maintained throughout the entirety of evolutionary history.
5. Divergence of life forms is at its maximum today, meaning that all past forms must either be aligned with or fall in-between extant forms.
6. The history of life is locked up in a rigid pan-divergent view.
7. So-called ancestors are best envisioned as "shadows," that is, pale reflections or miniatures of extant forms (preformationism).

[32]Herschel (1831: 149) writes: "If the analogy of two phenomena be very close and striking, while, at the same time, the cause of one is very obvious, it becomes scarcely possible to refuse to admit the action of an analogous cause in the other, though not so obvious in itself."

8. No evolutionary discontinuities exist between extant life forms and their extinct ancestral forms; they are all part of a united and homogeneous whole held together in a tightly knit fabric under the dual principle of gradualism-gradation (at least within classes and phyla).

Of course, Darwin's commitment toward the uniformitarian doctrine in the *Origin of Species* (1859–1872) could not have been as integral as Charles Lyell's was in the first edition of the *Principles of Geology* (1830–1833). Something had to bend, and it did.[33] But it did not break. Darwin allowed for a degree of evolutionary motion by *confining* it to lower taxonomic entities, as T.H. Huxley and R. Owen had also done. This is unsurprising, considering that the doctrine was under increasing pressure from what would eventually become an irresistible push toward directionalism. For Darwin, however, the stability of the world—its completeness—was preserved: (1) at the highest taxonomic levels; (2) in the continuous fabric binding the living world throughout time; and (3) in its perpetual re-equilibration. To Darwin's epistemological package already presented above, a few complementary notions should be added. Each is a mere methodological extension of an ontological commitment to a fully constituted world:

Actualism This method consists in interpreting the past in light of the present. We hold that Darwin's actualism is intertwined with his uniformitarian ontology: the present can legitimately be used to probe the past precisely because that past is believed to be similar enough to the present. If this is the case, why should one be deprived of observing directly what the past conceals?

Gradualism-Gradation Darwin holds that the evolutionary process is gradual, as can be seen everywhere in traces of gradation. This idea include the notion of a single, unified, homogeneous, and tightly knit world deprived of discontinuities, at least within distinct classes/phyla, as postulated in the uniformitarian doctrine.

Extrapolationism-Interpolationism "To travel in geographical space, is to travel in geological time" is surely an unspoken motto of the *Origin of Species*. Projecting himself forward in time (extrapolation), he assumed that the future would be similar to the pattern-process of divergence he thought he could find among extant forms. Projecting himself backward in time (interpolation), he used presumed homologies to reverse the flow of time through convergence in search of common ancestors.

---

[33]Even Charles Lyell went through a partial retreat from his early and nearly integral uniformitarian or steady-state view in biology between the 1850s and early 1870s. See Gould (1987: 167–173). Darwin and Lyell therefore offered two different ways of preserving uniformitarianism as defined in this book, to which two more should be added with the views of T.H. Huxley and R. Owen. On Huxley and Owen, see Chaps. 2 and 7 of this book. In a real sense, Huxley also modified his uniformitarian view between the 1850s and the 1870s, giving it more evolutionary motion through the recognition of the pattern-process of specialization at lower taxonomic levels. See Ruse (1979: 220–221) and Bowler (1996: 263–267, 331–335). Taken collectively, Lyell, Darwin, Huxley, and Owen represent four different ways of transitioning between a static worldview and a truly modern evolutionary worldview, among a long list of other scholars.

Darwin's commitment to uniformity strikes at the very heart of the aforementioned motto, even penetrating deeply into his rhetoric concerning the "tree of life." Indeed, Darwin tried to encapsulate his entire theory in the visual form of a Diagram (see Fig. 2.1). For a modern reader, a key feature of Darwin's Diagram is explained by its analogy with "fractal geometry": just as patterns are similar at all levels in the fractal object, Darwin depicts a pattern-process identical at all taxonomic levels. For Darwin, then, the world is perfectly identical and homogeneous: at lower taxonomic levels, at higher taxonomic levels, in the geological past, and in the expected future. Of course, whereas the rhetorical Darwin held that the pan-divergent view in the Diagram constitutes a truly open evolutionary process, the uniformitarian Darwin denied it.

8.4 Conclusion of Part IV

The epistemological connections between Newton and Darwin are significant: a science of *vera causa*, the use of analogy, empiricism, inductivism, and, above all, uniformity. During the pre-1859 period, as Darwin was working on the material that was eventually integrated into the *Origin of Species*, the name of Newton in England was invoked by epistemologists like John Herschel, William Whewell, and John Stuart Mill with increasing frequency. In the *Critique of Judgment* (1790), Immanuel Kant had wondered about the feasibility of one day seeing a Newton of a blade of grass, that is, a scholar who would do for biology what Newton had done for physics. Ever since Newton, the model of an ideal science based on a comprehensive unification never stopped captivating the esthetic and explanatory tastes of scholars. Consciously or unconsciously, Darwin worked toward fulfilling this epistemological goal, at least to a certain extent.

Regrettably for Darwin, he mostly failed where Newton had amply succeeded. Why is that? Let us mention two main reasons. First, the posterity of their common commitment to uniformitarianism turned out to be quite different. Newton assumed that the whole universe was uniform: in space, in time, and at the various levels of matter. The twentieth century showed that he was wrong on several major counts: at the atomic and subatomic levels (quantum mechanics), at the super-macroscopic level (general theory of relativity), and in the time dimension (an expanding universe). Nonetheless, Newtonian physics was very successful at the scale of our own solar system, allowing us to make remarkable predictions about the positioning of the celestial bodies, such that we can now successfully send probes to them years in advance. At that scale, it is still heuristically sound to exploit a steady-state view of the world. The reasoning could be reversed: it is because our solar system is stable enough that Newton encountered the success he did. For his part, Darwin did not share Newton's good fortune. His ontological choice proved unsuitable for an object of study inherently in transformation. His choice was founded on the idea of motion within a closed system, thus depriving him of the necessary explanatory flexibility. This can be seen in many instances, among which the following figure prominently:

(1) life being trapped in distinct static classes/phyla; (2) life lacking evolutionary freedom under the weak contingency thesis; (3) evolutionary novelties and discontinuities being prevented from emerging under a fabric of life too tightly knit (the gradualism-gradation principle); (4) the lack of evolutionary flexibility in life's motion uniformly envisioned as a pan-divergent pattern-process; (5) evolutionary directionality being impaired by a series of systems in equilibria.

Second, Newton was fortunate enough to encounter a physical world that was complex, but not to the point of preventing the human intellect from seeing order in it. This allowed Newton's successors, including Clairaut, Euler, d'Alembert, Lagrange, and Laplace, to bring the Newtonian system to its successful completion. On the other hand, Darwin faced a biological world so complex and multidimensional that the ideal of a unification of knowledge has largely evaded biologists ever since. One may fault Darwin for having applied an ill-designed interpretative framework, but surely one cannot blame him for the inherent complexity of the biological world. A glimpse of that complexity is reflected in the *Origin of Species*, one that overwhelmed Darwin beyond the possibility of unifying that knowledge.

References

Bachelard G (2002[1938]) The formation of the scientific mind. Clinamen Press, Manchester
Bellon R (2014) There is grandeur in this view of newton: Charles Darwin, Isaac Newton and Victorian conceptions of scientific virtue. Endeavour 38:222–234
Bowler PJ (1996) Life's splendid drama: evolutionary biology and the reconstruction of life's ancestry. University of Chicago Press, Chicago, pp 1860–1940
Brooke JH (1991) Science and religion: some historical perspectives. Cambridge University Press, Cambridge
Cain J (2009) Rethinking the Synthesis period in evolutionary studies. J Hist Biol 42:621–648
Camardi G (1999) Charles Lyell and the uniformity principle. Biol Philos 14:537–560
Chalmers AF (1999) What is this thing called science? 3rd edn. Hackett, Indianapolis
Cohen HF (1994) The Scientific Revolution: a historiographical inquiry. University of Chicago Press, Chicago
Cohen IB (1985a) The birth of a new physics, 2nd edn. W.W. Norton, New York
Cohen IB (1985b) Revolution in science. Belknap Press, Cambridge
Conner CD (2005) A people's history of science: miners, midwives, and 'low mechaniks'. Nation Books, New York
Darwin C (1859) On the origin of species. John Murray, London
Darwin C (1868) The variation of animals and plants under domestication, vol 1. John Murray, London
Darwin C (1872) The origin of species, 6th edn, with additions and corrections [1876]. John Murray, London
Deason GB (1986) Reformation theology and the mechanistic conception of nature. In: Lindberg DC, Numbers RL (eds) God and nature: historical essays on the encounter between Christianity and science. University of California Press, Berkeley, pp 167–191
Delisle RG (2008) Expanding the framework of the holism/reductionism debate in neo-Darwinism: the case of Theodosius Dobzhansky and Bernard Rensch. Hist Philos Life Sci 30:207–226
Delisle RG (2009a) Les philosophies du néo-darwinisme. Presses Universitaires de France, Paris

Delisle RG (2009b) The uncertain foundation of neo-Darwinism: metaphysical and epistemological pluralism in the Evolutionary Synthesis. Stud Hist Phil Biol Biomed Sci 40:119–132
Delisle RG (2012a) The disciplinary and epistemological structure of paleoanthropology: one hundred and fifty years of development. History and Philosophy of the Lifes Sciences 34: 283–330
Delisle RG (2012b) What was really synthesized during the Evolutionary Synthesis? a historiographic proposal. Stud Hist Phil Biol Biomed Sci 42:50–59
Delisle RG (2014) Evolution in a fully constituted world: Charles Darwin's debt towards a static world in the Origin of Species (1859). Endeavour 38:197–210
Delisle RG (2017a) Introduction: Darwinism or a kaleidoscope of research programs and ideas? In: Delisle RG (ed) The Darwinian tradition in context: research programs in evolutionary biology. Springer, Cham, pp 1–8
Delisle RG (2017b) From Charles Darwin to the Evolutionary Synthesis: weak and diffused connections only. In: Delisle RG (ed) The Darwinian tradition in context: research programs in evolutionary biology. Springer, Cham, pp 133–168
Di Fate VJ (2011) Arguing for uniformity: rethinking Lyell's Principles of Geology. Perspect Sci 19:136–153
Ellegard A (1990 [1958]) Darwin and the general reader: the reception of Darwin's theory of evolution in the British periodical press, 1859–1872. University of Chicago Press, Chicago
Evans LT (1984) Darwin's use of the analogy between artificial and natural Selection. J Hist Biol 17:113–140
Franklin A (1986) The neglect of experiment. Cambridge University Press, Cambridge
Galison P (1987) How experiments end. University of Chicago, Chicago
Ghiselin MT (1969) The triumph of the Darwinian method. Univesity of California Press, Berkeley
Gildenhuys P (2004) Darwin, Herschel, and the role of analogy in Darwin's Origin. Stud Hist Phil Biol Biomed Sci 35:593–611
Gillispie CC (1951) Genesis and geology. Harvard University Press, Cambridge
Gillispie CC (1958) Lamarck and Darwin in the history of science. Am Sci 46(4):388–409
Gooding D (1990) Experiment and the making of meaning. Kluwer, Dordrecht
Gould SJ (1980) Is a new and general theory of evolution emerging? Paleobiology 6:119–130
Gould SJ (1987) Time's arrow, time's cycle: myth and metaphor in the discovery of geological time. Harvard University Press, Cambridge
Greene JC (1981) Science, ideology, and world view. University of California Press, Berkeley
Hacking I (1983) Representing and intervening. Cambridge University Press, Cambridge
Herschel JFW (1831) A preliminary discourse on the study of natural philosophy. Longman, Rees, Orme, Brown and Green, London
Hodge MJS (1977) The structure and strategy of Darwin's 'long argument'. Br J Hist Sci 10: 237–246
Hodge MJS (1982) Darwin and the laws of the animate part of the terrestrial system (1835–1837): on the Lyellian origins of his zoonomical explanatory program. Studies in History of Biology 6:1–106
Hodge MJS (2000) Knowing about evolution: Darwin and his theory of natural selection. In: Creath R, Maienschen J (eds) Biology and epistemology. Cambridge University Press, Cambridge, pp 27–47
Hooykaas R (1963) The principle of uniformity in geology, biology and theology. E.J. Brill, Leiden
Hooykaas R (1975) Catastrophism in geology, its scientific character in relation to actualism and uniformitarianism. In: Albritton C (ed) Philosophy of geohistory, 1785–1970. Dowden, Hutchinson & Ross. Penn, Stroudsburg, pp 310–356
Hooykaas R (1987) The rise of modern science: when and why? Br J Hist Sci 20:453–473
Hopkins W (1861) Physical Theories of the Phenomena of Life Part II. Fraser's Magazine 62:74–90
Hull DL (1973) Darwin and his critics: the reception of Darwin's theory of evolution by the scientific community. Harvard University Press, Cambridge
Hull DL (1989) The metaphysics of evolution. State University of New York Press, Albany

Hull DL (2000) Why did Darwin fail? the role of John Stuart Mill. In: Creath R, Maienschen J (eds) Biology and epistemology. Cambridge University Press, Cambridge, pp 48–63

Hull DL (2009) Darwin's science and Victorian philosophy of science. In: Hodge MJS, Radick G (eds) The Cambridge companion to Darwin, 2nd edn. Cambridge University Press, Cambridge, pp 173–196

Huxley TH (1860) Darwin on the Origin of Species. Westminster Review 17:541–570

Huxley TH (1896) Darwiniana. Appleton, New York

Koyré A (1965) Newtonian studies. Chapman & Hall, London

Kubrin D (1967) Newton and the cyclical cosmos: providence and the mechanical philosophy. J Hist Ideas 28:325–346

Kuhn TS (1970) The structure of scientific revolutions, 2nd edn. University of Chicago Press, Chicago

Kuhn TS (1977) The essentiel tension: selected studies in scientific tradition and change. University of Chocago Press, Chicago

Largent M (2009) The so-called eclipse of Darwinism. Trans Am Philos Soc 99(1):3–21

Laudan L (1977) Progress and its Problems: towards a theory of scientific growth. University of California Press, Berkeley

Laudan L (1981) Science and hypothesis: historical essays on scientific methodology. D. Reidel, Dordrecht

Laudan R (1982) The role of methodology in Lyell's science. Stud Hist Phil Sci 13:215–249

Laudan R (1987) From mineralogy to geology: the foundations of a science, 1650–1830. University of Chicago Press, Chicago

Levine JM (1983) Natural history and the history of the Scientific Revolution. Cliometrica 13(1): 57–73

Levit GS, Simunek M, Hossfeld U (2008) Psychoontogeny and psychophylogeny: Bernhard Rensch's (1900–1990) selectionist turn through the prism of panpsychistic identism. Theory Biosci 127:297–322

Lloyd EA (1983) The nature of Darwin's support for the theory of natural selection. Philos Sci 50:112–129

Lyell C (1830) Principles of geology, vol 1. John Murray, London

Lyell C (1832) Principles of geology, vol 2. John Murray, London

Malherbe M (1996) Bacon's method of science. In: Peltonen M (ed) The Cambridge companion to Bacon. Cambridge University Press, Cambridge, pp 75–98

Mayo D (1996) Error and the growth of experimental knowledge. University of Chicago Press, Chicago

Mayr E (1980) Prologue: some thoughts on the history of the Evolutionary Synthesis. In: Mayr E, Provine WB (eds) The Evolutionary Synthesis: perspectives on the unification of biology. Harvard University Press, Cambridge, pp 1–48

Newton I (1726[1952]) Mathematical principles of natural philosophy, 3rd edn. William Benton Publisher, Chicago

Newton I (1730) Optiks: or a treatise of the reflections, refractions, inflections, and colour of light, 4th edn. William Innys, London

Ray J (1701) The wisdom of God manifested in the works of the creation, 3rd edn, London

Richards RA (1997) Darwin and the inefficacy of artificial selection. Stud Hist Phil Sci 28:75–97

Rossi P (1999) Aux origines de la science moderne. Seuil, Paris

Rossi P (2001) The birth of modern science. Blackwell, Oxford

Rudwick MJS (1970) The strategy of Lyell's Principles of Geology. Isis 61:5–33

Ruse M (1973) The philosophy of biology. Hutchinson, London

Ruse M (1975) Darwin's debt to philosophy: an examination of the influence of the philosophical ideas of John F.W. Herschel and William Whewell on the development of Charles Darwin's theory of evolution. Stud Hist Phil Sci 6:159–181

Ruse M (1976) Charles Lyell and the philosophers of science. Br J Hist Sci 9:121–131

Ruse M (1979) The Darwinian Revolution. University of Chicago Press, Chicago

Ruse M (2000) Darwin and the philosophers: epistemological factors in the development and reception of the theory of the Origin of Species. In: Creath R, Maienschein J (eds) Biology and epistemology. Cambridge University Press, Cambridge, pp 3–26
Sedgwick A (1860) Objections to Mr. Darwin's theory of the Origin of Species. The Spectator (April 7th), pp 334–335
Smith GE (2002) The methodology of the Principia. In: Cohen IB, Smith GE (eds) The Cambridge companion to Newton. Cambridge University Press, Cambridge, pp 138–173
Sterrett SG (2002) Darwin's analogy between artificial and natural selection: how does it go? Stud Hist Phil Biol Biomed Sci 33:151–168
Todhunter I (1876) William Whewell, D.D. Master of Trinity College Cambridge, vol 2. Macmillan, London
Westfall RS (1993) The life of Isaac Newton. Cambridge University Press, Cambridge
Whewell W (1832) [Review of volume 2 of Lyell's] Principles of Geology. Q Rev 47:103–132
Whewell W (1840) The philosophy of the inductive sciences founded upon their history, vol 2. J.W. Parker, London
Wilson DB (1974) Herschel and Whewell's version of Newtonianism. J Hist Ideas 35:79–97

Chapter 9
Conclusion: A Restored Unity in the *Origin of Species*?

Abstract Among scholars, Charles Darwin has often played the mythical role of a founding figure, one who must be either invoked or repudiated. In light of the analysis provided in this book, we argue that many of the notions Darwin employs—common ancestry, divergence, a self-regulated economy, evolutionary contingency, adaptation, natural selection, etc.—are incompatible with their uses in modern biology. Not only does modern evolutionary biology not originate with the *Origin of Species*, it also seems that the historiographical label "Darwinian Revolution" is largely without historical basis.

Keywords Darwin's Diagram · Darwinian Revolution · Common ancestry · Principle of divergence · Self-regulated economy of nature · Evolutionary contingency · Competition · Adaptation · Variation · Evolutionary taxonomy · Natural selection · Multilevel model of science · Reticulate model of science

The picture that emerges from our analysis has little to do with the mythical figure "Charles Darwin," one often encounters today. Among the larger public, Darwin is believed to have single-handedly erected the doctrine of evolutionism. There is no need for us to debunk this misconception. It seems, however, that Darwin also fills a mythical role among scholars. For instance, some see themselves as part of a central intellectual movement beginning with Darwin, culminating in an Evolutionary Synthesis promoting an even more inclusive Darwinism. Others propose alternative evolutionary theories to standard Darwinism, retrospectively seeking Darwin's sanction by claiming that some of their own ideas can be found in Darwin. Others, still, erect a straw man called "Darwinism" to better attack it and advance their own research agenda. One thing is certain: Darwin has taken on a mythical role as a presumed "founding figure," one that must be either invoked or repudiated.

We have been asking a great deal of Darwin, perhaps too much. The *Origin of Species* is a major piece of work, but one that belongs squarely to the nineteenth century. Our image of Darwin's work has been modernized to excess. The *Origin* offers an overall explanatory apparatus too rigid to capture the complexities of biological evolution, one centered on a steady-state view. No surprise that a

R. G. Delisle, *Charles Darwin's Incomplete Revolution*, Evolutionary Biology – New Perspectives on Its Development 1, https://doi.org/10.1007/978-3-030-17203-9_9

significant part of that evolutionary complexity could not be handled within his system. As if the situation was not already challenging enough, Darwin's empirico-inductivism forced him to confront a wealth of material typical of newly emerging scientific areas. The *Origin* is both an incomplete transition to evolutionism and a failed attempt at a unified theory of evolution. On both issues, Darwin's commitment to empirico-inductivism was strong enough to prevent him from entirely concealing a part of evolution's complexity, although it largely eluded his explanations (see Table 8.1).

One might want to object to the characterization proposed here on the grounds that many of Darwin's ideas are actively exploited today. Let us briefly review the most common of them in order to see how "Darwinian" these ideas really are today:

Common Ancestry and the Principle of Divergence These two concepts are central to evolutionary biology today, but Darwin exploited them in a very specific way no longer accepted. He treated them conjointly as an expression of the same reality: a universal and rigid pan-divergent view constraining the manifestations of other evolutionary patterns and processes.

A Self-regulated Economy of Nature Whereas Linnaeus held that the economy of nature was regulated by pre-established principles, Darwin failed to propose an alternative entirely free of pre-established notions. Such a complete liberation of these principles would have jeopardized Darwin's pan-divergent view.

Evolutionary Contingency For reasons just enumerated, Darwin could not fully embrace evolutionary contingency and make the full turn toward historical narrative. He therefore subscribed only to the weak contingency thesis, leaving the conceptualization of a strong contingency thesis to the twentieth century.

Competition and Adaptation Although the notions of competition and adaptation are instrumental for Darwin's thinking about the evolutionary drive, these are never embraced to the point of erasing the main evolutionary events. Darwin would not take stock of alternative evolutionary patterns-processes—polyphyletism, reticulate evolution, parallel evolution, convergent evolution, and the deletion of affiliative traces—alternatives that would blur his overall evolutionary outlook.

Variation Variations are for Darwin the material upon which the evolutionary drive feeds. Yet, no genuine new variations arise, under Darwin's view, if only because he conceives them as a mere expression of those already encountered among extant forms but projected backward in time.

Evolutionary Taxonomy Darwin's reinterpretation of the Linnean taxonomy to make it serve the purpose of indicating evolutionary relationships was deficient. Establishing an equation between taxonomy and phylogeny, Darwin wrongly assumes that taxonomy among extant forms captures the main evolutionary events of the past.

Natural Selection as an Explanatory Core In contradistinction to what is often believed today, Darwin's theory did not put the notion of natural selection in a central explanatory position. Rather, Darwin treated natural selection as a neutral auxiliary component with sufficient flexibility, having the main explanatory function of protecting the real immutable and irrefutable hard core of his theory: divergence-gradation. This is seen in the way Darwin modulates its application throughout the *Origin of Species* between different interpretative contexts. In Chaps. 1–5 and 8, natural selection is weak enough to have difficulty segregating closely related forms united in common reproductive networks. In Chaps. 6 and 7, natural selection is powerful enough to allow the linear ascent of independent evolutionary lines across taxonomic levels and evolutionary grades, at the price of blurring the crucial distinction for his theory between homologies and analogies. In Chaps. 9, 10, and 13, the power of natural selection is reduced to such an extent that its role is confined to the mere deployment of classes/phyla known today. The explanatory status of natural selection is most clearly revealed in Chaps. 11 and 12: in order to save the core of his theory (divergence-gradation), Darwin modulates the power of natural selection at will for the purpose of making its action fit theoretical predictions, with selective pressure being interpreted along a range from totally lacking to fairly strong.

A Unified Theory of Evolution Although Darwin is often assumed to have proposed a unified theory of evolution, the *Origin of Species* in fact displays a series of independent or competing pictures: (a) life forms caught in complex reproductive networks from which it is difficult to depart; (b) life forms rushing forward across taxonomic levels and evolutionary grades; (c) evolution at a standstill and lacking in novelties; (d) Evolution organized around a pan-divergent view; and (e) the exhaustion of the evolutionary drive following the encounter of major gaps or discontinuities in the knit of life. Darwin truly fought his intellectual way throughout the *Origin* in order to try to cover up evolution's complexity, keeping it within a pan-divergent view that assumes life to be deprived of any significant discontinuities.

The least that can to be said is that the ideas proposed by Darwin are heavily loaded, at the service of a unique and somewhat peculiar view of evolution. Of course, these ideas have been recycled by other evolutionary biologists since the nineteenth century, but rarely (if ever) as Darwin understood them. We argue that once removed from Darwin's own interpretative framework, these ideas are no longer properly called "Darwinian." Within new epistemological contexts, their meanings are profoundly modified. Among a range of possible reinterpretations encountered today, the following come to mind:

1. Whereas Darwin fused the notions of common ancestry and divergence together, these have since been uncoupled to serve, independently, explanatory functions other than Darwin's.
2. Darwin's weak contingency thesis has been disentangled from its original pan-divergent view, in order to allow for the establishment of a strong contingency thesis.

3. Whereas Darwin failed at entirely liberating the pre-established notions accompanying the economy of nature, this was eventually done to provide the evolutionary process the full adaptive flexibility required.
4. Natural selection can be placed at the irrefutable core of a causal theory of biological evolution if desired, something Darwin did not do.
5. Linnaean taxonomy was disconnected from Darwin's actualist and steady-state (uniformitarian) view in order to recognize that taxonomy practiced with extant forms constitutes a very imperfect reflection of the evolutionary past's complexity, under the assumption that taxonomy and phylogeny cannot be equated.
6. Variability (disparity) among extant forms has been reconceptualized from Darwin's use under a new interpretation, according to which it represents an incomplete sample of all the variability that have ever existed. Common ancestors, for instance, are no longer merely "shadows" of Darwin's preconception.
7. The concept of adaptation has been liberated from the constraints of keeping genealogical tracks of lineages over time, owing to blurred affiliative traces or the rise of identical features.
8. A unified evolutionary theory might be sought that incorporates, this time, a significant amount of evolutionary realities and complexities—polyphyletism, reticulate evolution, parallel evolution, and convergent evolution—all of which Darwin left out in their full expression.

Darwin was truly an inspiring scholar with respect to all these matters, yet all or most of them have since been reinterpreted in a piecemeal fashion and incorporated into a range of more or less incommensurable theories or research programs, even within the so-called "Darwinian tradition" (Delisle 2017). The epithet "Darwinism" has enjoyed widespread usage in evolutionary biology. In light of recent investigations, it is suggested that the label might best be abandoned altogether, considering that it now hides issues more than it reveals them. It was, perhaps, legitimate to maintain the appellation "Darwinism" when supported under two main historical events: Charles Darwin's unified theory in the nineteenth century and the Evolutionary Synthesis in the twentieth century.[1] As the historiographical reality of these two unification events is now questioned, an alternative historiography is needed. It is increasingly clear that during the last decades evolutionary biologists, philosophers, and historians have taken for granted that this research area is organized around a "multilevel model" of science, that is, a hierarchical explanatory structure postulating a causal core (mechanisms of evolution), itself explaining descriptive and peripheral phenomena from a host of related disciplines (see Fig. 9.1). An alternative historiography consists in promoting a "reticulate model" of science: a diffused and flexible explanatory structure with no privileged role given to causal components (i.e., natural selection) over descriptive ones, and no formal distinction between central and peripheral areas or topics (see Fig. 9.2). We have argued elsewhere that the development of evolutionary biology since Darwin conforms

[1] See our discussion in Chap. 8 of this book. References are given therein.

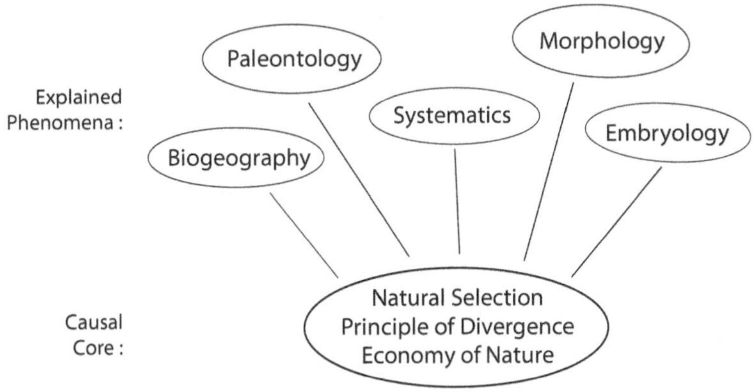

Fig. 9.1 The *Origin of Species*: a multilevel model of science. This model often implicitly associated with Darwin's theory of evolution assumes a hierarchical explanatory structure postulating a causal core (mechanisms of evolution), which in turn explains descriptive and peripheral phenomena from a host of disciplines. The well-foundedness of this model is based on Darwin's rhetorics and is often combined with the projection in time of the explanatory structure accompanying the so-called Evolutionary Synthesis in the twentieth century

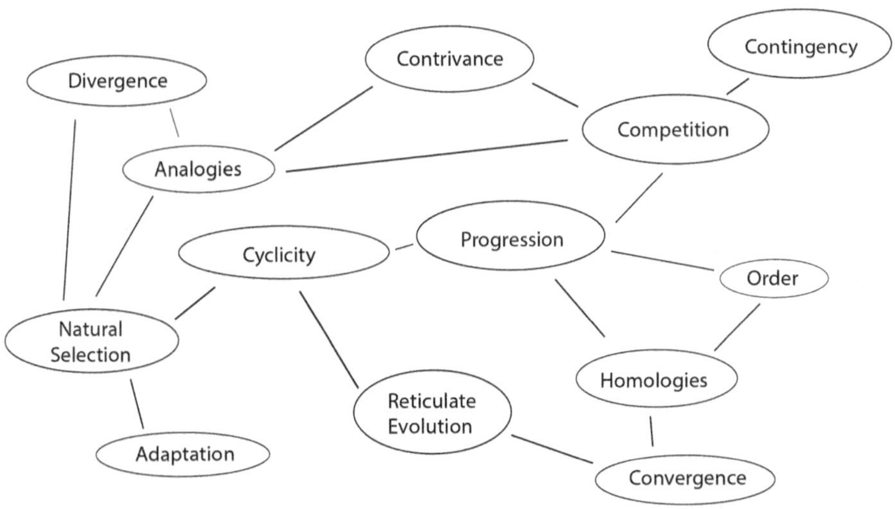

Fig. 9.2 The *Origin of Species*: a reticulate model of science. From a close reading of the *Origin*, one gathers that the book appeals to a model whereby a diffuse and flexible explanatory structure gives no privileged role to causal components (i.e., natural selection) over descriptive ones, with no formal distinction being made between central and peripheral areas or topics

more closely to the reticulate model than the multilevel model.[2] This area is best envisioned as a kaleidoscope of overlapping research programs and ideas running in

[2]This topic has been treated more fully in Delisle (2017).

parallel and interacting in complex ways. Charles Darwin's failed attempt at erecting a unified theory of biological evolution in the *Origin of Species* constitutes another expression of a science organized around a reticulate model, in light of the competing views contained in the same book.

Is it ultimately possible to save Darwin's unified theory by focusing exclusively on the rhetorical Darwin described in the previous chapters of this book? We think not. It should be remembered that the rhetorical Darwin tried to put natural selection at the causal core of a theory reuniting a whole range of descriptive disciplines. Even though Darwin did not succeed, it is perfectly legitimate to be inspired by the *spirit* of his attempt, leaving aside its *letter*. Darwin was pointing in a direction that has since appealed to the esthetic and epistemological tastes of many. However, the moment one recognizes that the essence of Darwin's synthetic effort is encapsulated in his Diagram (see Fig. 2.1), he can no longer be seen as a scholar of the future. Darwin instantly becomes a scholar of the past, in light of a Diagram visually representing his commitment to a steady-state view. Here's why:

1. Darwin's Diagram is not a representation of the entire tree of life. For him, evolution is confined within classes/phyla, that is, entities presenting themselves as independent, separate, and nearly static. Such high taxonomic level entities are, for Darwin, nearly permanent, given that no extinction of classes or phyla is believed to have ever occurred. With classes/phyla being unmovable, cyclical evolutionary motion is created through: (a) the rise and fall of lower taxonomic entities merely replacing each other (individuals, varieties, species, genera, families, and orders) and (b) a series of devices in equilibrium and locked up by counterweights (demographic expansion at the expense of a proportional decrease, evolutionary divergence occurring in reversed proportion to taxonomic ascent, and so on).
2. Darwin's integral uniformitarian stance is captured in the homogeneity of his Diagram. Life forms are characterized throughout by a unique and repetitive figure—a pan-divergent view—whether in the past, present, and future, and whether at low or high taxonomic levels. We have already insisted on its analogy with fractal geometry. In addition, the perfect homogeneity in the Diagram is guaranteed by the principle of gradation. The postulated homogeneity in Darwin's view is such that it annihilates the temporal dimension of evolution but putting it in an explanatory straitjacket wherein the past, the present, and the future are but more of the same.
3. Darwin's actualistic method finds its epistemological rationale in a uniformitarian ontology: if the past and present are believed to be essentially identical, it becomes obvious why the investigation of a present entirely open to our gaze will always be more informative than the investigation of its past. On this understanding, the world is explored by Darwin under three complementary assumptions, as clearly reflected in his Diagram: (a) all main evolutionary events are registered in today's taxonomy, since the arrangement has been maintained throughout the entire evolutionary history; (b) variability or disparity of the past is either aligned with or falls in-between the one observed today; and (c) divergence

of life forms is at its maximum today relative to its own past. Unsurprisingly, Darwin's so-called common ancestors are best envisioned as "shadows," that is, pale reflections or miniatures of extant forms (preformationism).

If one feels the need to identify a unity in the *Origin of Species*, one may find it in Darwin's steady-statism of the brand inherited from the tradition of natural theology. Of all the *Origin*'s competing perspectives, it seems that Darwin's steady-statism generates the greatest possible level of coherence. Of course, this unity will always be imperfect in light of Darwin's attempt at embracing evolutionism with what essentially is a static worldview, especially considering that he was overwhelmed from all sides with phenomena not fitting under his narrow conceptual purview (see Table 9.1). As already alluded to, Darwin is more properly classified among the last uniformitarians than among the first truly modern evolutionists. The net result is an incomplete transition to evolutionism, one too partial in our view to justify the historiographical notion of a "Darwinian Revolution."[3] Darwin's *Origin of Species* stopped at the threshold of the modern evolutionistic door without crossing it, although he was able to take a good peek at some of the issues that would eventually be developed beyond it.

If one is prepared to accept the theses proposed here, Darwin becomes a fascinating figure, one of many transitional figures inhabiting an intellectual space between two worldviews: the static and the evolutionary. If Darwin is no longer considered to be the absolute reference point about the rise of evolutionism—the first truly modern evolutionist—a reappraisal of this whole transitional period in the nineteenth century and beyond seems inescapable.

[3]A clearer view is likely to emerge about the so-called Darwinian Revolution when the rest of the Darwinian corpus (Darwin's other publications, etc.) is reevaluated.

Table 9.1 The *Origin of Species* confronts the careful reader with an array of evolutionary realities and explanatory components that, when taken collectively, reveal why Darwin was unable to complete his transition to modern and universal evolutionism, and why he failed to deliver a theory of evolution even remotely approaching unification

[A] *Evolutionary phenomena that undermine Darwin's theory when recognized in their full expression*	
• Hybridization	• Reproductive networks
• Polyphyletism	• Reticulate evolution
• Parallel evolution	• Convergent evolution
• Evolutionary stagnation	• Random evolutionary walk
• Evolution as an engine that erases its genealogical traces as it runs	• Wild power of natural selection generating analogies and erasing affiliations
• Gaps or discontinuities in nature	• When Darwin's "pressure cooker" leaks, losing evolutionary drive
• Exhaustion of the evolutionary drive among related forms in reverse proportion to their taxonomic ascent	• Decimation model hiding significant amounts of diversity or disparity in the evolutionary past
[B] *Persistent explanatory tensions often explicitly or implicitly acknowledged by Darwin*	
• Paleontology versus neontology	• A unified tree of life versus a segmented tree of life
• A static worldview versus an evolutionary worldview	• Homogeneity and contiguity in life's fabric versus gaps in life's fabric
• Gaps in nature impacting the evolutionary drive 1: a fully stocked world versus a world not fully stocked	• Gaps in nature impacting the evolutionary drive 2: when going from the sympatric phase to the allopatric phase of speciation
• Gaps in nature impacting the evolutionary drive 3: when related forms move too far apart from each other, taxonomically speaking	• Gaps in nature impacting the evolutionary drive 4: when a preadapted form conquers a new territory
• Homologies versus analogies	• Allopatric speciation versus sympatric speciation
• Selective contexts: under the competitive exclusion principle versus under the entangled bank	• Monophyletism versus polyphyletism
• A liberated balance of nature versus a balance of nature in a straitjacket	• Weak contingency thesis versus strong contingency thesis
• Fixed and entrenched taxonomic categories (classes, phyla) versus cycling (rising/declining) taxonomic categories (varieties, species, genera, families, and orders)	• Correlation of growth as an amplifier of change or as restrainer of change
• Ancestors versus shadows	• Directional drive across evolutionary grades and taxonomic levels versus divergent evolution
• The wild power of natural selection versus a pan-divergent view	• Fixed and irrefutable variables (core: divergence + gradation) versus flexible and auxiliary variables (taxonomic level, geological time, geographical space, and amount of selective pressure)
• Common ancestry versus divergence	

(continued)

Table 9.1 (continued)

• Inconsistent relationship between the power of artificial selection and the power of natural selection		• Life forms as abstract entities freely flowing in evolution versus life forms as interconnected entities impeding the evolutionary march (reproduction, symbiosis, parasitism, predator–prey relationship)
• Biological progress versus evolutionary opportunism		
• Species realism versus species nominalism		• Progression versus retrogression
[C] *Darwin's debt to natural theology and the ontology accompanying a static worldview*		
• Empirico-inductivism	• Uniformitarianism	• *Vera causa*
• Consilience of inductions	• Steady-statism	• Contrivance
• A designed world	• Interconnection	• Order
• Harmony	• Coadaptations	• Symbiosis
• Unmovable and entrenched taxonomic groups (classes and phyla) • Divergence at maximum today relative to the evolutionary past • Symmetry-equilibrium 1: ideal model of proportions between divergence, selective pressure, geological time, distance traveled, and taxonomic level • Symmetry-equilibrium 2: demographic expansion at the proportional expense of demographic decrease • Symmetry-equilibrium 3: exhaustion of the evolutionary drive among related forms in reverse proportion to their taxonomic ascent • Symmetry-equilibrium 4: evolutionary stalemate of species interlocked in a mutual adaptive entrenchment • Symmetry-equilibrium 5: the progression/retrogression duality • Evolution in a straitjacket 1: a rigid pan-divergent outlook throughout all taxonomic levels in perfect homogeneity ("fractal geometry") • Evolution in a straitjacket 2: Darwin's two embryological laws • Life at a standstill in the annals of life • Preformationism (shadows)		
[D] *Darwin tried to tackle the complexity of the world with a set of explanatory tools and simplifying assumptions, using rhetorical arguments as a cover-up*		
• Common ancestry	• Principle of divergence	• Selection of the most divergent forms
• Principle of specialization	• Principle of gradation	• Gradualism
• Natural selection	• Past forms being aligned or falling in between extant forms	
• Current taxonomic arrangement maintained throughout the entire evolutionary history		
• Competitive exclusion principle	• Entangled bank	• Economy of nature
• A fully liberated balance of nature		
• There is no law of development (strong evolutionary contingency)		
• Competition	• Adaptation	• Demographic pressure
• Variation	• Monophyletism	• Evolutionary taxonomy
• Unique centers of dispersal	• Extrapolationism	• Interpolationism
• Uniformitarianism	• *Vera causa*	• Consilience of inductions

Reference

Delisle RG (2017) From Charles Darwin to the Evolutionary Synthesis: weak and diffused connections only. In: Delisle RG (ed) The Darwinian tradition in context: research programs in evolutionary biology. Springer, Cham, pp 133–167

Index

R. G. Delisle, *Charles Darwin's Incomplete Revolution*, Evolutionary Biology – New Perspectives on Its Development 1, https://doi.org/10.1007/978-3-030-17203-9

MIX
Papier | Fördert gute Waldnutzung
FSC® C083411

Kolibri 360 GmbH
Ferdinand-Jühlke-Straße 7
99095 Erfurt, Deutschland
produktsicherheit@kolibri360.de